Stem Cell Biology and Regenerative Medicine

Volume 73

Series Editor

Kursad Turksen, Ottawa Hospital Research Institute, Ottawa, ON, Canada

Our understanding of stem cells has grown rapidly over the last decade. While the apparently tremendous therapeutic potential of stem cells has not yet been realized, their routine use in regeneration and restoration of tissue and organ function is greatly anticipated. To this end, many investigators continue to push the boundaries in areas such as the reprogramming, the stem cell niche, nanotechnology, biomimetics and 3D bioprinting, to name just a few. The objective of the volumes in the Stem Cell Biology and Regenerative Medicine series is to capture and consolidate these developments in a timely way. Each volume is thought-provoking in identifying problems, offering solutions, and providing ideas to excite further innovation in the stem cell and regenerative medicine fields.

Bhupendra V. Shravage · Kursad Turksen
Editors

Autophagy in Stem Cell Maintenance and Differentiation

Springer

Editors
Bhupendra V. Shravage
Developmental Biology Group
Agharkar Research Institute
Pune, Maharashtra, India

Kursad Turksen
Ottawa Hospital
Ottawa, ON, Canada

ISSN 2196-8985 ISSN 2196-8993 (electronic)
Stem Cell Biology and Regenerative Medicine
ISBN 978-3-031-17361-5 ISBN 978-3-031-17362-2 (eBook)
https://doi.org/10.1007/978-3-031-17362-2

This Springer imprint is published by the registered company Springer Nature Switzerland AG
The registered company address is: Gewerbestrasse 11, 6330 Cham, Switzerland

Preface

Autophagy is a conserved process that is essential for maintaining cellular homeostasis in all kinds of cells including stem cells. Stem cells have the ability to self-renew as well as differentiate into one or more kinds of cells as demanded by their external environment which has immense therapeutic potential in regenerative medicine. The regulation and role of autophagy in stem cells are poorly understood. This book offers insights into the mechanisms by which autophagy can regulate stem cell self-renewal and facilitate specific differentiation programs. Additionally, it provides a glimpse into modulating autophagy in stem cells as a therapeutic option in diseases such as cancer. We hope that researchers, teachers and students alike find this book useful. We greatly appreciate everyone who contributed or helped towards making this book possible that too during the difficult times of global COVID-19 pandemic. Thank you.

We remain very grateful to Gonzalo Cordova, the Editor of the series, and wish to acknowledge his continued support. A special thank you goes to the production team at Springer Nature Switzerland AG for their outstanding efforts in the production of this volume.

Finally, sincere thanks to the contributors not only for their support of the series but also for their willingness to share their insights and all their efforts to capture both the advances and the remaining obstacles in their areas of research. We trust readers will find their contributions as interesting and helpful as we have.

Pune, Maharashtra, India — Bhupendra V. Shravage
Ottawa, ON, Canada — Kursad Turksen

Contents

1 **Assays for Monitoring Autophagy in Stem Cells** 1
Aishwarya Chhatre and Bhupendra V. Shravage

2 **Autophagy in Stem Cell Maintenance and Differentiation** 35
Anirudha K. Sahu, Propanna Bandyopadhyay,
Rajdeep Chowdhury, and Sudeshna Mukherjee

3 **Autophagy in Embryonic Stem Cells and Neural Stem Cells** 59
Deepika Puri, Shalmali Bivalkar-Mehla, and Deepa Subramanyam

4 **Autophagy in Germline Stem Cells** 85
Kaitlin E. Kosinski and Alicia Meléndez

5 **The Role of Autophagy in the Regulation of Hematopoietic Stem Cells** 107
Pallavi Budgude, Prajakta Teli, Anuradha Vaidya,
and Vaijayanti Kale

6 **Autophagy in Muscle Stem Cells** 137
Shulei Li, Romina L. Filippelli, Alice Jisoo Nam,
and Natasha C. Chang

7 **Autophagy in the Intestinal Stem Cells** 169
Ebru Goncu

8 **The Autophagy Lysosomal Pathway: Friend or Foe in Adult Neural Stem Cells?** 193
Isabel Calatayud-Baselga and Helena Mira

9 **Autophagy in Mesenchymal Stem Cell-Based Therapy** 219
Carl Randall Harrell, Dragica Pavlovic, and Vladislav Volarevic

10 **Autophagic Control of Stem Cells Differentiation into Osteogenic Lineage—Implications in Bone Disorders** 235
Shalmoli Bhattacharyya and Aditi Mahajan

11 Autophagy in Cancer Metastasis 259
Ruhi Polara, Daphni van Rinsum, and Nirmal Robinson

Index 287

Chapter 1
Assays for Monitoring Autophagy in Stem Cells

Aishwarya Chhatre and Bhupendra V. Shravage

Abstract Macroautophagy (Autophagy hereafter) mediates degradation of cytoplasmic components and organelles via the lysosome in all types of cells including stem cells. It is an evolutionarily conserved process which involves formation of a double membrane, the autophagosome, and is crucial for maintaining homeostasis within the cells. Numerous assays for measuring autophagy (autophagic flux) have been designed and described in detail in Klionsky et al. [72]. However, not all may be suitable for a particular stem cell type, and interpreting the assays is key to understanding autophagy in stem cells. Several researchers use drugs or transcription factors to induce stemness or in differentiation protocols, and it would be prudent to know if the treatments affect autophagy and in what way. This review is an attempt to put together the relevant assays that can be used to monitor autophagy along with the advantages and limitations of each assay. In addition, we also discuss autophagy assays that have been successfully used by researchers in a particular stem cell type.

Keywords Stem cells · Autophagy assays · LC3/GABARAP · Flux · p62

Abbreviations

AKT	A serine/threonine protein kinase
AMPK	Adenosine 5'monophosphate activated protein kinase
ASC	Adult stem cells
Atg	Autophagy-related gene

A. Chhatre · B. V. Shravage (✉)
Developmental Biology Group, MACS-Agharkar Research Institute, Pune, India
e-mail: bvshravage@aripune.org

B. V. Shravage
Department of Zoology, Savitribai Phule Pune University, Pune, India

Department of Biotechnology, Savitribai Phule Pune University, Pune, India

B. V. Shravage and K. Turksen (eds.), *Autophagy in Stem Cell Maintenance and Differentiation*, Stem Cell Biology and Regenerative Medicine 73,
https://doi.org/10.1007/978-3-031-17362-2_1

ATP2A/SERCA	Sarcoplasmic/endoplasmic reticulum calcium ATPase
CCCP	Carbonyl cyanide m-chlorophenyl hydrazone
CLEM	Corelative light electron microscopy
CMA	Chaperone-mediated autophagy
CQ	Chloroquine
ESC	Embryonic stem cells
ESCRTs	Escort complexes
FIB-SEM	Focused ion beam scanning electron microscopy
FIP200	Focal adhesion kinase family interacting protein of 200 kDa
FOXO3	Forkhead box transcription factors
FUNDC1	Fun14 domain containing 1
FYVE	(Cysteine-rich proteins) Fab1 YOTB Vac1 and EEA1
GABARAP	Gamma amino butyric acid receptor-associated protein
GFP	Green fluorescent protein
HSC	Hematopoietic stem cells
Hsp	Heat shock protein
LAMP	Lysosome-associated membrane protein
LC3	Microtubule-associated protein light chain 3
LIR	LC3 interacting region
MAP	Mannose-associated protein
MAPK14	Mitogen-activated protein kinase
mCherry	MFruits family of monomeric red fluorescent proteins
MSC	Mesenchymal stem cells
mtHsp	Mitochondrial Heat shock protein
mTORC	Mammalian target of Rapamycin complex
MuSC	Epithelial/Muscle stem cells
NBR1	Neighbour of BRCA1 gene protein
P62	Protein of 62 kDa
PI3K	Phosphatidyl inositol 3 kinase
Pink1	PTEN-induced kinase1
RB1CC1	RB1 inducible coiled-coil protein 1
RFP	Red fluorescent protein
SEM	Scanning Electron Microscopy
SNARE	Soluble NSF attachment protein receptor
SOD2	Superoxide dismutase2
SQSTM1	Sequestrosome
STX	Syntaxin protein
TAX1BP1	Tax1 binding protein1
TBK1	TANK binding kinase1
TEM	Transmission Electron Microscopy
TOMM	Translocase of outer mitochondrial membrane
TSC	Tuberous sclerosis protein
Ulk1	Unc51-like autophagy activating kinase1
UVRAG	UV Radiation resistance-associated gene
VDAC	Voltage-dependent anion channel

VPS	Vacuolar protein sorting
WIPI2	WD repeat domain phosphoinositide interacting protein

Introduction

Stem cells essentially have two unique properties: they can differentiate into one or more types of cells and have the ability to self-renew to conserve their population. Based on their functional aspects, stem cells can be broadly divided into embryonic (ESC) and non-embryonic stem cells, which include Adult stem cells (ASC) like adult haematopoietic (HSC), mesenchymal (MSC), epithelial/muscle (MuSC) and neural stem cells (NSC) [1]. To maintain cellular homeostasis, stem cells possess certain intrinsic regulatory mechanisms and additionally extrinsic regulatory mechanisms in the case of niche-regulated stem cells. Autophagy is one of the intrinsic regulatory mechanisms involved in stem cell maintenance and homeostasis. Autophagy has been demonstrated to control mitochondrial ROS levels, premature ageing, prevent cell death, protect against DNA damage and promote stem cell longevity in long-lived stem cells [2]. Thus, autophagy is an important cellular process and has been shown to function in various kinds of stem cells and is vital for regulating stemness and differentiation.

Many stem cell researchers wish to focus on understanding the role of autophagy in regulating stem cell behaviour or as an end-point analysis. Hence, it is crucial to have knowledge of current acceptable standards for conducting and interpreting autophagy assays. For example, induction of pluripotent stem cells can be achieved by several methods including the addition of drugs to the growth medium. In such instances, it becomes imperative to know if these drugs, growth factors and transcription factors modulate autophagy and to what extent. Monitoring autophagy during quiescence, mitotic activity and differentiation of stem cells is key to understanding the potential role and influence of autophagy on several aspects of stem cells. An array of methodologies and assays have been employed to monitor autophagy in somatic cells and these have been adapted to study autophagy in stem cells. However, it is important to select the appropriate method to assay autophagy for a particular stem cell type. Whether a peculiar assay fits a particular stem cell type of interest or not has to be experimentally determined. A table is provided as a quick guide for your ready reference highlighting assays published so far (Table 1.1). A detailed summary of each assay described below is provided in the recently published "Guidelines for the use and interpretation of assays for monitoring autophagy" by Daniel Klionsky [3]. This review provides a comprehensive guide for assays that can be employed to monitor autophagy explicitly in stem cells.

Autophagy Process and Mechanism

The term autophagy was coined in 1963, by Christian de Duve as a lysosomal-related nomenclature on the occasion of *Ciba Foundation Symposium on Lysosomes* [4, 5]. It is an evolutionarily conserved catabolic process and is regulated in a complex manner in different organisms and within different kinds of cellular environments. Cells retain a basal level of autophagic activity for maintaining homeostasis, however, it is upregulated during stress including DNA damage, nutrient depletion, ROS, hypoxia and pathogen invasion. Cellular components including misfolded proteins, pathogenic bacteria, depolarized mitochondria, dysfunctional organelles and several others are recycled by lysosomal degradation during autophagy [6]. The components are sequestered in a double-membrane vesicle called the autophagosome.

Several forms of autophagy have been described in the past 50 years of research and are primarily categorized into 3 major types—macroautophagy, microautophagy and chaperone-mediated autophagy (CMA). Macroautophagy is a bulk degradation process involving the engulfment of cargo into a double-membraned vesicle, the autophagosome. Figure 1.1 depicts autophagy and the core proteins involved in the process. For simplicity, the process is divided into sequestration of cargo, formation of initiation complex, recruitment of phagophore membrane (e.g. endoplasmic reticulum), formation of autophagosomes, sequestration and engulfment of materials, and fusion with lysosome and degradation [7]. Recycled cellular material is transported out via permeases and efflux pumps back into the cytoplasm. Microautophagy involves the direct delivery of cytoplasmic cargo to the lysosome which involves complex lysosomal membrane dynamics. The cargo to be delivered is either taken up by protrusions or invaginations of the lysosomal membrane. Several proteins which function in membrane trafficking including ESCRTs, Vac 8, etc. are required for this kind of autophagy. CMA involves the selective degradation of KFERQ-like motif-bearing proteins delivered to the lysosome via chaperone Hsc70 and other co-chaperones via the lysosome-associated membrane protein-2A [8, 9].

Autophagy is negatively regulated by the mTORC1 complex, which receives inputs from several upstream signalling molecules including MAP kinase, AMP kinases and AKT kinase. AMPK can inhibit mTORC1 activity by phosphorylating Raptor and TSC2 [10–12]. AMPK is also responsible for initiating the Ulk1 complex formation [12]. The autophagy process initiates as a function of Ulk1/Atg1 phosphorylation by AMPK and its further activation of Atg13 and Atg101, with FIP200 acting as a scaffold protein in the formation of Ulk1-Atg13-Atg101-FIP200 initiation complex (in metazoans) [10, 11, 13, 14]. This complex is responsible for the activation of phagophore nucleation at endoplasmic reticulum sites, trans-Golgi apparatus or plasma membrane. Ulk1 activation is directly linked to Atg9 recruitment along with another complex formation of VPS34 (Class III PI3K), Beclin1 (Atg6), Atg14 and p150 (Vps15) on the phagophore membrane [10, 15]. This complex is responsible for vesicle membrane nucleation, whereas Atg14 a binding partner in this complex is involved in membrane stabilization [14, 15]. Other proteins and lipids involved in phagophore nucleation and membrane formation are explained in detail

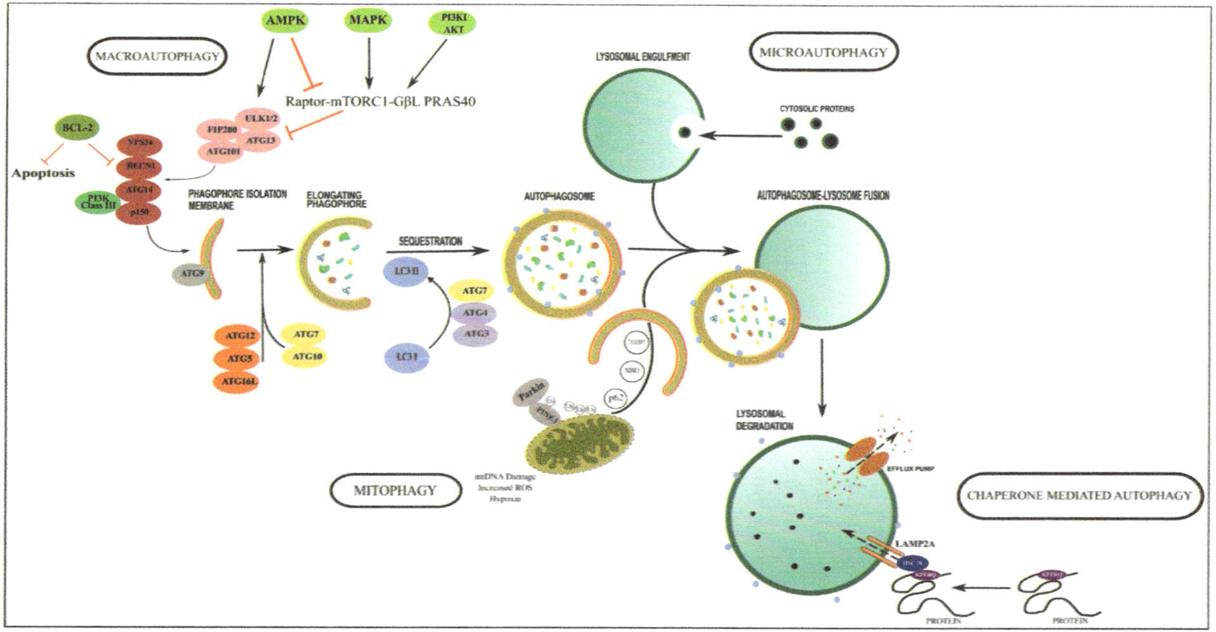

Fig. 1.1 Schematic representation of macroautophagy, microautophagy, chaperone-mediated autophagy and mitophagy processes: The figure depicts core autophagy proteins involved in nucleation, elongation and completion of phagophore to form the autophagosome. Sequestration of cargo into the autophagosome is coupled with autophagosome formation and fully formed autophagic vesicles and lysosomes fuse to facilitate subsequent degradation of cargo. AMPK, MAPK, mTOR complex and PI3K, Akt modulate autophagy depending on the upstream inputs received by intracellular and extracellular cues. The membrane formation requires VPS34, PI3K Class III, Becn1, Atg14 and p150 complex formation. Atg9 is attached to the nucleation membrane and cargo is sequestered for engulfment. Elongation is mediated by Atg7, Atg3, Atg10, Atg12-5-16L complex and the lipidated LC3-II (blue dots on autophagosome). Lysosome is fused with the autophagosome to form the autolysosome. Sequestered material is degraded in the lysosome and recycled by efflux pumps. Depolarized mitochondria are sequestered in a similar fashion to the autophagosome by an orchestrated mechanism involving kinases, phosphatases and ubiquitination. p62, TAX1BP1 and NBR1 cargo receptors are usually involved in the sequestration of mitochondria. Parkin ubiquitinates PINK1 and other OMM proteins, giving out a signal to the cargo receptors. Mitochondria and other bulk autophagy cargo are degraded via lysosomal hydrolases (black dots)

by Melia and colleagues [7, 14, 15]. Elongation of the phagophore membrane into autophagosome is mediated by two ubiquitin-like conjugation systems. These are facilitated by Atg7, Atg3, Atg10 that act like E1, E2 like enzymes and Atg4 a protease that is necessary for activation of LC3 family proteins [16–19]. Atg12 is activated by El like Atg7 and, transferred to E2 like Atg10 to form a complex with Atg5. This tethers with Atg16L to form the Atg12-Atg5-Atg16L complex which acts as an E3-like enzyme for LC3/GABARAP/Atg8 [20]. LC3/GABARAP are involved in the sequestration and tethering of the autophagosome with sequestered cargo by its LIR motifs. LC3-I/GABARAP-I/Atg8-I is cleaved to form LC3-II/GABARAP-II/Atg8-II by Atg4 [13, 18, 21, 22]. LC3II/GABARAP-II/Atg8-II, the active form, attaches to the phagophore membrane with its LIR motif that interacts with the sequestered cargo via autophagy receptors like p62, NBR1, OPTN, NDP52, TAX1BP1 and TOLLIP, and tethers the autophagosome membrane. LC3II/Atg8-II is tethered to phosphatidylethanolamine present in the autophagosome membrane covalently through activities of Atg7, Atg3 and the Atg12-5-16L complex. LC3-I/GABARAP-I/Atg8-I are the only proteins that are lipidated during autophagy [23]. Tethering of LC3II/GABARAP-II/Atg8-II to the autophagosome has been a widely used biomarker to monitor autophagy flux [24–26]. Autophagy receptors like p62, attach to LC3II/GABARAP-II/Atg8-II via LIR and simultaneously interact with poly-ubiquitinated cargo thus facilitating their recruitment to the autophagosome. Lysosome fuses with cargo-laden autophagosome forming autolysosome during macroautophagy. Lysosomal hydrolases and cathepsins degrade the cargo, and degradation products such as amino acids, simple sugars and fatty acids are transported to the cytoplasm [7, 27]. Effects of autophagy proteins in maintaining stem cell homeostasis are described in detail in many reviews [1, 28–30] (see Table 1.1).

Techniques for Assessing Autophagy Orchestration

Autophagy flux is a total measure of cargo sequestration from uptake to degradation within the lysosomes in cells per unit time. One of the ways to measure the flux is to quantify the amount of protein that is degraded via autophagy. This can be done using several methods from conventional radiometric turnover assays of long-lived proteins to fluorescently labelled proteins assaying for their degradation. The fundamental guidelines to perform assays for monitoring autophagy components are discussed briefly in Refs. [71, 72]. In the following sections, we describe the tools, and discuss and compare the techniques used for monitoring autophagy exclusively in stem cells.

Table 1.1 Representative figure of immunofluorescence assays (A) LC3-II (blue dots) and SQSTM1/p62 (red dots) antibody tagging with respective antibodies with their mechanism of action. (B) mCherry and GFP tagging of protein LC3. (C) Tandem reporter mRFP/mCherry-GFP-tfLC3 is a dual fluorescent autophagosome and autolysosome tagging probe. GFP-mRFP combination emits yellow fluorescence, whereas RFP emits red. (D) Improvement over the mCherry/RFP-GFP-tfLC3 reporter, a reporter with internal control GFP-LC3-RFP-LC3ΔG. (E) Bimodal excitation providing Keima (pink dots) dual fluorescent probe, with excitation at 480 and 586 nm with 620 nm emission spectra. (F) Tracking lysosomes in stem cells, by using Lysotracker, LAMP1/2 and CathepsinL biomarkers. (G) Tracking mitophagy by variants of Mitotrackers, MitoSOX Red and Orp1

S. no.	Type of assay	Advantages	Limitations	Stem cell types	References
1	LC3 and p62/SQSTM1 tagging 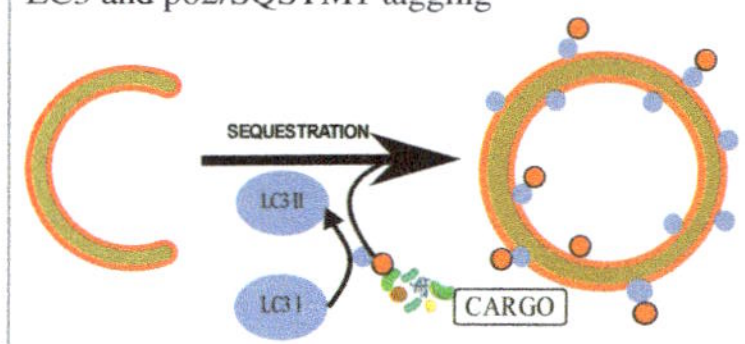	Antibody specific for core autophagy proteins tagged along with fluorophores. Easy estimate of autophagy-related proteins like LC3-I and p62	Lower resolution of images. Non-specific binding of secondary fluorophore antibodies. Qualitative estimation and localization of protein, no quantitative determination of the protein	Muscle stem cells Haematopoietic stem cells Mouse germ cells Germline stem cells (female) Human embryonic stem cells Neural stem cells Skeletal Muscle stem cells	[31, 32–36]

(continued)

Table 1.1 (continued)

S. no.	Type of assay	Advantages	Limitations	Stem cell types	References
2	mCherry and GFP tagging 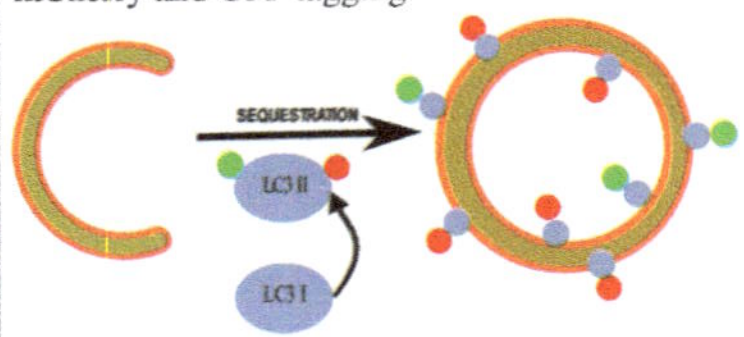	Bioluminescent protein tagged with autophagy-related gene. Endogenous expression and estimation of protein localization. No requirement of externally tagged fluorophore	Single protein expression can be observed at a time. For example,: mCherry/GFP attached to LC3-I or p62 cannot be observed simultaneously due to the limitation of similar emission spectral range and the possibility of bleed through	Mouse embryonic stem cells Human embryonic stem cells Induced pluripotent stem cells Haematopoietic stem cells Neural stem cells Muscle stem cells Mesenchymal stem cells Germline stem cells	[28, 29, 37, 38, 39, 40, 41, 42, 43, 44, 45 46–]

(continued)

Table 1.1 (continued)

S. no.	Type of assay	Advantages	Limitations	Stem cell types	References
3	mRFP/mCherry-GFP-tandem reporter 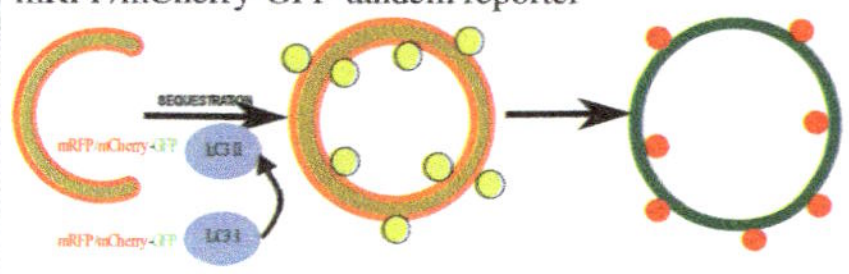	Dual fluorescence. Yellow puncta formed indicates autophagosome formation due to the mRFP and GFP colocalization. GFP fluorescence quenches in acidic pH, an advantage to distinguish between autophagosome and autolysosome	Does not allow distinguishing between LC3-I and LC3-II localization. GFP sensitive to lysosomal pH. Cannot be coupled with other cellular biomarkers	Cancer stem cells Mesenchymal stem cells (BM) Germ cells Neural stem cells/induced neural stem cells	[47, 48, 34, 49, 50–52]
4	GFP-LC3-RFP-LC3ΔG reporter 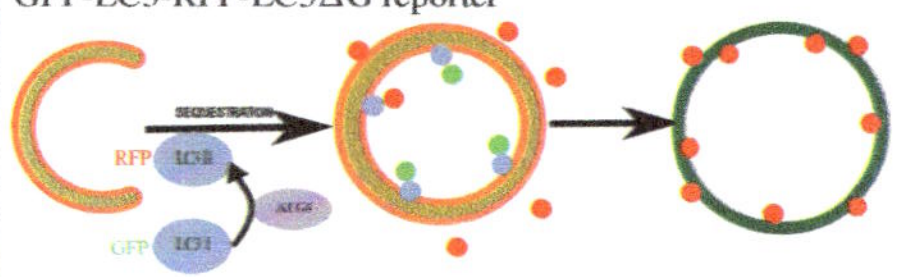	Advanced dual coupled fluorescence reporters with RFP-LC3-I as internal control. Provides a quantitative estimate of LC3-I to LC3-II. Tracks LC3-II localization in autophagosome and autolysosome	Allows only observation of LC3-II tagged autophagosomes or autolysosomes. Does not take into account other autophagy markers	–	[26]

(continued)

Table 1.1 (continued)

S. no.	Type of assay	Advantages	Limitations	Stem cell types	References
5	Keima dual fluorescent reporter 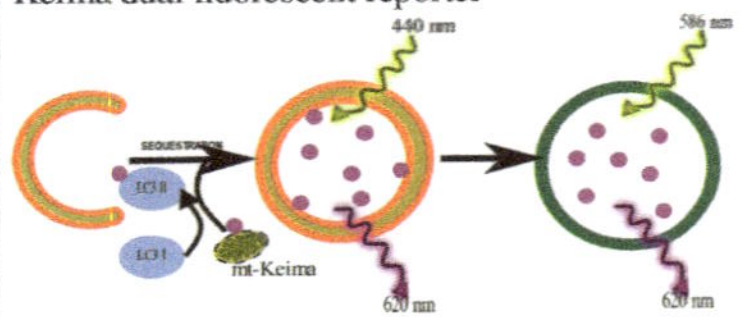	Single protein with bimodal excitation and dual emission spectrum. Insensitive to lysosomal degradation. Provides accurate localization of tagged autophagy-related protein/sequestered material (e.g. Mitochondria)	Autophagy substrates cannot be directly observed. Quantitative estimation of proteins involved in the autophagy process is not possible. Can be utilized only in live imaging set-ups.	HCN stem cells Induced pluripotent stem cells	[40, 53]
6	Tracking lysosomes 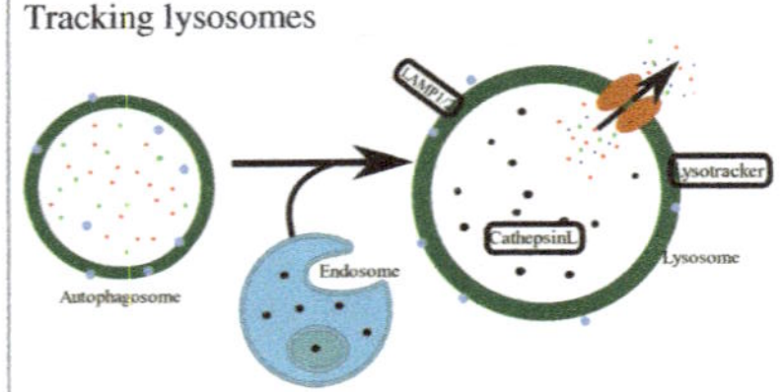	Employs fluorescence-emitting dyes, compounds and proteins to track lysosomes, the crucial vesicle of autophagy and cellular degradation processes. Insensitive to the lower lysosomal pH	Requires a static imaging sample. Cannot specifically estimate autophagic degradation turnover since lysosomal degradation cannot be specific to autophagy cargo	Induced pluripotent stem cells Quiescent Neural stem cells Follicle stem cells Germline stem cells	[40, 48, 43, 54]

(continued)

Table 1.1 (continued)

S. no.	Type of assay	Advantages	Limitations	Stem cell types	References
7	Tracking mitophagy 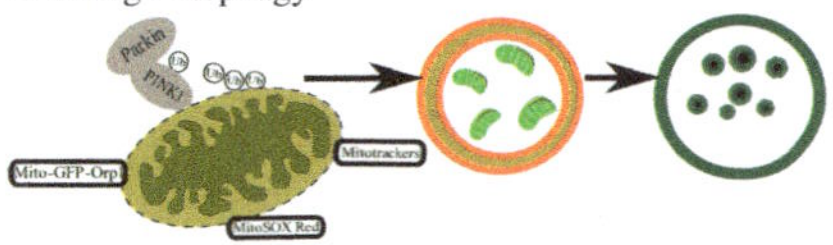	Tags proteins coupled with membrane or localized inside mitochondria. Fluorescent reporters, dyes and endogenous coupled fluorescent markers are the popular choice	Quality of mitochondrial visualization depends on the quality of the compound, bioluminescent protein or dyes employed. Allows static visual observations. Requires coupling with autophagy biomarker to observe the possible mitophagy	Haematopoietic stem cells Neural stem cells Induced Pluripotent stem cells Germline stem cells	[39, 55, 40, 56, 57, 5843 53,]

(continued)

Table 1.1 (continued)

S. no.	Type of assay	Advantages	Limitations	Stem cell types	References
8	Western blotting/LC3 turnover assay 	Quantitative estimation of proteins involved in autophagy process. Protein–protein interactions can be estimated with immunoblots and pull-down assays. Popular choice is to quantitate LC3-I and LC3-II ratio, indirect estimation of autophagic degradation turnover	Does not allow visual localization of endogenous protein. Requires cell lysate preparation, relatively tedious and time-consuming to quantitatively estimate protein expression levels	Mouse embryonic fibroblast/induced pluripotent stem cells Muscle stem cells Neural stem cells Haematopoietic stem/progenitor cells Mesenchymal stem cells Germline stem cells	[29, 59, 40, 60, 62, 33–35, 46–43, 60, 62, 44, 60, 62–6566 41,]

(continued)

Table 1.1 (continued)

S. no.	Type of assay	Advantages	Limitations	Stem cell types	References
9	Flow cytometry 	Allows autophagy estimation in a heterogenous cell population. Distinguishes and sorts specific cell population with fluorescent-tagged autophagy biomarkers Provides a rough estimate of the amount of autophagic degradation in a cellular population. Streamlined quantitative protein estimation evaluations	Does not allow visual localization of endogenous protein. Quantitative estimation of mRNA levels or protein expression is difficult	Haematopoietic stem cells Neural stem cells Mesenchymal stem cells HCN stem cells Cancer stem cells	[32, 67, 53, 68, 62, 44, 60, 64, 69]
10	Electron microscopy	Higher magnification and greater resolution imaging and analysis. Visualization of autophagosome structures and cargo sequestered for degradation. Estimation of protein interactions and localization using immune-gold electron microscopy technique	Fails to quantitatively estimate protein levels. Time-consuming sample preparation and only allows observation of autophagic structures in static environment	Embryonic stem cells Mesenchymal stem cells Haematopoietic stem cells Neural stem cells Induced pluripotent stem cells	[37–55, 69, 51, 69]

Electron Microscopy Imaging

A solid inference is based on concrete visual observation of any biological entity as it provides evidence in support of a hypothesis. Biological research is vastly based on observing molecules using several probes and biomarkers by microscopy. Electron microscopy is a robust, high-resolution imaging technique required to assess the morphological construction and appearance of the autophagy process and its constituent structures in vivo. Eventually, after the discovery of lysosomes by Christian de Duve in 1940, electron microscopy was employed to image lysosomes, endosome and autophagosome structures to understand their physical nature [5]. Electron microscopy imaging has been successful in deciphering the structures of initial autophagic vacuoles and their differentiation with degradative autophagy vacuoles later renamed as autophagosomes and autolysosomes, respectively [37]. Advances in Scanning (SEM) and transmission electron microscopy (TEM) have been successful so far in stem cell research, and it has immensely contributed to our understanding of the autophagy process in stem cells [59, 47, 38].

TEM is crucial for autophagy research as it is the only technique that allows to visualize autophagic vesicles in the nm range. Most importantly, TEM along with appropriate sampling strategies can be used for quantifying autophagy [39]. TEM can reveal sequential changes in the morphology of autophagic vacuole in both selective and non-selective autophagy. This is usually achieved by following the size and shape of autophagic vacuoles as well as the stepwise degradation of cytoplasmic structures [55, 31]. Typically, the destruction of cytoplasmic organelles leads to an increase in electron density, then to vacuoles with moderate heterogenous density which finally become more homogenous and amorphous electron-dense vacuoles [40, 60]. Using fusion inhibitors like Bafilomycin A1, it is possible to chase the formation of autophagosomes. It is possible to identify cargo within the autophagic vacuoles using immuno-TEM followed by gold labelling [37, 69]. This technique uses antibodies specific to LC3 (or anti-GFP antibodies marking GFP-LC3) to mark the autophagosomes and antibodies specific to proteins that are suspected as cargo [62, 48]. One can more specifically identify autophagic bodies as compared to conventional TEM alone, however, the immune-TEM with gold labelling needs extensive standardization.

For quantification of data using TEM, it is important to use proper volumetric analysis with the limitation that such studies reflect steady-state autophagy levels. However, it is still possible to measure autophagy flux if the experiments are done in the presence and absence of autophagy inhibitors at different time intervals [73]. To get a better insight, it may be useful to perform whole cell quantification of autophagosomes using fluorescence microscopy along with qualitative verification with TEM [74]. New advances in electron microscopy like STEM, Cryo-EM, CLEM and other potential techniques used for high-resolution imaging of autophagy degradation have been discussed extensively elsewhere and have to be employed to understand autophagy involvement in stem cells. EM although efficient in providing high-quality images requires huge setups, constant maintenance of vacuum in the tube, time-consuming specimen preparation and cannot be used to understand the localization of multiple target proteins [75]. Some of the limitations of TEM could be overcome by focused ion beam scanning electron microscopy (FIB-SEM) which eliminates the sample processing treatments involved in TEM [76]. It is important to identify autophagosomes correctly, not all double-membraned structures are autophagosomes. For example, autophagosomes have a characteristic distended empty space between the two membranes. Measuring autophagic structures per cell area is a more reliable technique and shows a better correlation with measurements done using other techniques. Tomographic reconstruction of TEM images can confirm the spherical nature of autophagosomes thus allowing to rule out other possible structures like inner membrane cisternae or damaged mitochondria within cells. To overcome these limitations of TEMs, researchers utilize fluorescence-based microscopy techniques to visualize endogenous protein localization.

Fluorescence Microscopy

This technique utilizes fluorophores directly or indirectly to monitor Atg8, LC3, GABARAP and associated proteins including Atg-related proteins during autophagy. LC3 family proteins (LC3/GABARAP/Atg8) can be detected directly using antibodies that include both monoclonal and polyclonal. LC3 version fused with fluorescent proteins can be easily detected in fixed cells and tissues as well as in live imaging.

LC3 Antibody Tagging

LC3/GABARAP family of proteins (Atg8s) consists of 6 proteins in mammals, LC3A, LC3B and LC3C variants proteins, whereas the latter consists of GABARAP, GABARAP1 and GABARAP2 (also known as GATE-16) in eukaryotes [3]. Lower order eukaryotes contain Atg8 homologues (Atg8a and Atg8b) that are involved in cargo recognition, engulfment and sequestration of autophagosomes to lysosomes. LC3 being the only facilitator of cargo to vesicles is widely used as a biomarker to monitor autophagy flux [24, 72]. Several monoclonal and polyclonal antibodies are available for immunofluorescence detection of LC3 using confocal microscopy, and it offers several advantages. It obviates the need for transfection or generation of transgenic animals expressing tagged LC3 and also avoids artefacts associated with overexpression of the protein [32–34]. LC3-II appears as punctate structures in confocal microscopy, and changes in the number of LC3-II puncta can be used as an indicator of induction or inhibition of autophagy [24, 35, 36]. However, in some cases a ratio of LC3-II signal intensity to the total LC3-I intensity may be used. Quantification of LC3-II puncta can be done using several image processing and analysis software (ImageJ and Imaris). LC3 variants being the only proteins attached to autophagosomes and autolysosomes are successful in tracking their localization, however, they also have limitations. LC3 antibodies provide an estimate only of the membranes required for autophagy-mediated degradation asthey do not label the sequestered cargo. To observe the cargo localization in vivo, it is necessary to observe the localization of autophagy receptor proteins attached to the proteins targetted for degradation. Hence, autophagy sequestration cannot be only studied by LC3-II localization [32]. For example, tracking cargo sequestered to the autophagosome and estimating colocalization of LC3 and p62 would facilitate relatively accurate measurements [37, 77]. One more disadvantage is that endogenous proteins may not be detected if the expression is very low (below the level of detection). To overcome this, LC3 proteins are typically fused with fluorescent proteins and expressed in cells and transgenic animals.

Fluorescent Protein Tagging of LC3

LC3 tagged with fluorescent proteins like GFP (RFP, mCherry, BFP, etc.; GFP-LC3, GFP-Atg8 or GFP-GABARAP) at the N-terminal can be used to monitor autophagy by immunofluorescence or direct fluorescence microscopy [78]. Such an approach is desired when endogenous LC3 levels are below detectable levels. Autophagosomes can be detected as green fluorescent dots (in case of GFP-LC3) localized in distinct areas of the cells/tissues. GFP-LC3 (RFP/mCherry/BFP) can be used for live imaging of cells/tissue and monitoring autophagy in real time [28, 37]. This is very useful when autophagy is to be monitored under different stressors or during the differentiation of cells such as iPSC or ESC differentiation [38, 46]. With advancements in knock-in approaches and genome editing recombinant technologies, GFP-LC3 constructs have been employed in varied types of stem cells, to understand cross-talks, autophagy activators, repressors and regulators in stem cells [46,49, 66, 79, 67].

There are several caveats that are associated when using GFP (RFP/mCherry/BFP) tagged LC3 for tracking autophagy. When generating cells expressing GFP-LC3, it should be noted that transfection reagents increase autophagy [41, 80]. Transfection should be performed with low levels of constructs. It is crucial to achieving uniform levels of GFP-LC3 expression within cells and tissues. This may require optimization of the duration to allow optimal expression of GFP-LC3 in cells. In the case of differential GFP-LC3 expression, it is recommended to normalize the intensity of GFP-LC3-II present in the puncta to the total GFP-LC3 intensity. High levels of GFP-LC3 can result in artefacts such as nuclear localization. GFP-LC3-II puncta counts need to be expressed per cell or on a per cell area basis. Automated counting approaches may be misleading, hence, it is recommended to perform manual scoring of LC3-II puncta. GFP-LC3 may associate with protein aggregates and can bias the counts. Typically, GFP-LC3 puncta represent a mix of ubiquitinated protein aggregates in the cytosol, ubiquitinated protein aggregates in the autophagosomes and phagophores [81]. GFP is sensitive to acidic pH and hence when the autophagosome and lysosome fusion occurs, GFP is quenched due to which it becomes difficult to monitor autolysosomes [82]. mRFP/mCherry fluorescent proteins are resistant to acidic pH and hence the fluorescence emitted by mRFP/mCherry-LC3 fusion proteins is not quenched in autolysosomes [78, 83, 84], thus offering a better alternative to GFP-LC3. This property of mRFP/mCherry-LC3 allows for differential counts of autophagosomes and autolysosomes [29, 40, 42, 43]. mCherry is an efficient and reliable fluorescent protein marker, however, it fails to track the complete autophagy degradation from cargo engulfment to recycling from lysosomes. It fails to present the difference between autophagosomes and lysosomes and lacks in revealing the complete process. To examine the process in a comprehensive approach, it is obligatory to visualize autophagosomes and autolysosomes in the same milieu [32]. To achieve this, the employment of two separate fluorophores in a simultaneous routine is a necessity.

mRFP/mCherry-GFP-LC3 Tandem Reporter Constructs and Their Advancements

To measure autophagy flux, Yoshimori lab took advantage of dual fluorescent labelling comprehending pH-sensitive GFP and pH-stable RFP construct, i.e. mRFP-GFP-tfLC3 (tandem fluorescence tagged LC3). Colocalization of GFP and RFP produces yellow puncta which indicate the formation of autophagosomes, whereas individual RFP red puncta indicate autolysosome/lysosomal compartments [26]. This reporter allows for the estimation of both induction of autophagy and the flux via autophagy simultaneously. However, the actual degradation of the cargo has to be monitored in the presence of lysosomal protease inhibitors such as E64d. mRFP/RFP can be replaced by mCherry. An added advantage of these reporters is that they could be used in high throughput screens and are amenable to automation necessary for large image-based screens [34, 49, 79]. Chloroquine and Bafilomycin A1 can be used to prevent the formation of autolysosomes which provides additional control for measuring the extent of autophagosome formation [47, 60, 48, 50]. Rosella pH biosensor can be used as an alternative to complement the mRFP/mCherry-GFP-LC3 reporter [51, 52]. Rosella is a fusion between pH-stable fast-maturing RFP variant and pH-sensitive GFP variant and has been successfully used in yeast as well as mammalian cells. These sensors can be very well adapted for flow cytometry-based measurement of autophagy flux [85].

A few precautions need to be taken when using these reporters for flux assays. Overexpression of these proteins can themselves target the proteins for degradation. Optimization of these expression levels or the use of different fluorescent proteins such as TagmRFP and mWasabi is useful [25]. When using fixed tissues or cells, it is important to maintain an acidic environment as the use of pH neutral or basic fixatives may lead to restoration of the GFP signal thus complicating measurement of flux. An alternative to this is to use a more acid-sensitive version of GFP line pHluorin [84]. When performing live imaging, photobleaching becomes a major issue and needs to be appropriately addressed. In cases when the drug or the gene of interest affects proteolytic degradation, the colocalization of GFP and mRFP/mCherry increases which can lead to misinterpretation. In some instances, the overexpression of these reporters can lead to toxicity or cell death. In such cases, appropriate controls need to be included and alternative fluorescent protein fusion reporters should be used.

GFP-LC3-RFP-LC3ΔG

Despite the coupling of GFP with RFP in the tandem reporter, it does not take into account the activation of LC3-I to LC3-II. To overcome this conundrum, **GFP-LC3-RFP-LC3ΔG** is one of the finest probes developed by Mizushima lab that provides an alternative approach to measuring the flux. Atg4 (Atg4A, Atg4B, Atg4C and Atg4D in mammals) are endopetidases that cleave the terminal amino acid(s) residue from LC3/GABARAP/Atg8 which exposes reactive glycine and facilitates transfer to E1-like enzyme, Atg7. Based on this property of Atg4, cleavage at C terminal of GFP-LC3-RFP-LC3ΔG fusion protein releases equal amount of RFP-LC3ΔG and GFP-LC3 [86]. RFP-LC3ΔG acts as an internal control for autophagosome localization (GFP-LC3-RFP works equally well). GFP-LC3 gets lipidated onto the phagophore, and autophagosomes are degraded. GFP/RFP signal ratio provides an estimate of autophagy flux [86]. Tandem fluorescent proteins fused to LC3 provide a convenient way of measuring autophagy flux. However, it is best to perform an independent autophagy flux assay in addition to the tandem fluorescent proteins-LC3-based assays.

Keima

Keima, a *Montipora* sp. Coral-derived fluorescent protein is resistant to lysosomal degradation, and has a bimodal excitation of 440 and 586 nm depending on its ionized and neutral state. It has a single emission peak at 620 nm. At pH 7.0, Keima has an excitation peak at 440 nm, and in acidic conditions (i.e. lysosomal compartment) its excitation peak shifts to 586 nm. A ratio of fluorescence at 550 nm to that at 438 nm provides a measure of degradation of cargo through the lysosomal system [87]. Keima can be expressed in the cytosol, and its delivery to the lysosomes can be tracked to measure non-selective (bulk) autophagy [88, 89]. Keima can be fused to a specific protein (such as mitochondria-targeted Keima) which is indicative of selective autophagy [40, 53]. Keima cannot be used when cells/tissues are fixed since the shift in excitation of Keima relies on the acidification of lysosomes. Keima is advantageous for developing strong probes for live monitoring of turnover of autophagy-related proteins. For further reading regarding Keima and its modulations, refer to Ding and Hong [90].

SQSTM/p62/Ref(2)P Tagging

SQSTM1/p62 (Ref(2)P in Drosophila) is a common autophagy cargo protein involved in sequestering and recognition of damaged, unfolded and accumulated bulk proteins/organelles to autophagosomes [72]. Monitoring p62 localization in vivo along with LC3/GABARAP/Atg8 can serve as a measure for autophagic degradation of cargo. SQSTM1/p62 has the ability to bind several polyubiquitinated substrates including RNA and can serve as an indicator for RNAphagy. Importantly, steady-state levels of SQSTM1/p62 reflect the autophagic status as disruption of autophagy leads to the accumulation of this protein within the cytosol [91, 92, 93]. Conversely, upregulation of autophagy typically leads to a decrease in SQSTM1/p62 levels. Together with CQ treatment, SQSTM1/p62 can be used to measure autophagy flux. Stem cell studies have utilized p62 biomarker for understanding degradation turnover of autophagy (Table 1.1).

SQSTM1/p62 levels are different in different cell types, and changes in the levels of this protein are context-dependent [67, 42, 43, 54]. However, in some cases transgenic expression of SQSTM1/p62 does not reflect the changes of the endogenous protein. Another complication is that SQSTM1/p62 is transcriptionally upregulated under certain conditions of stress. In certain cases of aneuploidy, both mRNA and protein levels of SQSTM1/p62 are altered, however, the flux remains unaltered. It is therefore recommended to monitor mRNA as well as protein levels of SQSTM1/p62 along with appropriate controls that include transcriptional inhibitors. SQSTM1/p62 may be chased using radioactive pulse-chase experiments when other methods can't be reliably used. SQSTM1/p62 is a large molecule with multiple domains, and these have multiple functions in several processes within the cell. For instance, SQSTM1/p62 can interact with Keap1 through its Keap1 interacting domain during oxidative stress. One important aspect to note is that overexpression of SQSTM1/p62 can lead to the formation of inclusion bodies and this could create problems in measurement of flux [94]. SQSTM1/p62 is also degraded via the proteasome, and this must be accounted for when performing autophagy assays that include disruption of the proteasome [95]. While performing western blots for SQSTM1/p62, it is important to assure that lysates are prepared in a buffer containing 1% SDS. To conclusively establish SQSTM1/p62 degradation by autophagy, a time course of SQSTM1/p62 levels at 6 and 24 h (and up to 48 h) after autophagy induction is advisable [96]. In cases where SQSTM1/p62 fails to be a reliable marker, it may be prudent to track other autophagy receptors such as NBR1, Optineurin, TAXBP1, CUET and NDP52. Cargo recognition proteins like TAX1BP1 and NBR1 can also be recruited as biomarkers to observe their localization to selective autophagy cargo degradation. Their localization in adult NSCs is an estimate of the cargo sequestered for degradation [92]. Several other cargo protein biomarkers are involved in selective autophagy processes. Although not being studied in stem cells, they are extensively described by Klionsky et al. [72]. Thus, it is advisable to conduct SQSTM1/p62 assays in combination with LC3/GABARAP/Atg8 assay and not solely depend on SQSTM1/p62.

So far, we have discussed tagging the crucial autophagy proteins involved in autophagosome sequestration, maturation and cargo recognition. However, tracking LC3/GABARAP/Atg8 and p62 with respective antibodies does not always provide accurate resolution for imaging. Fluorescence quenching due to drastic changes in pH/temperature during experimentation and non-specific binding of the antibodies can lead to false results. Hence, several fluorescent protein tagging techniques have been developed to ameliorate the problem of lower resolution. So far we have discussed monitoring global autophagy within cells. While monitoring global autophagy levels within cells is an important step to understand autophagy, certain situations may require evaluation of specific steps/stages of autophagy.

Evaluation of Specific Steps During Autophagy

Several years of research has suggested that the Ulk1 complex is directly or indirectly regulated by several upstream kinases. One of the most important kinases is mTORC1 which negatively regulates the ULK1 complex by phosphorylating Ser 757. This inhibitory phosphorylation can be tracked on western blots using phospho-specific (Ser 757) antibody of Ulk1. It should be noted that mTORC1 itself is regulated by multiple upstream kinases that are involved in nutrient sensing, energy levels and growth of cells/tissues [12]. Ulk1 complex can phosphorylate several downstream effectors that can also be monitored for autophagy induction. These downstream effectors include Ulk1 complex proteins like Atg13, RB1CC1, Atg101, Atg6 complex proteins including Vps34, Becn1, Atg14, AMBRA1, core autophagy proteins such as Atg4, Atg9 and Atg16, and several other non-autophagy proteins such as AMPK, p62/SQSTM1, FUNDC1, MAPK14, TBK1, and Sec16A (an exhaustive list is available in Klionsky et al. [72]). In certain cases, relocalization of certain proteins may be tracked upon induction of autophagy. For instance, Atg13 relocates to the omegasome when ULK1 is activated. It is important to note that decreased mTORC1 activity may not lead to the induction of autophagy *per se*. Also, autophagy induction can occur independently of Ulk1. A low abundance of Ulk1 may make it difficult to analyse phosphorylation levels [97].

Vps34 activity can be tracked to monitor autophagy induction in its initial stages. One way to test Vps34 activity is to immunoprecipitate the protein from cells and test for in vitro phosphorylation using Phosphatidyl-inositol and radiolabelled ATP. However, due to the limitations of this assay, it is recommended to use liposomes that mimic the lipid composition of the autophagosome. Atg6/Beclin1 is in a complex with Vps34, Vps15 and Atg14 and the activity of this complex can be monitored by in vitro phosphorylation of Phosphatidylinositol [98]. To achieve this, Atg6 and its partner proteins can be immunoprecipitated as a complex. In addition to this, Atg6-GFP puncta formation can also be used to monitor autophagy induction. However, Atg6 can localize to endosomes, and hence caution should be exercised when using this assay for monitoring autophagy.

Atg9 is a membrane protein that relocalizes to phagophore assembly sites from peripheral reservoirs upon Ulk1 and Atg13 activation. GFP-Atg9 can be used to track changes in its localization which can be used as a measure for autophagosome formation [99].

Atg12-Atg5 and Atg16 localize to the phagophore from the cytosol upon autophagy induction and can be detected by fluorescence (if tagged to fluorescent proteins like GFP, RFP etc.) and immunofluorescence. The conjugation of Atg12 to Atg5 can be monitored and used to measure autophagy. Tracking the formation of the Atg12-Atg5-Atg16 complex is an indirect way of evaluating autophagy [100]. Atg16 is a substrate of Atg1 and is phosphorylated at Ser 278 under autophagy-inducing conditions in cells and can be detected using a phospho-Atg16 antibody in immunofluorescence assay or using Western Blotting technique. This method is recommended as phosphorylation can be monitored and is a reliable way to identify autophagy induction. This is the recommended method as the phosphorylation of Ser 278 by Ulk1 is not affected by late-stage autophagy block which is one of the limitations of the LC3-based assay. However, like other proteins it may be altered independent of autophagy, and this assay should be performed in addition to other assays to monitor autophagy [101].

Atg14 localizes to phagophores and can be tracked using antibodies or a GFP-tagged version of Atg14. However, Atg14 can also localize to ER and autophagosomes. Hence, an autophagosome marker should be included to identify a pool of Atg14 positive phagophores [102].

WIPI (Atg18 in lower organisms) is required for autophagy and can form complex with Atg2 at the ER-phagophore junction during omegasome formation. IN general WIPI proteins bind PI3P and accumulate on membranes that build autophagosomes and omegasome. Quantitative fluorescence microscopy is routinely used to track Atg18/WIPI proteins to monitor autophagy. It is possible to measure punctate Atg18/WIPI under autophagy stimulating conditions where the numbers of such puncta are elevated. Atg18/WIPI puncta are reduced in conditions of autophagy inhibition and correlate well with LC3-II levels. Atg18/WIPI levels tend to be very low to undetectable and in some instances, overexpression of GFP/RFP/mCherry tagged Atg18/WIPI may lead to artefacts such as mislocalization [101].

STX17 is a SNARE protein that participates in autophagosome-lysosome fusion and can be detected on complete autophagosomes in most cases. In the case of starvation-induced autophagy, SXT17 has been shown to participate in omegasome formation [103].

Autophagy Visualization Dyes

Staining with acidotropic dyes like MDC, acridine orange, neutral red, lysosensor blue and lysotracker red can be used when tagging core autophagy proteins using fluorescent proteins is not possible or antibodies that recognize Atg proteins are unavailable. These fluorescent dyes are useful in identifying an increase in acidic compartments (lysosomal, endosomal) within cells, however, they should not be relied upon altogether. Lysotracker Red (DND-99) does not quench upon fixation and hence can be used where fixation by paraformaldehyde is necessary. The other lysotracker series of dyes (e.g. Lysosensor Blue. Lysotracker green) can only be used in live cell imaging. Standardization of concentration and incubation time of the dye is necessary for individual cell types and tissue types. Additional control is to monitor the incorporation of these acidotropic dyes in the presence and absence of CQ or HCQ. A commercially available dye for monitoring autophagy CytoID may be used. This dye has been validated in cell lines and localizes to vesicles which are RFP-LC3 positive upon amino acid derivation. It exhibits low background fluorescence and responds well to autophagy modulators [104, 105].

Lysosome-associated membrane proteins (LAMP) are predominantly found on lysosomes and are crucial for their biogenesis, hence tracking LAMP proteins can be used to monitor lysosome numbers and localization [40, 48, 54]. LAMP1/LAMP2 antibodies can be used to track lysosomes using immunofluorescence and flow cytometry techniques. Alternatively, GFP (mCherry, BFP)-LAMP1/LAMP2 fusion protein can be used to track lysosomes in IF, flow cytometry and live imaging studies.

Tracking Mitophagy in Stem Cells

Mitochondrial dysfunction in stem cells causes either massive loss of stem cells or accumulation leading to premature ageing and senescence. Further loss in mitochondrial homeostasis may lead to mtDNA mutagenesis, reactive oxygen species turmoil and increased hypoxia responsible for developing severe pathogenesis [106, 107]. Maintaining mitochondrial homeostasis is crucial for the normal functioning of the cell. To understand the molecular mechanisms involved in the accumulation of dysfunctional mitochondria, it is essential to decipher the mitophagy process. Several mitochondrial biomarkers have been employed in diversified stem cell types as probes for studying mitophagy. Based on in vivo data, 3 types of mitophagy have been described. Type 1 mitophagy involves the formation of phagophore, requires mitochondrial fission to occur and requires Vps34 complex including Becn1. Type 2 mitophagy is independent of Becn1 and is induced when mitochondria are damaged and requires LC3. Type 3 mitophagy is independent of Atg5 and LC3 but requires Pink1 and Parkin [108]. However, additional types of mitophagy have also been recently described that involve participation from FUNDCI and BNIP3.

TEM is a useful technique to track mitophagy as it can reveal details of mitochondrial membrane and cristae at nm levels allowing researchers to determine several aspects of mitochondrial health. TEM can be coupled with Bafilomycin or CQ treatment to get measurements of mitophagy flux [39]. Like Lysotrackers dyes, Mitotracker Red, Mitotracker Green and Mitotracker Deep Red can be used to track mitochondrial morphology as well as a number of different cell types [39, 53, 56]. Dyes which exhibit dual fluorescence in different redox states such as JC-1 and Mitotracker CMSRox can also be used to track mitochondrial redox potential [40]. These dyes in combination with appropriate lysotracker dyes can be used to track mitophagy using confocal microscopy and flow cytometry in live cells. However, it is important to note that autophagosomes form at the ER-mitochondrial contact sites, and hence this aspect needs to be accounted for when interpreting results. Antibodies that specifically bind mitochondrial proteins such as VDAC, TOMM20, SOD2, Hsp60 and mtHsp70 can be used to visualize mitochondria in primary cells or tissues [60]. To measure autophagic sequestration of mitochondria, researchers can use mitotracker dyes/mito-specific antibodies together with lysosomal dyes/lysosome specific antibodies. Additionally, GFP/RFP-LC3/GABARAP can be employed with mitochondrially targeted GFP (e.g. mito-GFP and mitochondrial resident protein fused to GFP/RFP/BFP) to measure colocalization [57, 58]. Mitophagy can also be quantified using mt-Keima using fluorescence microscopy or alternatively with flow cytometry. The peak of excitation of Keima shifts from 440 nm shifts to 586 nm when mitochondria containing autophagosomes fuse with the lysosomes. mito-QC reporter is another construct to efficiently track mitochondria undergoing fission and sequestration for degradation [40, 53, 90]. The mito-QC reporter comprises mCherry-GFP-Fis1 construct that localizes to mitochondria, allowing dual tagging of mitochondria in autophagosomes and autolysosomes [60].

It is also possible to monitor mitochondrial mass in the presence and absence of a mitophagy inducer. This can be achieved using flow cytometry coupled with mitotracker dyes and by immunofluorescence techniques using mitochondria-specific antibodies [109]. qPCR to measure mitochondrial DNA is yet another technique that allows for the measurement of mitochondrial mass. qPCR can be performed for MT-ND1 (mitochondrially encoded NADH dehydrogenase 1 or 2) and normalizing expression levels to nuclear DNA encoded pyruvate kinase M1/2 [110]. In some cases, mitophagy inducers may be necessary to measure mitochondrial degradation capacity and in such cases it may be essential to optimize the type and concentration of the inducer [111]. CCCP/FCCP are stronger inducers of mitophagy, while Antimycin A and oligomycin are milder mitophagy inducers. Mitotimer is a time-sensitive fluorescent protein that shifts fluorescence from green to red and can be used to track mitochondrial turnover, mitophagy and mitochondrial oxidative stress. Pink1-Parkin-dependent mitophagy could be detected using markers such as anti-parkin antibody and anti p-S65-Ub antibody. Translocation of Parkin from the cytoplasm to the mitochondrial membrane occurs upon induction of, in Pink1-Parkin dependent mitophagy S65 residue of ubiquitin is phosphorylated upon mitophagy induction, and this can be detected using anti-P-S65-Ub antibody [112]. It is important to use the appropriate maker when following mitophagy in cells and tissues. It is important to choose an inner membrane protein or a mitochondrial matrix protein when labelling mitochondria as this provides a more precise measurement of mitochondrial numbers [113]. Localization of Parkin to the mitochondrial outer membrane alone should not be used as a sign of mitophagy. Also, the disappearance of outer membrane proteins alone should not be considered as induction of mitophagy as these proteins VDAC1 and TOMM20 are also degraded by the proteosomal systems. PINK-Parkin-dependent mitophagy can occur independently of Atg-proteins. Thus, it is advisable to use several methods such as TEM, Atg-family proteins that colocalize with mitochondria and western blotting that can complement each other in deciphering mitophagy status in stem cells [37].

Atg8s/LC3 and p62 Protein Expression in Stem Cells

Western blot has been widely engaged to measure the expression levels of two of the most important autophagy proteins, LC3/GABARAP/Atg8s and p62 over the years. LC3/GABARAP/Atg8 and p62 levels are quantifiable using western blot, and their accurate analysis provides the closest clues to measuring autophagy flux. Different stages of autophagosome formation can be depicted from distinct LC3 levels in stem cells [68]. Since autophagy has been widely studied in stem cells, western blots have provided concrete means to understand expression levels of LC3 in a variety of stem cells. Atg8/LC3/GABARAP are expressed as precursor proteins and need to be activated by Atg4 (Atg4A, Atg4B, Atg4C and Atg4D) which cleaves terminal cysteine residue (in most cases) exposing glycine to be conjugated to phosphatidyl ethanolamine (PE). The PE-conjugated form /LC3-II/GABARAP-II/Atg8-PE (also called Atg8-II) shows faster mobility on SDS-PAGE as compared to the unconjugated form. The positions of unconjugated and conjugated Atg8/LC3/GABARAP are approximately 16–18 kDa and 14–16 kDa, respectively. LC3-II/GABARAP-II/Atg8-PE is the only protein that is reliably associated with completed autophagosomes [114].

It is important to note that both LC3-I and LC3-II levels display dynamic behaviour depending on the cell type and context. This is important as the basal autophagy may be different in different kinds of stem cells. Upon upregulation of autophagy, the levels of LC-I and LC3-II might change non-uniformly. An increase in LC3-II levels usually indicates an increased autophagic degradation, however, an increase in LC3-I and LC3-II, simultaneously, should be interpreted as accumulated autophagosomes and hindered autophagy [71]. Western blotting is able to detect the accurate levels of LC3-I and LC3-II in stem cells. It is recommended to use 4–20% gradient gels or 4–125 Bis–Tris gel using MES buffer as these allow for proper separation of LC3-I/LC3-II bands. It should also be taken into account that the levels of LC3-I and LC3-II are subject to change in different stem cell type, stage of differentiation and the kind of stress the cells are subjected to [40, 32, 46]. LC3-II can accumulate when autophagosome-lysosome fusion is disrupted by using vinblastine, by disrupting Calcium pump ATP2A/SERCA with thisparagon, by inhibiting V-ATPase by Bafilomycin A_1 or by raising lysosomal pH using CQ. However, stem cell cultures are often treated with modulators that affect signalling pathways which may affect autophagy flux (LC3-II levels) [60, 62, 44]. There are several limitations when following LC3/GABARAP/Atg8 transitions using Western blotting for measuring autophagy. Changes in LC3-II are cell type and context-dependent and in some cases, they may be independent of core Atg-proteins such as Atg6. LC3-II can associate with membranes that do not participate in autophagy. Standardization with appropriate controls is essential for evaluating LC3 by westerns. It may be necessary to directly process the cells or tissue in Laemmli buffer when preparing protein samples for proper extraction of LC3-I and LC3-II forms. Levels of LC3-II should not be compared with LC3-I, but compared to housekeeping proteins. Since mammalian stem cells have multiple isoforms of LC3, following LC3-II (LC3B-II) levels alone may not be an accurate way to measure autophagy. It is necessary to identify if anti-LC3 antibodies recognize one or more isoforms. This will enable to report specific isoforms that show differential expression during autophagy induction.

LC3 Turnover Assays

The assay is usually estimated as the difference between LC3-II levels in the presence and absence of autophagy modulators. Lipidated (LC3-II) and non-lipidated forms (LC3-I) of LC3 provide a more accurate estimate of the autophagosome formation/autophagy flux. A greater ratio should be interpreted as higher autophagy turnover, whereas lower numbers indicate basal or decreased autophagy turnover. However, as mentioned previously both conditions should also be subjected to the same concentration of autophagy modulators. The time of incubation and concentration of the autophagy inhibitor needs to be empirically determined for each kind of stem cell/stem cell population. However, it is recommnded to expose stem cells to Bafilomycin A1 and CQ for shorter duration [115, 116].

Time lapse microscopy with photo-switchable fluorescent proteins (such as PA-GFP) can be monitored by assessing the half-life of LC3 protein after photoactivation. This can be achieved by measuring autophagosome pool size and its transition time upon fusion with lysosomes. Alternatively, the formation of autophagosomes can be quantified. Autophagic flux can be monitored by following the turnover of LC3-II by utilizing a *Renilla sp.* Luciferase (Rluc). Autophagy-dependent turnover can be monitored by stimulating a change in the ratio of luminescence between cells expressing wild type LC3 fused to Rluc and cells expressing non-lipidable form of LC3 ($LC3^{G120A}$ mutant)[117].

Flow Cytometry and Fluorescence-Associated Cell Sorting (FACS)

For high throughput analyses involving cell populations, it may be advantageous to use multispectral flow cytometry. This is particularly advantageous when studying stem cell populations in vitro. Commercial kits and several methods for analysing autophagy using flow cytometry have been developed and are capable of simultaneous analysis in a varied range of specifications. Automated flow cytometric analysis can be coupled with imaging for distinguishing the desired population of cells expressing autophagy markers from a heterogenous group and sorting the cell types accordingly [118]. Several autophagy markers tagged with fluorophores in stem cells have been employed for their analysis using flow cytometry, especially for monitoring cell death rate. Additionally, fluorescent dyes such as lysotracker, lysosensor as well as genetically coded fluorescence proteins like EGFP/BFP/mRFP-LC3 can be used in flow cytometry assays. LC3 degradation can be monitored using total cellular fluorescence before and after induction of autophagy. It is also possible to deplete the LC3-I form by treating cell populations with saponin and measuring LC3-II form alone. In addition to these, tandem mCherry-EGFP-LC3 can be used to measure flux using flow cytometry [67]. GFP-SQSTM1 exhibits the largest magnitude of change upon autophagy induction and could be a better marker for assaying autophagy flux in cells. It is important to keep appropriate controls when assaying autophagy in adherent cells using flow cytometry. This is very important as embryonic stem cell is culture tend to form embryoid bodies. Alternatively, autophagic vesicles can be purified and can be used for measuring autophagy flux. Flow cytometry with multispectral imaging should be used with additional autophagy assays to determine flux.

Potential Probes for Monitoring Autophagy

So far, we have discussed and mentioned the methods employed to monitor autophagy using fluorescence-based techniques in stem cells, although there have been certain advancements in building novel constructs that could be potentially used in stem cells as well. The approach to quantifying long-lived protein degradation was isotope labelling, an in vitro assay. The study used Atg5-deficient human primary fibroblast, with a proteome-wide approach of isotope labelling by L-arginine and L-lysine by stable isotope labelling with amino acids in cell culture (SILAC). Protease degradation was observed using mass spectrometry and flow cytometry. Autophagy activity was found important for the degradation of long-lived proteins during quiescent to activated state change. CathepsinD enzyme activity increase during activation also suggested a high degradation rate of long-lived proteins by autophagy [119].

Although protein biomarker LC3 family attached fluorescent probes are efficient, withstanding lysosomal pH and producing a bright outcome is not possible. Miyawaki lab improved upon their previous Keima reporter to develop mito-SRAI (signal retaining autophagy indicator), a recent tool for monitoring mitophagy. The tool takes advantage of CFP tolerance of lysosomal environment (TOLLES) with GFP (mAG) monomeric Azami Green [88]. Although initially developed for mitochondria, this construct can be attached to the desired protein to observe its potential localization in the autophagosome and lysosome.

Jiang lab recently developed a novel clicking of organelle-enriched probes for tracking lysosomes, endosomes and autophagosomes (mitochondria). Tetrazine and trans-cyclo-octenol (TCO) undergo Deils-Alder reaction forming a bio-orthogonal compound. TCO is targeted to the lysosome via Lyso-BODIPY (boron di-pyrromethane), whereas tetrazine is targeted to mitochondria via Mito-rhodamine and endosomes via cholesterol units. Lyso-BODIPY-TCO reaction with tetrazine re-establishes the fluorescence by forming a pyrazine product with tetrazine quencher, i.e. Forster resonance energy transfer (FRET). This allows for a dual fluorescence as Lyso-BODIPY-TCO has greater emission in acidic conditions [120]. Several probes have been developed using fluorescent proteins and compounds for monitoring autophagy, however, their use in stem cells is sparse. Recent studies have shown constructing fluorescent molecule based probes specifically designed for monitoring selective autophagy, and are detailed in reviews [for more information please refer 90].

Gene Expression Assays

Fluorescence imaging, quantitative protein expression of autophagy process and morphological analysis of autophagosomes provide a useful measure of autophagy at the protein level. For a thorough examination of protein regulation, structural changes and modification, genomics should be coupled with these methodologies. Induction of autophagy also leads to an increase in mRNA levels of several Atg genes such as Atg1, Atg6, Atg7, Atg8, Atg9, Atg12, Atg13, Atg14 and p62 [49, 121]. Several of these genes are transcribed at higher levels upon autophagy induction, however, this is both cell type and context-dependent. Quantitative real-time polymerase chain reaction (qRT-PCR) and northern blots for assessing mRNA levels can provide data about autophagy induction. However, it is highly recommended to assess the mRNA levels at different time points during autophagy induction. This is due to the fact that mRNA levels are highly dynamic and change over course of time. Promoter analysis studies, chromatin immunoprecipitation and Atg-gene promoter luciferase assays can provide additional information on the transcriptional regulation of Atg genes. Certain transcriptional activators and repressors like TFEB, ZKSCAN3 and FOXO regulate several Atg genes and lysosomal biogenesis genes simultaneously [29, 60, 36]. Tracking such gene regulatory networks and systems biology approach (e.g. CLEAR—coordinated lysosomal expression and regulation, RNA-seq and microarrays) is useful in monitoring autophagy. It is important to note that not all Atg genes show altered mRNA levels and, that the changes in mRNA levels may be used as indicators of change in autophagy status but only when strongly supported by more robust assays.

Protein Interaction Databases

Over the years of biological research, molecular interactions have been studied from a large number of varied cell populations and whole organisms. A huge amount of data has been generated from these studies along with their genome sequences; annotated, curated and stored in distinct databanks or datasets. Data from these publicly available databases can be extracted and benefited from to understand biological systems as a whole in silico. One of the studies in stem cells used the STRING database to analyse potential FOXO3 binding partners in the autophagy process. FOXO3 was found to bind with WIPI1/2, ZFYVE, Ambra1, UVRAG, Atg9a, Atg2a/b, RB1CC1, Pink1, Atg13, LC3/GABARAPs, NBR1 and Ulk2, a wide range of proteins involved in autophagy regulation and orchestration [48]. Kyoto Encyclopaedia of Genes and Genomes (KEGG) pathway analysis was used to study the interrelation between different metabolic pathways as well as the upregulation and downregulation of proteins. In a distinct study by Maycotte et al., an analysis of quiescent NSCs in KEGG showed higher levels of lysosomes than activated NSCs, indicating a higher amount of cellular degradation prior to NSCs activation [60]. The aforementioned databases and several others coupled with computational in silico techniques of homology modelling, sequence similarity, protein interaction and modelling can be employed to predict the possibility of autophagy components in the desired species. These in silico genomic approaches are described in detail in Klionsky et al. [72].

Conclusion

This review discusses the techniques used to monitor autophagy in stem cells over the years. From the review, it may be inferred that Confocal microscopy has dominated over other immunofluorescence methods, due to its ability to visualize the autophagy process in a direct first-hand proof. Apart from the in vivo confocal imaging techniques, western blots and flow cytometry are powerful and equally reliable techniques that could be employed as an integrated approach. High-resolution imaging provided by electron microscopy is eminent to understand the morphological changes in the autophagy involved vesicles. Genetic analysis is crucial to understand the impact of transcriptional and translation regulators on autophagy and concomitantly on stem cells. This becomes important when using stem cell differentiation protocols. Emphasis on the effects of loss of autophagy in stem cells using different approaches underlines the cruciality of the process.

Stem cell research in itself is a huge arena. Autophagy has a concrete foundation of more than 40-years of research, however, its importance in stem cell maintenance, longevity and homeostasis has been recently gained importance. Regardless of the diversified stem cell populations, autophagy functions as a part of the anabolic and catabolic regulatory system within these cells. As has been discussed, impairment in autophagic degradation has devastating effects on stem cell populations. Not only macroautophagy but also other types of autophagy like chaperone-mediated autophagy modulation has also shown to affect stem cell metabolic regulation. Mitophagy has been especially emphasized due to the effects of impaired mitochondrial clearance on stem cell longevity, and in particular on stem cell maintenance and differentiation.

Effects of autophagy modulating drugs in certain cancer stem cell populations such as CML have shown promising drug targeting strategies. Whether similar effects will be observed on other cancer stem cell populations will be decided by further research in the future. Studies so far have marked the importance of autophagy in stem cells, and their pharmacological significance in engineering drugs to modulate autophagy to advance regenerative medicine and stem cell based therapies. Contemporary medical approaches like elevating rapamycin levels to induce autophagy have been employed in tissue grafts, organs and stem cell transplants to improve prognosis. These approaches can be improvised and advanced by targeted and precision therapies that are directed specifically towards stem cell populations.

There is no doubt investigations into autophagy have increased exponentially. Most of the methodologies and techniques used so far have been developed over a decade. Macroautophagy has been vastly explored in stem cells, whereas selective autophagy in animal models and tissue cultures has shown to be of great importance, perhaps its modulation in stem cells would uncover novel mechanisms involved in metabolic regulation. The review highlights techniques that can be employed to understand the autophagy process in stem cells. However, employment of each assay is dependent on the autophagic degradation stage under study, i.e. autophagosome membrane formation, autophagy flux and autophagosome-lysosome fusion; it is wise to integrate the approaches rather than isolate them on a single one for results. An approach that provides an accurate measurement of the whole process in a single assay has not been developed yet. Given the supreme influence of autophagy on stem cells understanding its diversified role in stem cell mechanisms, using advanced methodologies is pivotal. The future holds new possibilities for investigating autophagy processes in stem cells and regenerative medicine.

Acknowledgements This work was supported by grants from DBT-Ramalingaswami Fellowship, DST-SERB grant number CRG/2020/002716 and DBT grant number BT/PR/12718/MED31/298/2015 to BS. BVS is affiliated to Savitribai Phule Pune University (SPPU), Pune, India, and is recognized by SPPU as a Ph.D. guide (Biotechnology and Zoology). The authors would also like to thank Director, ARI, and the entire ARI fraternity for facilitating our research activities in the Developmental Biology Group.

Compliance with Ethical Standards

Disclosure of Interests All the authors declare they have no conflict of interest.

Ethical Approval This article does not contain any studies with human participants or animals performed by any of the authors.

References

1. Guan JL, Simon AK, Prescott M et al (2013) Autophagy in stem cells. 9:830–849
2. Chang NC (2020) Autophagy and stem cells: self-eating for self-renewal 8
3. Klionsky DJ (2011) For the last time, it is GFP-Atg8, not Atg8-GFP (and the same goes for LC3) 7

4. Klionsky DJ (2008) Autophagy revisited: a conversation with Christian de Duve 4
5. Ohsumi Y (2014), Historical landmarks of autophagy research
6. Tanida I (2011) Autophagosome formation and molecular mechanism of autophagy 14
7. Melia TJ, Lystad AH, Simonsen A (2020) Autophagosome biogenesis: from membrane growth to closure
8. Majeski AE, Fred Dice J (2004), Mechanisms of chaperone-mediated autophagy
9. Dice JF (2007) Chaperone-mediated autophagy
10. Jung CH, Jun CB, Ro SH et al (2009) ULK-Atg13-FIP200 complexes mediate mTOR signaling to the autophagy machinery 20:1992–2003
11. Russell RC, Tian Y, Yuan H et al (2013) ULK1 induces autophagy by phosphorylating Beclin-1 and activating VPS34 lipid kinase 15
12. Kim J, Kundu M, Viollet B et al (2011) AMPK and mTOR regulate autophagy through direct phosphorylation of Ulk1 13:132–141
13. Pengo N, Agrotis A, Prak K et al (2017) A reversible phospho-switch mediated by ULK1 regulates the activity of autophagy protease ATG4B 8
14. Park JM, Jung CH, Seo M et al (2016) The ULK1 complex mediates MTORC1 signaling to the autophagy initiation machinery via binding and phosphorylating ATG14 12
15. Park JM, Seo M, Jung CH et al (2018) ULK1 phosphorylates Ser30 of BECN1 in association with ATG14 to stimulate autophagy induction 14
16. Hosokawa N, Sasaki T, Iemura SI et al (2009) Atg101, a novel mammalian autophagy protein interacting with Atg13 5
17. Papinski D, Schuschnig M, Reiter W et al (2014) Early steps in autophagy depend on direct phosphorylation of Atg9 by the Atg1 Kinase
18. Satoo K, Noda NN, Kumeta H et al (2009) The structure of Atg4B-LC3 complex reveals the mechanism of LC3 processing and delipidation during autophagy 53
19. Mizushima N (2020) The ATG conjugation systems in autophagy 63:1–10
20. Tanida I, Ueno T, Kominami E (2004) LC3 conjugation system in mammalian autophagy 36:2503–2518
21. Sánchez-Wandelmer J, Kriegenburg F, Rohringer S et al (2017) Atg4 proteolytic activity can be inhibited by Atg1 phosphorylation 28
22. Maruyama T, Noda NN (2018) Autophagy-regulating protease Atg4: structure, function, regulation and inhibition
23. Yu L, Chen Y, and Tooze SA (2018) Autophagy pathway: Cellular and molecular mechanisms 14
24. Schaaf MBE, Keulers TG, Vooijs MA et al (2016) LC3/GABARAP family proteins: autophagy-(un)related functions 30
25. Zhou C, Zhong W, Zhou J et al (2012) Monitoring autophagic flux by an improved tandem fluorescent-tagged LC3 (mTagRFP-mWasabi-LC3) reveals that high-dose rapamycin impairs autophagic flux in cancer cells 8:1215–1226
26. Kimura S, Noda T, Yoshimori T (2007) Dissection of the autophagosome maturation process by a Novel Reporter Protein, Tandem Fluorescent-Tagged LC3 3
27. Erskine D, Koss D, Korolchuk VI et al (2021) Lipids, lysosomes and mitochondria: insights into Lewy body formation from rare monogenic disorders 141
28. Tra T, Gong L, Kao LP et al (2011) Autophagy in human embryonic stem cells 6
29. Xu Y, Zhang Y, García-Cañaveras JC et al Chaperone-mediated autophagy regulates the pluripotency of embryonic stem cells
30. Chen X, He Y, Lu F (2018), Autophagy in stem cell biology: a perspective on stem cell self-renewal and differentiation
31. Nguyen TD, Shaid S, Vakhrusheva O et al (2019) Loss of the selective autophagy receptor p62 impairs murine myeloid leukemia progression and mitophagy 133
32. Tang AH, Rando TA (2014) Induction of autophagy supports the bioenergetic demands of quiescent muscle stem cell activation 33
33. Fiacco E, Castagnetti F, Bianconi V et al (2016) Autophagy regulates satellite cell ability to regenerate normal and dystrophic muscles 23

34. Zhang Q, Gao M, Zhang Y et al (2016) The germline-enriched Ppp1r36 promotes autophagy 6
35. Chen P, Zhao X, Tian GG et al (2021) C28 induced autophagy of female germline stem cells in vitro with changes of H3K27 acetylation and transcriptomics 766
36. Li X, Hu X, Tian GG et al (2019) C89 induces autophagy of female germline stem cells via inhibition of the PI3K-Akt pathway in vitro 8:606
37. Mizushima N, Yamamoto A, Hatano M et al (2001) Dissection of autophagosome formation using Apg5-deficient mouse embryonic stem cells 7
38. Warr MR, Binnewies M, Flach J et al (2013) FOXO3A directs a protective autophagy program in haematopoietic stem cells 494
39. Liu K, Zhao Q, Liu P et al (2016) ATG3-dependent autophagy mediates mitochondrial homeostasis in pluripotency acquirement and maintenance 12
40. Xiang G, Yang L, Long Q et al (2017) BNIP3L-dependent mitophagy accounts for mitochondrial clearance during 3 factors-induced somatic cell reprogramming 13
41. Zhang Z, Yang M, Wang Y et al (2016) Autophagy regulates the apoptosis of bone marrow-derived mesenchymal stem cells under hypoxic condition via AMP-activated protein kinase/mammalian target of rapamycin pathway 40
42. Zhao S, Fortier TM, Baehrecke EH (2018) Autophagy promotes tumor-like stem cell niche occupancy 28:3056–3064.e3
43. Nilangekar K, Murmu N, Sahu G et al (2019) Generation and characterization of germline-specific autophagy and mitochondrial reactive oxygen species reporters in Drosophila 7
44. Chung H, Choi J, Park S (2018) Ghrelin protects adult rat hippocampal neural stem cells from excessive autophagy during oxygen-glucose deprivation 65
45. Orsini M, Chateauvieux S, Rhim J et al (2019) Sphingolipid-mediated inflammatory signaling leading to autophagy inhibition converts erythropoiesis to myelopoiesis in human hematopoietic stem/progenitor cells 26
46. Wang S, Xia P, Ye B et al (2013) Transient activation of autophagy via Sox2-mediated suppression of mTOR is an important early step in reprogramming to pluripotency 13
47. Nuschke A, Rodrigues M, Stolz DB et al (2014) Human mesenchymal stem cells/multipotent stromal cells consume accumulated autophagosomes early in differentiation 5
48. Leeman DS, Hebestreit K, Ruetz T et al (2018) Lysosome activation clears aggregates and enhances quiescent neural stem cell activation during aging 359
49. Gong C, Bauvy C, Tonelli G et al (2013) Beclin 1 and autophagy are required for the tumorigenicity of breast cancer stem-like/progenitor cells 32
50. Ha S, Jeong S-H, Yi K et al (2019) Autophagy mediates astrogenesis in adult hippocampal neural stem cells 28
51. Jung S, Choe S, Woo H et al (2020) Autophagic death of neural stem cells mediates chronic stress-induced decline of adult hippocampal neurogenesis and cognitive deficits 16
52. Sênos Demarco R, Uyemura BS, Jones DL (2020) EGFR signaling stimulates autophagy to regulate stem cell maintenance and lipid homeostasis in the drosophila testis 30:1101–1116.e5
53. Ha S, Jeong S-H, Yi K et al (2017) Phosphorylation of p62 by AMP-activated protein kinase mediates autophagic cell death in adult hippocampal neural stem cells 292
54. Singh T, Lee EH, Hartman TR et al (2018) Opposing action of hedgehog and insulin signaling balances proliferation and autophagy to determine follicle stem cell lifespan 46:720–734.e6
55. Vazquez-Martin A, Haute C van den, Cufí S et al (2016) Mitophagy-driven mitochondrial rejuvenation regulates stem cell fate 8
56. Xue F, Hu L, Ge R et al (2016) Autophagy-deficiency in hepatic progenitor cells leads to the defects of stemness and enhances susceptibility to neoplastic transformation 371
57. Lieber T, Jeedigunta SP, Palozzi JM et al (2019) Mitochondrial fragmentation drives selective removal of deleterious mtDNA in the germline 570:380–384
58. Patterson GH (2002) A photoactivatable GFP for selective photolabeling of proteins and cells 297
59. Sanchez CG, Penfornis P, Oskowitz AZ et al (2011) Activation of autophagy in mesenchymal stem cells provides tumor stromal support 32

60. Audesse AJ, Dhakal S, Hassell L-A et al (2019) FOXO3 directly regulates an autophagy network to functionally regulate proteostasis in adult neural stem cells 15
61. Maycotte P, Jones KL, Goodall ML et al (2015) Autophagy supports breast cancer stem cell maintenance by regulating IL6 secretion 13
62. Zhang Q, Yang Y-J, Wang H et al (2012) Autophagy activation: a novel mechanism of atorvastatin to protect mesenchymal stem cells from hypoxia and serum deprivation via AMP-activated protein kinase/mammalian target of rapamycin pathway 21
63. Liu P, Liu K, Gu H et al (2017) High autophagic flux guards ESC identity through coordinating autophagy machinery gene program by FOXO1 24
64. Rothe K, Lin H, Lin KBL et al (2014) The core autophagy protein ATG4B is a potential biomarker and therapeutic target in CML stem/progenitor cells 123
65. Nazio F, Bordi M, Cianfanelli V et al (2019) Autophagy and cancer stem cells: molecular mechanisms and therapeutic applications
66. Yazdankhah M, Farioli-Vecchioli S, Tonchev AB et al (2014) The autophagy regulators Ambra1 and Beclin 1 are required for adult neurogenesis in the brain subventricular zone 5
67. García-Prat L, Martínez-Vicente M, Perdiguero E et al (2016) Autophagy maintains stemness by preventing senescence 529
68. Gómez-Sánchez R, Pizarro-Estrella E, Yakhine-Diop SMS et al (2015) Routine Western blot to check autophagic flux: cautions and recommendations 477
69. Jung M, Choi H, Mun JY (2019) The autophagy research in electron microscopy 49
70. Yue Z, Jin S, Yang C et al (2003) Beclin 1, an autophagy gene essential for early embryonic development, is a haploinsufficient tumor suppressor 100
71. Ueno T, Komatsu M (2020) Monitoring autophagy flux and activity: principles and applications 42
72. Klionsky DJ, Abdel-Aziz AK, Abdelfatah S et al (2021) Guidelines for the use and interpretation of assays for monitoring autophagy, 4 ed. 1
73. Chu CT (2010) A pivotal role for PINK1 and autophagy in mitochondrial quality control: Implications for Parkinson disease 19
74. Zhu J-H, Horbinski C, Guo F et al (2007) Regulation of autophagy by extracellular signal-regulated protein kinases during 1-methyl-4-phenylpyridinium-induced cell death 170:75–86
75. Razi M and Tooze SA (2009), Chapter 17 Correlative Light and Electron Microscopy, In Methods in Molecular Biology
76. Colasuonno F, Borghi R, Niceforo A et al (2017) Senescence-associated ultrastructural features of long-term cultures of induced pluripotent stem cells (iPSCs) 9:2209–2222
77. Al-Younes HM, Al-Zeer MA, Khalil H et al (2011) Autophagy-independent function of MAP-LC3 during intracellular propagation of Chlamydia trachomatis 7:814–28
78. Costantini LM, Baloban M, Markwardt ML et al (2015) A palette of fluorescent proteins optimized for diverse cellular environments 6
79. Gomes LC, Odedra D, Dikic I et al (2016) Autophagy and modular restructuring of metabolism control germline tumor differentiation and proliferation in *C. elegans* 12:529–546
80. Ames K, Cunha DS da, Gonzalez B et al (2017) A non-cell-autonomous role of BEC-1/BECN1/Beclin1 in coordinating cell-cycle progression and stem cell proliferation during germline development 27:905–913
81. Li M, Hou Y, Wang J et al (2011) Kinetics comparisons of mammalian Atg4 homologues indicate selective preferences toward diverse Atg8 substrates 286:7327–38
82. Szeto J, Kaniuk NA, Canadien V et al ALIS are stress-induced protein storage compartments for substrates of the proteasome and autophagy 2:189–99
83. Shu X, Shaner NC, Yarbrough CA et al (2006) Novel chromophores and buried charges control color in mFruits 45
84. Tanida I, Ueno T, Uchiyama Y (2014) A super-ecliptic, phluorin-mkate2, tandem fluorescent protein-tagged human LC3 for the monitoring of mammalian autophagy 9
85. Rosado C, Mijaljica D, Hatzinisiriou I et al (2008) Rosella: a fluorescent pH-biosensor for reporting vacuolar turnover of cytosol and organelles in yeast 4

86. Kaizuka T, Morishita H, Hama Y et al (2016) An autophagic flux probe that releases an internal control 64
87. Kogure T, Karasawa S, Araki T et al (2006) A fluorescent variant of a protein from the stony coral Montipora facilitates dual-color single-laser fluorescence cross-correlation spectroscopy 24
88. Katayama H, Hama H, Nagasawa K et al (2020) Visualizing and modulating mitophagy for therapeutic studies of neurodegeneration 181
89. Katayama H, Kogure T, Mizushima N et al (2011) A sensitive and quantitative technique for detecting autophagic events based on lysosomal delivery 18
90. Ding S, Hong Y (2020) The fluorescence toolbox for visualizing autophagy
91. Cho Y-H, Han K-M, Kim D et al (2014) Autophagy regulates homeostasis of pluripotency-associated proteins in hESCs 32
92. Wang C, Chen S, Yeo S et al (2016) Elevated p62/SQSTM1 determines the fate of autophagy-deficient neural stem cells by increasing superoxide 212
93. Mulakkal NC, Nagy P, Takats S et al (2014) Autophagy in *Drosophila*: from historical studies to current knowledge 2014
94. Komatsu M, Kurokawa H, Waguri S et al (2010) The selective autophagy substrate p62 activates the stress responsive transcription factor Nrf2 through inactivation of Keap1 12:213–223
95. Lerner C, Bitto A, Pulliam D et al (2013) Reduced mammalian target of rapamycin activity facilitates mitochondrial retrograde signaling and increases life span in normal human fibroblasts 12:966–977
96. Komatsu M, Waguri S, Koike M et al (2007) Homeostatic levels of p62 control cytoplasmic inclusion body formation in autophagy-deficient mice 131:1149–1163
97. Karanasios E, Stapleton E, Manifava M et al (2013) Dynamic association of the ULK1 complex with omegasomes during autophagy induction 126:5224–38
98. Itakura E, Kishi C, Inoue K, et al (2008) Beclin 1 forms two distinct phosphatidylinositol 3-kinase complexes with mammalian Atg14 and UVRAG 19:5360–72
99. Barve G, Sridhar S, Aher A et al (2018) Septins are involved at the early stages of macroautophagy in *S. cerevisiae*
100. Mastorci K, Montico B, Faè DA et al (2106) Phospholipid scramblase 1 as a critical node at the crossroad between autophagy and apoptosis in mantle cell lymphoma 7:41913–41928
101. Proikas-Cezanne T, Ruckerbauer S, Stierhof Y-D et al (2007) Human WIPI-1 puncta-formation: a novel assay to assess mammalian autophagy 581:3396–404
102. Fan W, Nassiri A, Zhong Q (2011) Autophagosome targeting and membrane curvature sensing by Barkor/Atg14(L) 108:7769–74
103. Hamasaki M, Furuta N, Matsuda A et al (2013) Autophagosomes form at ER–mitochondria contact sites 495:389–393
104. Guo S, Liang Y, Murphy SF et al (2015) A rapid and high content assay that measures cyto-ID-stained autophagic compartments and estimates autophagy flux with potential clinical applications 11
105. Ozeki N, Hase N, Hiyama T et al (2017) MicroRNA-211 and autophagy-related gene 14 signaling regulate osteoblast-like cell differentiation of human induced pluripotent stem cells 352
106. Bavister BD (2006) The mitochondrial contribution to stem cell biology 18
107. Ye G, Xie Z, Zeng H et al (2020) Oxidative stress-mediated mitochondrial dysfunction facilitates mesenchymal stem cell senescence in ankylosing spondylitis 11
108. Wang Y, Zhang Y, Chen L et al (2016) Kinetics and specificity of paternal mitochondrial elimination in Caenorhabditis elegans 7:12569
109. Geisler S, Holmström KM, Skujat D et al (2010) PINK1/Parkin-mediated mitophagy is dependent on VDAC1 and p62/SQSTM1 12:119–131
110. Djavaheri-Mergny M, Amelotti M, Mathieu J et al (2006) NF-kappaB activation represses tumor necrosis factor-alpha-induced autophagy 281:30373–82

111. Wiemer EA, Wenzel T, Deerinck TJ et al (1997) Visualization of the peroxisomal compartment in living mammalian cells: dynamic behavior and association with microtubules 136:71–80
112. Farré J-C, Krick R, Subramani S et al (2009) Turnover of organelles by autophagy in yeast 21:522–530
113. Shiba-Fukushima K, Ishikawa K-I, Inoshita T et al (2017) Evidence that phosphorylated ubiquitin signaling is involved in the etiology of Parkinson's disease
114. Hemelaar J, Lelyveld VS, Kessler BM et al (2003) A single protease, Apg4B, Is specific for the autophagy-related ubiquitin-like proteins GATE-16, MAP1-LC3, GABARAP, and Apg8L 278:51841–51850
115. Sarkar S, Korolchuk V, Renna M et al (2009) Methodological considerations for assessing autophagy modulators: a study with calcium phosphate precipitates 5:307–13
116. Tanida I, Minematsu-Ikeguchi N, Ueno T et al (2005) Lysosomal turnover, but not a cellular level, of endogenous LC3 is a marker for autophagy 1:84–91
117. Farkas T, Høyer-Hansen M, Jäättelä M (2009) Identification of novel autophagy regulators by a luciferase-based assay for the kinetics of autophagic flux 5:1018–25
118. Zappavigna S, Lombardi A, Misso G, et al (2017) Measurement of autophagy by flow cytometry In: Methods in Molecular Biology
119. Zhang T, Wolfe C, Pierle A et al (2017) Proteome-wide modulation of degradation dynamics in response to growth arrest 114
120. Liu X, Xiang M-H, Zhou W-J et al (2021) Clicking of organelle-enriched probes for fluorogenic imaging of autophagic and endocytic fluxes
121. Bernard A, Jin M, Xu Z et al (2015) A large-scale analysis of autophagy-related gene expression identifies new regulators of autophagy 11
122. Zhang Y, Xu X, Hu M et al (2021) SPATA33 is an autophagy mediator for cargo selectivity in germline mitophagy 28:1076–1090
123. Wang C, Yeo S, Haas MA et al (2017) Autophagy gene FIP200 in neural progenitors non–cell autonomously controls differentiation by regulating microglia 216
124. Mortensen M, Soilleux EJ, Djordjevic G et al (2011) The autophagy protein Atg7 is essential for hematopoietic stem cell maintenance 208
125. Mikhaylova O, Stratton Y, Hall D et al (2012) VHL-regulated MiR-204 suppresses tumor growth through inhibition of LC3B-mediated autophagy in renal clear cell carcinoma 21:532–546

Chapter 2
Autophagy in Stem Cell Maintenance and Differentiation

Anirudha K. Sahu, Propanna Bandyopadhyay, Rajdeep Chowdhury, and Sudeshna Mukherjee

Abstract Autophagy is one of the intracellular machinery for maintaining organelle as well as physiological homeostasis in cells by clearance of cellular debris and recycling of essential raw materials. It is different from other cellular processes like apoptosis and necrosis in the sense that it acts as a double-edged sword that might lead either to survival or death based on the stimuli. There are broadly three different types of autophagy: macroautophagy, microautophagy, and chaperone mediated autophagy. Macroautophagy is one of the commonly understood forms of autophagy and has been discussed simply as autophagy throughout the chapter. The role of autophagy in stem cell maintenance and differentiation is essential as both the processes require intensive intracellular remodeling which involves a continuous cycle of synthesis and degradation of event-specific proteins. Several pathways are involved in the regulation of autophagy and vice versa in stem cells. Among them, there are master proteins mandatory for stem cell maintenance and/or differentiation reported to be directly regulating autophagy. The current chapter discusses the different signaling pathways in stem cells; regulating or being regulated by autophagy and its role in the maintenance and differentiation of various types of stem cells.

Keywords Autophagy · Cellular Homeostasis · Differentiation · Stem cells

Abbreviations

3-MA	3-Methyl adenine
ADSC	Adipose derived stem cell
ALT	Alanine Aminotransferase
Ambra1	Activating molecule in Beclin1-regulated autophagy protein 1

A. K. Sahu · P. Bandyopadhyay · R. Chowdhury · S. Mukherjee (✉)
Department of Biological Sciences, Birla Institute of Technology and Science (BITS), Pilani Campus, Pilani 333031, Rajasthan, India
e-mail: sudeshna@pilani.bits-pilani.ac.in

B. V. Shravage and K. Turksen (eds.), *Autophagy in Stem Cell Maintenance and Differentiation*, Stem Cell Biology and Regenerative Medicine 73,
https://doi.org/10.1007/978-3-031-17362-2_2

AMPK	AMP-activated protein kinase
APJ	Apelin receptor
ATG	Autophagy related gene
ATP	Adenosine triphosphate
BAT	Brown adipose tissue
BMSC	Bone marrow derived mesenchymal stem cells
BNIP3	BCL2/Adenovirus E1B 19 kDa Protein Interacting Protein 3
C/EBPβ	CCAAT-enhancer binding protein
CMA	Chaperone mediated autophagy
CNS	Central nervous system
CSC	Cardiac stem cells
CSF	Colony stimulating factor
CYLD	Cylindromatosis
DNA	Deoxyribo Nucleic Acid
ER	Endoplasmic Reticulum
ESCRT	Endosomal sorting complex required for transport
Eva1	Eva-1 homolog A
FGF	Fibroblast growth factor
FGFR	Fibroblast growth factor receptor
FIP200	Focal adhesion kinase family-interacting protein of 200kD
FOXO	Forkhead box transcription factor
FRS2α	FGF receptor substrate 2α
GATA 1	GATA-binding factor 1
GIT1	G-protein-coupled receptor kinase-interacting protein 1
GSK3	Glycogen Synthase Kinase 3
HIF-1α	Hypoxia Inducible Factor 1 alpha
HMPC	Hematopoietic mesenchymal progenitor cells
HPC	Hepatic progenitor cells
HSC	Hematopoietic stem cells
HSC-70	Heat shock Protein 70 kDa
IL-6	Interleukin-6
ISC	Intestinal stem cells
JNK	C-Jun N-terminal kinase
Klf	Kruppel-like factors
LC3	Microtubule-associated protein 1 light chain 3 protein
LPC	Liver Progenitor Cell
miRNA	Micro Ribo-nucleic acid
MMP	Matrix metalloproteinases
MSC	Mesenchymal stem cells
mTOR	Mammalian target of rapamycin
NBR1	Neighbor of BRCA1 gene 1 protein
NFATc1	Nuclear factor of activated T-cells 1
NF-κβ	Nuclear Factor kappa light chain enhancer of activated B cells
NSC	Neural stem cells
PE	Phosphatidylethanolamine

PI3K	Phosphatidylinositol—3—kinase
PIP2	Phosphatidylinositol 4,5-bisphosphate
PIP3	Phosphatidylinositol 3,4,5-triphosphate
PPARγ2	Peroxisome proliferator activated receptor gamma
RANKL	Receptor activator of nuclear factor kappa-B ligand
RBC	Red Blood cells
ROS	Reactive oxygen species
SC	Satellite cells
STAT-3	Signal transducer and activator of transcription 3
TAB2TGF-β	activates kinase 1 binding protein 2
TGs	Triglycerides
TIP60	60KDa HIV-Tat interactive protein
TNF-α	Tumor Necrosis Factor alpha
TRAF	Tumor necrosis factor receptor-associated factor
TUNEL	Terminal Deoxy-nucleotidyl Transferase dUTP Nick End Labeling
ULK1	Unc-51-like kinase 1
VPS	Vacuolar protein sorting
VZ/SVZ	Ventricular and subventricular zone
WAT	White Adipose Tissue
WIPI	WD repeat protein interacting with phosphoinositide
Wnt	Wingless-related Integration site

Introduction

Autophagy, which literally means 'Self eating' although might seem like an intimidating devouring process, actually refers to an evolutionarily conserved intracellular cleansing process for the recycling of damaged organelles, unwanted proteins, surplus storage nutrients (glycogen and lipids), reactive oxygen species (ROS), macro and micro-molecules (nucleotides and metabolic byproducts) in the form of essential intracellular raw materials like amino acids, sugars, fatty acids, nucleosides, etc., to maintain cellular homeostasis [1–4]. Even intracellular pathogens can be taken up by the autophagy program to break them down into antigenic peptides to further incite specific immune responses [5, 6]. Autophagy was first discovered in yeast by Yoshinori Ohsumi and his colleagues at the University of Tokyo, in 1992, although the terminology was given long back by Christian de Duve in 1963 [7]. Induction of autophagy was initially believed to be only under starvation [8], but later on, it was also found to be activated by DNA damage, cellular stress, infection, and hypoxia [4]. Autophagy has been implicated in the prognosis of various human diseases, cellular, and host immunity as well as in developmental processes like stem cell renewal, maintenance, and metabolic remodeling leading to differentiation and morphogenesis, especially (but not limited to) during embryo development [5, 9–14]. The cellular machinery for autophagy basically includes the formation of a double-membraned vesicular body called autophagosome which carries the intracellular junk and wholly moves toward the lysosome and fuses with it for degradation and recycling. The fused body is called auto-phagolysosome which finally completes the process of autophagy [1, 2].

Autophagy Compared to Necrosis and Apoptosis

Autophagy is far different from apoptosis and necrosis. Although apoptosis and necrosis themselves differ from each other in terms of the process and what they do to the cell; both are bound to cause cell death unlike autophagy. Both are incited by mostly similar but somewhat different intracellular or extracellular cues. In short, apoptosis is a caspase-mediated programmed cell death that is characterized by chromosome condensation, nuclear fragmentation, and membrane blebbing intended to basically get rid of excess, undesirable, and sometimes intracellularly damaged cells, whereas necrosis is considered to be an accidental cell death caused by nonspecific or nonphysiological stimuli, mostly extracellular stress inducers, and is characterized by the expansion of cellular organelles, disintegration of cell membrane, and further inflammatory responses mediated by the release of the intracellular contents. However, autophagy is basically a cleaning and recycling process associated with the formation of the autophagosome, which is a bilayer vesicle containing damaged organelles, proteins, and other cytoplasmic components [15–17]. Although autophagy is intended to be more of a survival mechanism maintaining cellular homeostasis by resource recycling and consuming cellular waste; its excessive triggering by single or multiple upstream signaling might lead to unplanned cell death [4].

The Autophagic Cascade

The whole process of autophagy is basically divided into 6 steps: Induction, Nucleation, Elongation, Maturation, Fusion, and 'Degradation and recycling'. Unless there is a signal; the whole process is in check and is negatively regulated by the mTOR complex. Induction of autophagy results in the dephosphorylation and activation of ULK1 which then forms a ULK1–ATG13–FIP200–ATG101 complex. The ULK1 complex translocates to sites on the Endoplasmic Reticulum where the nucleation of autophagosome takes place leading to pre-autophagosome formation. ULK1 further activates another effector protein called Beclin1 from the phosphatidylinositol 3-kinase (PI3K) complex by phosphorylation. Beclin1 is one of the most important proteins within the PI3 Kinase complex associated with other accessory proteins like VPS34 and VPS15. Now, this PI3 Kinase converts PIP2 (Phospho-inositol diphosphate) to PIP3 (Phospho-inositol triphosphate) on the pre-autophagosome membrane which leads to the recruitment of WIPI proteins on the surface. This happens during the elongation process. WIPI proteins help in the recruitment of P62 and NBR1 sequestered LC3-I proteins to PE (Phosphatidylethanolamine) via another complex called ATG16-ATG5-ATG12 complex and facilitate the conversion of LC3-I to LC3-II by lipidation. Now, P62 and NBR1 are cargo carrier proteins that carry the debris for degradation and recycling into the fully formed autophagosomes. The presence of LC3-II generally is an indicator of a fully matured autophagosome. Once the matured autophagosome is formed, it finally fuses with the lysosome for degradation of the cargo and recycling of resources [2, 6, 7].

Types of Autophagy

There are basically three different types of autophagy: macroautophagy, microautophagy, and chaperone mediated autophagy [18].

Macroautophagy is an umbrella term for several different sub-types of autophagy and essentially refers to bulk degradation. All different types of organellophagy (bulk degradation of organelles through autophagy) like mitophagy (mitochondria), reticulophagy (Endoplasmic reticulum), nucelophagy (nucleus) including lipidophagy (lipid droplets), xenophagy (foreign micro-organisms), etc., come under macroautophagy [19–21].

Microautophagy by far is understood to be a simpler form of autophagy where the lysosomes directly take up cytoplasmic contents. The different forms of microautophagy are general microautophagy (cytoplasmic contents), Micro-ERphagy (Endoplasmic reticulum), Micropexophagy (Peroxisome), Micronucleophagy (Nucleus), Micromitophagy (Mitochondria), and Microlipophagy (Lipid droplets). There is no need for ATG proteins or core autophagy machinery for general microautophagy as such though ESCRT (Endosomal Sorting Complexes Required for Transport) proteins mediate membrane budding and scission in many cellular processes and other proteins like VSP4 and lipidated LC3 (for cargo selection) might be required. However, special kinds of microautophagy like micropexophagy might require few other proteins from the core autophagy machinery as well [22].

Chaperone mediated autophagy (CMA) is a special kind of autophagy wherein a chaperone protein Hsc-70 is involved in the identification of the proteins destined for degradation. The cytosolic proteins targeted for degradation must contain a specific amino acid signature, namely, 'KFERQ peptide sequence'. In this process, first, the target protein is identified based on the signature sequence followed by its unfolding and later taken up by the lysosomes for degradation. Regulation may depend on whether the KFERQ motif present in the target protein is accessible to Hsc-70 or not [23].

Role of Autophagy in Stem Cell Maintenance

Stem cells are unique cells of the human body that have the ability to self-renew for pool maintenance and differentiate to develop into any cell of the physiological system. They are mainly classified into the following four types based on their potency to form the various cell types—totipotent stem cells, pluripotent stem cells, multipotent stem cells, and unipotent stem cells. Stem cells are found not only in growing embryos, but also in adults where they are preserved for a long period in order to give rise to progenitor cells in the hour of need. This is to be kept in mind that with each differentiation, the stem cells' potential to form a variety of cells becomes restricted; such as pluripotent stem cells can form any cell of the body but a multipotent stem cell can give rise to cells of a specific lineage [24].

Stem cell maintenance is one of the essential requirements both in case of embryonic, as well as adult stem cells, for replenishing the mother stem cell reserve. Maintenance of stem cells requires a lot of intracellular metabolic remodeling which involves periodic synthesis and degradation of proteins and thus requires a very active or multiple modes of protein degradation machinery [11].

The two major machineries to control the turnover of proteins in cells are autophagy and ubiquitin mediated proteasomal degradation [25]. If we consider autophagy as a tool that the cells can use as per their requirement; for each requirement, the cells might induce autophagy in a different way and the outcomes could be different depending on the upstream signaling induced by the cell responding to a particular stimulus. To understand whether autophagy plays a role in stem cell maintenance, several researchers have inhibited different key proteins in the pathway of autophagy and have searched for its effect on the renewal of various stem cells.

An insight into how autophagy helps in the maintenance of particular stem cells is discussed subsequently.

Role of Autophagy in Maintenance of Hematopoietic Stem Cells (HSCs)

The HSCs undergo hematopoiesis for generating blood cells of both myeloid and lymphoid lineages. For hematopoiesis to occur; a balance between hematopoietic stem cell (HSC) quiescence, activation, and differentiation is essential which is guided by autophagy [14].

In order to maintain the self-renewal capacity, intracellular reactive oxygen species (ROS) levels must be low, and the nuclear genome must be protected. Autophagy helps to maintain genome integrity and degrades defective mitochondria. This helps in the removal of excess ROS leading to the maintenance of HSC. There is also a close involvement of autophagy-related protein 7 (Atg7) whose deletion is coupled with the accumulation of damaged mitochondria, increased ROS levels, and DNA damage [26]. The absence of Atg 7 and FIP 200 genes inhibits regeneration activity and self-renewal capacity, and increases stress-induced apoptosis at an older age. Another autophagy gene, Atg12, deficiency also accompany impaired self-renewal and regenerative potential of HSCs [14, 27].

In order to maintain quiescence, when there is no need for multiplication or differentiation; HSCs show low levels of oxidative phosphorylation. Here in, autophagy plays a critical role in regulating intracellular oxidative metabolism by removing active healthy mitochondria which are fascinating, thus helping in energy conservation and maintenance of quiescence [7, 14].

Enduring metabolic stress is also important for the survival of any cell irrespective of whether it is a stem cell or somatic cell. HSCs maintain a very high level of FOXO3A which is a transcription factor that targets and induces pro-autophagy genes when the autophagy program has already been induced. In a way, higher levels of FOXO3A enhance autophagy rather than inducing it directly [7, 28].

Being an integral part of blood cell population maintenance, any disruption in hematopoiesis regulation leads to hematopoietic disorders like anemia and leukemia [7].

Role of Autophagy in Maintenance of Neural Stem Cells

Neural stem cells (NSCs) are the stem cells of the nervous system which are multipotent in nature and reside in discrete niches inside the subventricular zone of the lateral ventricles and subgranular zone of the hippocampal dentate gyrus of the adult brain.

Like HSCs; neural stem cells also depend on the FOXO family of proteins like FOXO1, FOXO3, and FOXO4 for enhancing autophagy to maintain cellular homeostasis. Inactivation of any of these proteins in NSCs has shown defective self-renewal and accumulation of protein aggregates [27]. In contrast, when compared with the quiescent NSCs, lysosome-related genes are surprisingly seen to be highly upregulated unlike activated NSCs; those instead harbor genes involved in proteasomal degradation pathway. The above is further supported by the fact that quiescent NSCs harbor a substantial amount of large lysosomes containing insoluble protein aggregates. The cells can reversibly get rid of these lysosomes whenever they are activated. This is probably a strange strategic way, how NSCs conserve energy during quiescence and stay in a docile state till activation.

Unlike HSCs, where autophagy is also known to degrade metabolically active mitochondria to maintain quiescence, NSCs only use autophagy to remove excess intracellular ROS via mitophagy [7].

Role of Autophagy in Maintenance of Muscle Stem Cells

Homeostasis and regeneration of skeletal muscles are maintained by muscle stem cells; also called Satellite cells. The nomenclature came up based on their position beneath the basal lamina of muscle fibers. These myoblasts are somite-derived that have not fused with other myoblasts and retained their stemness throughout adult life [29, 30].

The functioning of autophagy is quite interesting in case of muscle stem cells. A group of researchers has shown that starvation in the form of short-term caloric restriction could induce autophagy that could increase satellite cell number and muscle regeneration. Whereas, when talking about the basal level of autophagy, quiescent satellite cells from young mice display a higher level of autophagic flux than those isolated from aged mice. Intentional induction of autophagy by inhibition of mTOR in quiescent Satellite cells from aged mice shows enhanced regeneration. This suggests that the role of autophagy in muscle stem cells is majorly associated with delaying senescence.

On another note, the transition of Satellite cells from quiescence to active state also needs autophagy. In this case autophagy acts as a mechanism to recycle ATP to cater to the energy requirements of the cells during the transition [7].

Another unique phenomenon attributed to the muscle stem cells regarding autophagy is the expression of autophagic genes in an oscillatory fashion synchronizing with the circadian rhythm, i.e., more during the day compared to night. Such rhythm is absent in aged muscle stem cells [27]. Although the functional implication of this phenomenon is not yet attributed to anything, it might have something to do with the fluctuating energy requirements of the cells during day and night.

Role of Autophagy in Maintenance of Bone Marrow Derived Mesenchymal Stem Cells (BMSCs)

Bone Marrow Derived Mesenchymal stem cells are adult pluripotent stem cells that are capable of giving rise to diverse other cell types like adipocytes, endothelial cells, osteocytes, neurons, and cardiomyocytes in response to appropriate signal exposure.

The role of autophagy in the maintenance of Bone Marrow Derived Mesenchymal stem cells is controversial. A study made by Zhang et al. in 2016 interestingly suggests that hypoxia-induced autophagy can lead to apoptosis of BMSCs by activation of AMPK that inhibits mTOR and in turn activates autophagy. The group has shown that inhibition of autophagy by an autophagy inhibitor 3-MA has reduced apoptosis of BMSCs, while apoptosis is aggravated by Rapamycin: an autophagy inducer; in a time-dependent manner. The group checked for apoptosis by TUNEL assay and expression of cleaved caspase 3 in the presence of 3-MA or Rapamycin under hypoxic condition [31].

However, on a different note, another group of authors has established that hypoxia-based autophagy which is induced via Apelin/Apj signaling helps in the proliferation of BMSCs [32]. In this regard, both cases could be possible under different circumstances depending upon the severity of the hypoxia induced.

Besides, autophagy also plays an important role in precluding senescence in different types of BMSCs like mandible-derived BMSCs and tibia-derived BMSCs [13, 26, 33].

Role of Autophagy in Maintenance of Hepatic Progenitor Cells

Liver progenitor cells (LPCs) or Hepatic progenitor cells (HPCs) have the capacity to self-renew and differentiate to form hepatocytes and biliary epithelial cells [26]. The regenerative potential of hepatic cells after partial hepatectomy is immense and is known to be mediated by IL-6 and TNF-α along with their downstream targets STAT-3 and NF-κB. Very recently, a close association of autophagy has been found to play a role in hepatocyte differentiation and their maintenance [34]. On impairment of Atg 5, damaged mitochondria get accumulated along with the reduction in ATP, total protein, and albumin levels, and an increase in ALT (alanine aminotransferase) and glucose levels. Autophagy regulates their metabolic functions, and on ablation or impairment of this process, the intrinsic apoptotic pathway is activated, reducing proliferation and increasing the death rate of hepatocellular cells.

The abolition of autophagy proteins, Atg5 and Beclin 1, in liver progenitor cells reduces stemness and induces senescence. Upon aging, lipofuscin, which is an aging pigment gets accumulated in the lysosomes causing a reduction in the number and function of autophagosomes leading to decreased regenerative potential of the liver.

Thus, autophagy is required in both differentiation and maintenance of stemness of LPCs where it is protected during liver injury by this homeostatic mechanism.

Role of Autophagy in Maintenance of Intestinal Stem Cells (ISCs)

The intestinal epithelium, which acts as a physical, as well as a physiological barrier, to prevent the gut microbiota from entering the host system, confronts extensive wear and tear due to its protective and digestive role. Thus, it requires frequent repair and maintenance. This is where intestinal stem cells come into play.

Similar to HSCs; autophagy in ISCs is also implicated in preserving cellular homeostasis by removing excess mitochondria and intracellular ROS. Moreover, research has also shown that intentional deletion of ATG7 (an intermediary protein indirectly involved in autophagophore elongation) from ISCs leads to impaired antioxidant and DNA repair responses emphasizing the role of autophagy in the maintenance of intracellular homeostasis in ISCs [35].

The role of autophagy in the maintenance of all the above different types of stem cells has been schematically shown in Fig. 2.1.

Autophagy and Stem Cell Differentiation

Differentiation is a very essential phenomenon for the continuous replacement of damaged cells from tissues to maintain tissue-specific homeostasis and another very important factor controlling stem cell maintenance. There are unique ways in which stem cells practice autophagy for undergoing differentiation. The role of autophagy for inducing or enhancing differentiation could vary in a context specific depending on different types of stem cells.

Role of Autophagy in the Differentiation of HSCs

Previously, we discussed that Hematopoietic stem cells (HSCs) have the ability to differentiate into different kinds of blood cells both from lymphoid and myeloid lineages. Along with the maintenance of the HSC pool, it is also important for differentiation to happen as well; for the replacement of lost or damaged blood cells.

One of the major roles of autophagy specifically during the differentiation of HSCs to final mature erythrocytes is the clearance of mitochondria [36]. In this process of differentiation, HSCs first form erythroblasts which further mature into reticulocytes after losing the nucleus. Reticulocytes then get rid of organelles including mitochondria to finally mature into erythrocytes (RBCs). Elimination of mitochondria and nucleus provides accommodation for more hemoglobin and also helps in debarring oxygen being used by RBCs allowing RBCs to store oxygen for supply and transport. Therefore, clearance of mitochondria is essential for erythrocyte formation [37]. This is evidenced by the fact that GATA1, a transcription factor and master regulator of hematopoiesis, regulates several autophagy genes [18, 38].

However, interestingly, it has been observed that even ATG 5 deficient embryos harbor functional autophagic vacuoles in their reticulocytes. This has been confirmed by ultrastructural analysis [39]; however, in another independent study, it was shown that ATG7 is still necessary for the seamless removal of mitochondria in erythroid cells [18, 40]. Such observations from researchers suggest that an alternative mode of autophagy is operative in reticulocytes. The occurrence of an alternative mode of autophagy is not new and has been vividly discussed by Shimizu (2018) from the Dept. of Pathological Cell Biology, Tokyo Medical and Dental University, Japan, in a mini-review [41].

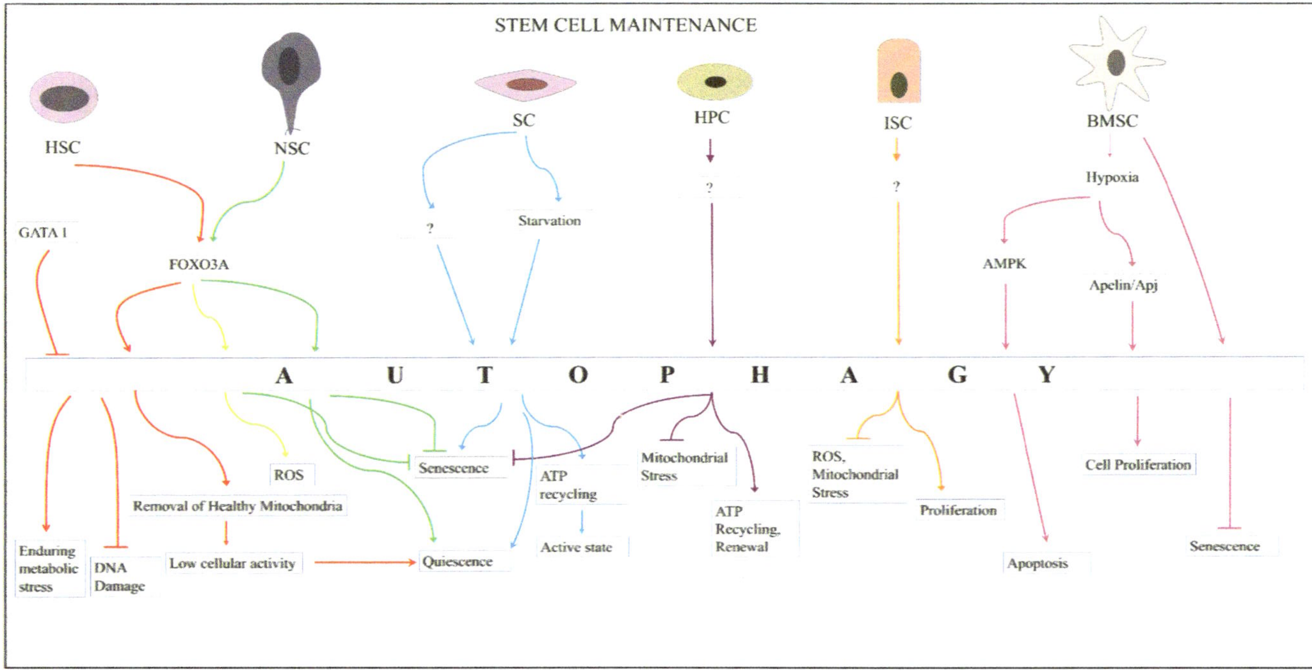

Fig. 2.1 *Role of autophagy in stem cell maintenance.* The maintenance of different stem cells is determined by diversified roles of autophagy which is solely context dependent and cell specific. Functions like clearing mitochondrial ROS and damaged DNA, circumventing through cell senescent pathways, ATP recycling, and saving energy by quiescence help in stem cell maintenance, survival, and proliferation. Sometimes, depending upon the upstream signaling, autophagy can directly encourage the process of stem cell proliferation by regulating energy metabolism, intracellular resource recycling, and cyclic synthesis and degradation of proteins involved in the cell proliferative pathways. HSC (Haematopoietic Stem cells), NSC (Neural Stem cells), SC (Satellite Cells), ISC (Intestinal Stem Cells), HPC (Hepatic Progenitor Cells), and BMSC (Bone marrow derived mesenchymal stem cells)

Role of Autophagy in the Differentiation of NSCs

Neural stem cells (NSC) as discussed earlier are multipotent self-renewing stem cells in the developing and adult mammalian central nervous system (CNS), those have the potential to give rise to either neurons or glial cells in the brain. As NSCs have the unique capability of repairing neural circuits, they tend to remain highly metabolically active under circumstances where cell replacement is necessary. Previously, we saw that autophagy is essential for the maintenance of NSCs.

Several researchers have also shown that neurogenesis also requires autophagy to happen during embryogenesis and also during adult tissue repairing. This is evident from the fact that key proteins from the autophagy cascade like Atg9a, Atg5, Atg7, Beclin1, Ambra1, Eva1a, and LC3-II go up during neurogenesis in the forebrain, olfactory bulb, and cerebral cortex derived NSCs [18]. Moreover, silencing of Vps34 also affects neurogenesis in the cerebral cortex by reducing excitatory neuron migration and axonal growth without affecting the cell cycle of NSCs at the Ventricular and subventricular zone (VZ/SVZ). Further, in cortical NSCs, intended deletion of ATG5 has shown to impair neuronal differentiation while in VZ/SVZ NSCs proliferation is induced instead of differentiation [42, 43].

In another interesting study miR-34a: an miRNA known to regulate Atg9a (a key regulator protein in autophagy induction which provides membrane for autophagosome formation); also negatively regulates neuronal differentiation suggesting autophagy is essential for neuronal differentiation [44].

On another note, if discussed about signaling known to cross-talk with autophagy in NSCs: Wnt3a and Notch signaling are reported.

The role of Wnt3a in regulating autophagy and, in turn, affecting neuronal differentiation is controversially discussed. On one hand, wnt3a is shown to decrease autophagy in mature neurons after traumatic brain injury while escalating hippocampal neurogenesis; whereas on the other hand, wnt3a increases autophagy in cells under cultured conditions in hippocampal neuronal cultures from embryonic rat through AMPK activation. Thus, whether wnt3a affects neuronal differentiation by either promoting or demoting autophagy could be highly context dependent and/or situation specific.

In case of Notch signaling, autophagy itself has been observed to regulate it in NSCs. In the process, the Notch1 receptor is degraded through its uptake into pre-autophagosomal vesicles in an Atg16L1-dependent manner. Increased Notch1 due to faulty autophagic machinery hinders neuronal differentiation and inflates the NSC pool [42, 45].

Role of Autophagy in the Differentiation of Cardiac Stem Cells

Human heart replaces its cardiomyocyte population completely around 18 times throughout life unless there is some form of cardiac damage where the cardiac stem cells (CSCs) come into play. The CSCs inhabit the stem cell niche in the adult heart and divide symmetrically/asymmetrically as per the shift in homeostasis. The asymmetric division gives rise to a CSC and another committed cell like a myocyte [46].

It has been reported that fibroblast growth factor (FGF) signaling regulates cardiac development. It inhibits 'premature differentiation' of CSCs. FGF attaches to fibroblast growth factor receptor (FGFR) tyrosine kinases and activates its downstream targets via FRS2α (FGF receptor substrate 2α) thereby down regulating autophagy. With the inhibition of autophagy, differentiation of myocardial stem cells also gets inhibited. Thus, it can be concluded that FRS2α is essential for autophagy suppression by FGF through the mTOR pathway. In the presence of a FGFR inhibitor, there has been an increased expression of Beclin 1 and p27 which marks the initiation of autophagy leading to CSCs differentiation. Interestingly, it has also been found that autophagy is not only a regulator of differentiation, but also is positively correlated with ectopic foci formation of the heart.

On the contrary, the Wnt pathway, which is an upstream regulator of FGF, inhibits CSCs differentiation via GSK3-TIP60-ULK1 pathway. It is even reported that cholesterol metabolism showed an increase in Atg5 proving that induction of autophagy triggers differentiation of CSCs. The other signalings which get activated are GSK3β/β-catenin and JNK/STAT-3, which cause an increase in cardiac transcriptional proteins, factors, and enhancers' expressions, mediating their movement inside the nucleus and enhancing differentiation [14].

Role of Autophagy in Adipocyte Formation from MSCs

Adipocytes are derived from mesenchymal stem cells (MSCs) through adipogenesis; those, in turn, give rise to adipose tissue, a process closely governed by autophagy. The adipose tissue is complex enough to be considered as a vital endocrine organ in mammals for storing energy and regulating inflammation, cell signaling, and metabolism by secreting endocrines. In general, white adipose tissue (WAT) is primarily known to store lipids during the fed state, whereas during starvation, it releases fatty acids into the bloodstream for muscle energy production by breaking down triglycerides (TGs). Instead, brown adipose tissue (BAT) stores a limited amount of TGs, does not secrete fatty acids and rather uses them for heat production in the body [13, 47].

Studies have shown that activation of autophagy is important to WAT differentiation; by attenuating the proteasome-dependent degradation of PPARγ2: a regulator of fatty acid storage and glucose metabolism and cutting down the number of mitochondria. This is evident from the fact that pharmacological inhibition or knockdown of ATG5 and ATG7 genes lead to browning of WAT, hence decreasing lipid accumulation. In detail, PPARγ2 under proliferating conditions is degraded by the proteasomal system unless an adipogenic differentiation signal is received. This is followed by induction of autophagy stabilizing PPARγ2 promoting differentiation into adipogenic fate. This establishes the fact that both ubiquitin-dependent proteasomal degradation system and autophagy cross talk with each other to regulate the level of PPARγ2, in turn, regulating adipogenesis [18, 48, 49].

Deficiency of autophagy in this tissue causes diminished differentiation and development of adipose stem cells and abnormal secretion of adipocytokine. It also leads to the conversion of white adipose tissue to brown adipose tissue, clearly demarcating calorie loss, storage of which is essential to provide energy according to body requirements [47, 50].

Along with differentiation, autophagy also helps in lipid droplet expansion. C/EBPβ is found to be responsible for autophagy activation via Atg 4b and it functions as an important adipogenic factor. Upon activation, p62 (an autophagy related protein) breaks down Klf 2 and Klf 3 which are the negative regulators of adipocyte differentiation so as to promote the formation of mature adipocytes [51].

Atg7 and Atg5 also play a keen role in the maintenance of differentiation, lipid accumulation, insulin sensitivity, and white adipose mass formation. The rate of energy production through beta oxidation of fatty acids is also controlled by autophagy-related proteins. They keep a check on the equilibrium of white and brown adipose tissue formation, suggesting a significant part of autophagy being involved in adipose stem cell differentiation and balancing its zeal in metabolic activities [11, 52].

Role of Autophagy in Osteoclastogenesis

Osteoclasts are cells involved in bone resorption which develop from hematopoietic myeloid progenitor cells under the influence of colony-stimulating factor 1 (CSF1) and RANKL signals. This differentiation of osteoclasts is also regulated by autophagy, the disruption of which causes changed osteoclast function leading to osteoporosis. Apart from differentiation, survival, and function of other bone cells like osteocytes, osteoblasts are also governed by autophagy [53].

Deletion of autophagy proteins like Atg5, Atg 7, Beclin 1, and LC3 in osteoclasts are associated with reduced bone cell differentiation and mineralization, impaired secretion, bone resorption, and decreased bone mass. On the other hand, ablation of autophagy increases endoplasmic reticulum stress and RANKL secretion causing osteoblast dysfunction; leading to the activation of osteoclasts and bone resorption. Beclin 1 ubiquitination is mediated by TRAF, which, in turn, is deubiquitinated by the CYLD (Cylindromatosis) gene under the influence of p62, an autophagy cargo protein, which also causes RANKL stimulated osteoclast differentiation [54].

Upon coming across a low oxygen tension, osteoclasts show increased differentiation and expression of BNIP3 via HIF-1α; stimulating Beclin 1, Atg5, and Atg 12 release and LC3 expression on autophagosomes clearly indicating increased autophagic flux. Alongside the expressions of RANKL, Cathepsin K, NFATc1, and MMPs are also increased leading to increased osteoclastogenesis. miR-155 microRNA targets TGF-β activated kinase 1 binding protein 2 (TAB2) to induce autophagy in osteoclasts and modulates their function and differentiation. Another protein, GIT1 (G-protein-coupled receptor kinase 2 interacting protein 1), triggers phosphorylation of Beclin 1 and its dissociation from Bcl2 leading to autophagy induction and osteoclastogenesis [54].

Role of Autophagy During Embryogenesis

Embryogenesis is a phenomenon post fertilization where the maternally obtained products like proteins and mRNA are degraded along with the simultaneous synthesis of new products from the 'zygotic genome'. This is a very rapid process with ubiquitin mediated proteasomal degradation playing an important role. But when this operation is not adequate, autophagy comes into play bringing about cellular remodeling [55].

Autophagy is activated within four hours of fertilization, regulated by E2 and progesterone which inhibits mTOR activation, a step found to have a very important role in embryo development. Atg5 null zygotes are reported to have arrested growth at the four to eight cell stage, whereas Atg5 positive zygotes proceed to the blastocyst stage. In zygotes with autophagy deficiency, there is a decreased rate of protein synthesis due to defects in the acquired protein clearance rate. Also, in such embryos, there is a reduced rate of implantation [56].

Other autophagy proteins like Beclin 1, FIP200, Ambra 1 deficiency leads to embryonic death because of pro-amniotic canal closing defect, heart failure and liver degeneration during mid or late gestational phases, defect in neural tube respectively. Atg 3, Atg5, Atg 7, Atg 16L ablation causes death of neonates just after birth due to defects in milk sucking and adjustment issues with the normoxic conditions outside the mother's womb [56].

Although there have been several publications, researchers are still trying to understand the complete involvement of autophagy in embryogenesis (Table 2.1).

The role of autophagy in all the above different types of stem cell differentiation has been schematically shown in Fig. 2.2.

Table 2.1 Functions of important proteins from the Autophagy cascade

Protein	Description	Function	References
ULK 1	Unc-51-like kinase 1	Translocates to sites on the Endoplasmic reticulum where nucleation of the autophagosome takes place. Phosphorylates ATG13 and FIP200 and forms the initiation complex Also activates Beclin 1	[2, 7]
FIP200	Focal adhesion kinase family-interacting protein of 200 kD	Part of the ULK1–ATG13–FIP200–ATG101 complex involved in nucleation and initiation Also interacts with ATG16L1 to regulate later stages of autophagophore formation	[57]
ATG 3	Autophagy-related Genes and proteins 3/4b/5/7/9a/12/13/16L1/101	An E2 enzyme responsible for lipidation of LC3, essential for elongation of the forming autophagosome	[58]
ATG 4b		Processing and delipidation of LC3-PE which is necessary for autophagosome formation	[59]
ATG 5		Together with ATG 16 and ATG 12, it helps to target particular substrates for developing autophagosome	[60]
ATG 7		It is an E1 enzyme that plays a role in conjoining ATG 5 with ATG 12. It is also involved in ATG 8 lipidation	[58]
ATG 9		It is involved in phagophore expansion by attaching itself to the lengthening part of it	[61, 62]
ATG 12		Together with ATG 5 and ATG 16, it helps to target particular substrates for developing autophagosome	[60]
ATG 13		In the initiation stage, ATG 13 interacts with ATG 101 and ULK1, and forms a complex which causes the pre-autophagosome formation	[63, 64]

(continued)

Table 2.1 (continued)

Protein	Description	Function	References
ATG16L		It is an E3-like enzyme formed in association with ATG 5 and ATG 12 during the elongation stage, responsible for LC3 lipidation required for autophagosome formation	[60]
ATG101		Stabilizes ATG13 and prevents its proteasomal degradation	[63, 64]
Beclin1		It binds to the forming pre-autophagosome and leads to its elongation, an essential event during nucleation	[65, 66]
PI3K	Phosphatidylinositol-3-kinase	It regulates Beclin 1 and initiates autophagy by getting phosphorylated during the nucleation stage	[2]
VPS34	Vacuolar protein sorting 34	Only protein identified in yeast which is conserved in mammals. It regulates the formation of PI3P, essential for elongation of the forming autophagosome	[67]
VPS15	Vacuolar protein sorting 15	It stabilizes and regulates the activity of VPS34 for phosphorylating PIP2 to PIP3	[68, 69]
WIPI	WD repeat protein interacting with phosphoinositides	It attaches to newly formed PIP3 and mediates phagophore formation. Individual proteins of this family have many other roles in regulating autophagy	[70, 71]
P62		It is also referred to as SQSTM1 (Sequestosome 1), a ubiquitin cargo binding protein	[72, 73]
LC3-I	Microtubule-associated protein 1 light chain 3 proteins I/II	It is the cytosolic form of the non-lipidated LC3 family protein and gets associated with p62 during the elongation stage	[74, 75]
LC3-II		It is the lipidated form of LC3-I and is a marker of mature autophagosomes	[74, 75]
Ambra1	Activating molecule in Beclin1-regulated autophagy protein 1	Essential regulator of cell death by apoptosis or cell survival via autophagy which upon phosphorylation due to activation of its upstream target, binds with Beclin1, initiating the functions of the VPS complex	[76, 77]

(continued)

Table 2.1 (continued)

Protein	Description	Function	References
Eva1a	Eva-1 homolog A	It is also known as TMEM 166 (transmembrane protein 166) or FAM176A (family with sequence similarity 176). It has a major role in autophagosome formation wherein it acts as a downstream target of Beclin1 and an upstream protein of ATG12-ATG5 complex	[78, 79]

Conclusion

As we come to the end of this chapter, we can say that the role of autophagy in stem cell maintenance and differentiation is quite diversified and purely context specific. It's necessitated in both embryonic as well as adult stem cells. Along with maintaining cellular and tissue-specific homeostasis, it is involved in various other functions like energy metabolism, regulating and dysregulating essential downstream pathways resulting in cell death, immune response, proliferation, and asymmetric division along with transcriptomic shift, ultimately leading to survival or differentiation. This role of autophagy is slightly different than its normal function of cell clearance during stress. Different forms of autophagy can simultaneously exist and function in the same cell. Even macroautophagy can perform differently based on the upstream signal received, i.e., the decision whether to go for debris clearance or maintenance or differentiation. Other than autophagy, ubiquitin mediated proteasomal degradation pathway also plays a crucial role to maintain cellular homeostasis under regular intracellular physiological conditions.

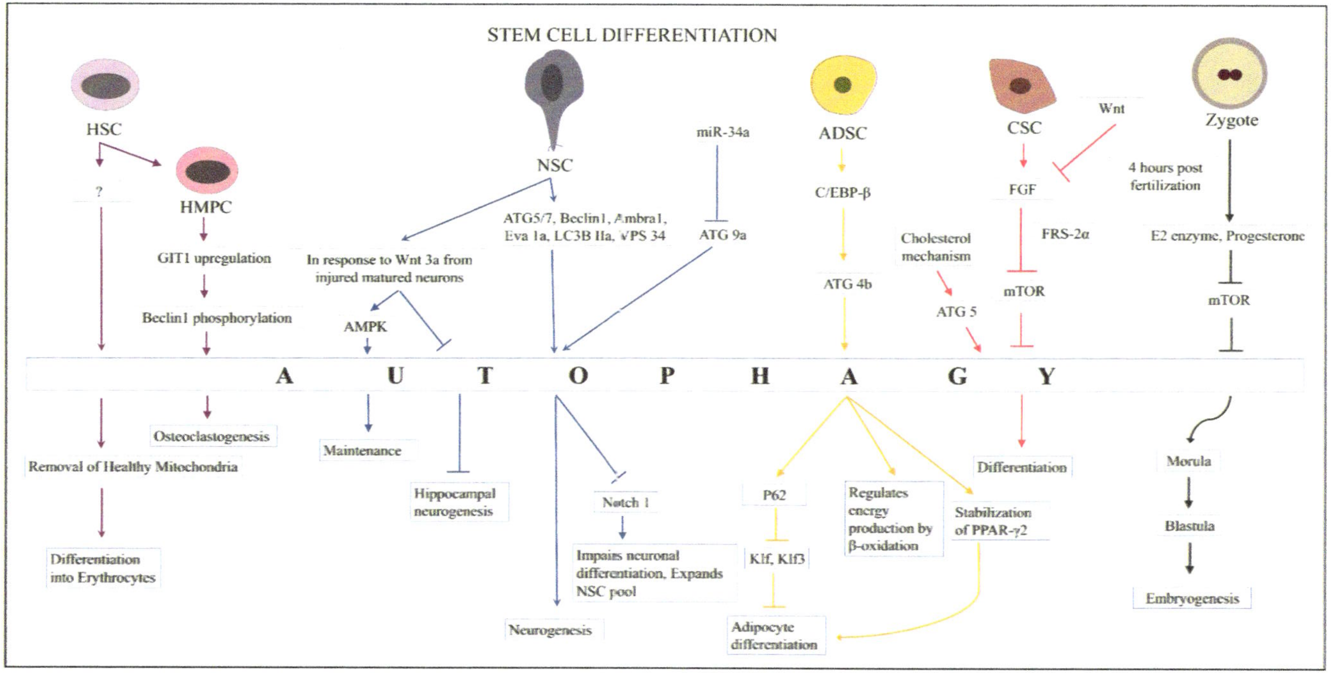

Fig. 2.2 *Role of autophagy in stem cell differentiation*. In case of stem cell differentiation, several upstream key proteins which are involved in deciding the fate and timing of stem cell differentiation regulate or dysregulate autophagy. As autophagy is basically involved in protein degradation followed by clearance; it might influence the half-life of several proteins which directly either enhance or inhibit stem cell differentiation. Thus, autophagy has an essential role to play in stem cell differentiation as well; along with stem cell maintenance and preserving cellular homeostasis. HSC (Hematopoietic stem cells), HMPC (Hematopoietic Mesenchymal Progenitor Cells), NSC (Neural stem cells), ADSC (Adipose Derived Mesenchymal Stem Cells), and CSC (Cardiac stem cells)

Acknowledgements We would like to acknowledge the DST-SERB grant (EMR/2017/004149) provided to SM and DRDO-LSRB grant (O/o DG(TM)/81/48222/LSRB-351/PEE&BS/2019) provided to RC as a funding source for our research. The authors acknowledge BITS Pilani, Pilani campus, for providing infrastructural support. AKS and PB acknowledge DST-SERB (EMR/2017/004149) and DRDO (O/o DG(TM)/81/48222/LSRB-351/PEE&BS/2019), respectively, for providing fellowship.

Compliance with Ethical Standards

Funding Source DST-SERB grant (EMR/2017/004149) provided to SM and DRDO-LSRB grant (O/o DG(TM)/81/48222/LSRB-351/PEE&BS/2019) provided to RC.

Disclosure of Interests All authors declare they have no conflict of interest.

Ethical Approval This article does not contain any studies with human participants or animals performed by any of the authors.

References

1. Khan I, Baig MH, Mahfooz S, Rahim M, Karacam B (2021) deciphering the role of autophagy in treatment of resistance mechanisms in Glioblastoma
2. Chun Y, Kim J (2018) autophagy: an essential degradation program for cellular homeostasis and life.https://doi.org/10.3390/cells7120278
3. El-gowily AH, Abosheasha MA (2020) Differential mechanisms of autophagy in cancer stem cells: emphasizing gastrointestinal cancers, 1–12. https://doi.org/10.1002/cbf.3552
4. Jung S, Jeong H, Yu S-W (2020) Autophagy as a decisive process for cell death. Exp Mol Med 526(52):921–930. https://doi.org/10.1038/s12276-020-0455-4
5. Levine B, Mizushima N (2011). Autophagy in immunity and inflammation, 1–5. https://doi.org/10.1038/nature09782
6. He C, Klionsky DJ (2009) Regulation mechanisms and signaling pathways of autophagy. Annu Rev Genet 43:67–93. https://doi.org/10.1146/annurev-genet-102808-114910.Regulation
7. Chang NC (2020) Autophagy and stem cells: self-eating for self-renewal 8:1–11. https://doi.org/10.3389/fcell.2020.00138
8. Castro-Obregon S (2010) The discovery of lysosomes and autophagy. Scitable by Nat Educ 3:49. https://www.nature.com/scitable/topicpage/the-discovery-of-lysosomes-and-autophagy-14199828/. Accessed August 10, 2021
9. Turksen K (2018) Autophagy in health and disease: potential therapeutic approaches. https://books.google.co.in/books?hl=en&lr=&id=Q3xxDwAAQBAJ&oi=fnd&pg=PR5&dq=autophagy+in+health+and+disease+Kursad+Turksen&ots=h4P592smya&sig=Bj20UB_RJAH5b8OQWa3ouCjwhDo. Accessed August 10, 2021
10. Simon H (n.d) Autophagy in myocardial differentiation and cardiac development. https://doi.org/10.1161/CIRCRESAHA.112.265157
11. Vessoni AT, Muotri AR, Okamoto OK (2012) Autophagy in stem cell maintenance and differentiation, 21. https://doi.org/10.1089/scd.2011.0526
12. Yang ZJ, Chee CE, Huang S, Sinicrope FA (2011) The role of autophagy in cancer: therapeutic implications. Mol Cancer Ther 10:1533–1541. https://doi.org/10.1158/1535-7163.MCT-11-0047
13. Sbrana FV, Cortini M, Avnet S, Perut F, Columbaro M, De Milito A, Baldini N (2016) The role of autophagy in the maintenance of stemness and differentiation of mesenchymal stem cells. Stem Cell Rev Reports, 621–633. https://doi.org/10.1007/s12015-016-9690-4

14. Guan J, Simon AK, Prescott M, Menendez JA, Wang F, Wang C, Wolvetang E, Vazquez-martin A, Zhang J (2013) Autophagy in stem cells. Taylor Fr. 9:830–849. https://doi.org/10.4161/auto.24132
15. Chen Q, Kang J, Fu C (2018) The independence of and associations among apoptosis, autophagy, and necrosis. Signal Transduct Target Ther. https://doi.org/10.1038/s41392-018-0018-5
16. Chaabane W, User SD, El-Gazzah M, Jaksik R, Sajjadi E, Rzeszowska-Wolny J, Łos, Autophagy MJ (2012) Apoptosis, Mitoptosis and necrosis: interdependence between those pathways and effects on cancer. Arch Immunol Ther Exp 611(61):43–58. https://doi.org/10.1007/S00005-012-0205-Y
17. Coleman J, Liu R, Wang K, Kumar A (2016) Detecting apoptosis, autophagy, and necrosis, 77–92. https://doi.org/10.1007/978-1-4939-3588-8_5
18. Rodolfo C, Di Bartolomeo S, Cecconi F (2016) Autophagy in stem and progenitor cells. Cell Mol Life Sci 73:475–496. https://doi.org/10.1007/s00018-015-2071-3
19. Jin M, Liu X, Klionsky DJ (2013) SnapShot: selective autophagy. Cell 152:368. https://doi.org/10.1016/J.CELL.2013.01.004
20. Li W, He P, Huang Y, Li YF, Lu J, Li M, Kurihara H, Luo Z, Meng T, Onishi M, Ma C, Jiang L, Hu Y, Gong Q, Zhu D, Xu Y, Liu R, Liu L, Yi C, Zhu Y, Ma N, Okamoto K, Xie Z, Liu J, He RR, Feng D (2020) Selective autophagy of intracellular organelles: recent research advances. Theranostics. 11:222–256. https://doi.org/10.7150/THNO.49860
21. Reggiori F, Komatsu M, Finley K, Simonsen A (2012) Autophagy: more than a nonselective pathway. Int J Cell Biol. https://doi.org/10.1155/2012/219625
22. Schuck S (2020) Microautophagy—distinct molecular mechanisms handle cargoes of many sizes. https://doi.org/10.1242/jcs.246322
23. Kaushik S, Cuervo AM (2012) Chaperone-mediated autophagy: a unique way to enter the lysosome world. Trends Cell Biol 22:407–417. https://doi.org/10.1016/j.tcb.2012.05.006.Chaperone-mediated
24. Zakrzewski W, Dobrzyński M, Szymonowicz M, Rybak Z (2019) Stem cells: past, present, and future. Stem Cell Res Ther 10. https://doi.org/10.1186/S13287-019-1165-5
25. Dikic I (2017) Proteasomal and Autophagic degradation systems. Annu Rev Biochem 86:193–224. https://doi.org/10.1146/ANNUREV-BIOCHEM-061516-044908
26. Chen X, He Y, Lu (2018) Review article autophagy in stem cell biology : a perspective on stem cell self-renewal and differentiation
27. Boya P, Codogno P, Rodriguez-muela N (2018) Autophagy in stem cells: repair, remodelling and metabolic reprogramming, 1–14. https://doi.org/10.1242/dev.146506
28. Warr MR, Binnewies M, Flach J, Reynaud D, Garg T, Malhotra R, Debnath J, Passegué E (2013) FOXO3A directs a protective autophagy program in haematopoietic stem cells. Nature 494:323–327. https://doi.org/10.1038/NATURE11895
29. Almeida CF, Fernandes SA, Ribeiro Junior AF, Keith Okamoto O, Vainzof M (2016) Muscle satellite cells: exploring the basic biology to rule them. Stem Cells Int. https://doi.org/10.1155/2016/1078686
30. Yin H, Price F, Rudnicki MA (2013) Satellite cells and the muscle stem cell niche. Physiol Rev 93:23. https://doi.org/10.1152/PHYSREV.00043.2011
31. Zhang Z, Yang M, Wang Y, Wang L, Jin Z, Ding L, Zhang L, Zhang L, Jiang W, Gao G, Yang J, Lu B, Cao F, Hu T (2016) Autophagy regulates the apoptosis of bone marrow-derived mesenchymal stem cells under hypoxic condition via AMP-activated protein kinase/mammalian target of rapamycin pathway 40:671–685. https://doi.org/10.1002/cbin.10604
32. Li L, Li L, Zhang Z, Jiang Z (2015) Hypoxia promotes bone marrow-derived mesenchymal stem cell proliferation through apelin/APJ/autophagy pathway 47:362–367. https://doi.org/10.1093/abbs/gmv014
33. Dong W, Zhang P, Fu Y, Ge J, Cheng J, Yuan H, Jiang H (2015) Roles of SATB2 in site-specific Stemness, autophagy and senescence of bone marrow mesenchymal stem cells. J Cell Physiol 230:680–690. https://doi.org/10.1002/JCP.24792

34. Xu F, Hua C, Tautenhahn H, Dirsch O, Dahmen U (2020) The role of autophagy for the regeneration of the aging liver
35. Trentesaux C, Fraudeau M, Luana C, Lemarchand J, Jacques S (2020) Essential role for autophagy protein ATG7 in the maintenance of intestinal stem cell integrity, 117. https://doi.org/10.1073/pnas.1917174117
36. Mortensen M, Ferguson DJ, Edelmann M, Kessler B, Morten KJ, Komatsu M, Simon AK (2010) Loss of autophagy in erythroid cells leads to defective removal of mitochondria and severe anemia in vivo. Proc Natl Acad Sci USA 107:832–837. https://doi.org/10.1073/PNAS.0913170107
37. Moras M, Lefevre SD, Ostuni MA (2017) From erythroblasts to mature red blood cells: organelle clearance in mammals. Front Physiol 8:1076. https://doi.org/10.3389/FPHYS.2017.01076
38. Kang YA, Sanalkumar R, O'geen H, Linnemann AK, Chang CJ, Bouhassira EE, Farnham PJ, Keles S, Bresnick EH (2012) Autophagy driven by a master regulator of hematopoiesis. Mol Cell Biol 32L:226–239. https://doi.org/10.1128/MCB.06166-11
39. Honda S, Arakawa S, Nishida Y, Yamaguchi H, Ishii E, Shimizu S (2014) Ulk1-mediated Atg5-independent macroautophagy mediates elimination of mitochondria from embryonic reticulocytes. Nat Commun 5:1–5. https://doi.org/10.1038/ncomms5004
40. Zhang J, Randall MS, Loyd MR, Dorsey FC, Kundu M, Cleveland JL, Ney PA (2009) Mitochondrial clearance is regulated by Atg7-dependent and independent mechanisms during reticulocyte maturation. Blood 114:157. https://doi.org/10.1182/BLOOD-2008-04-151639
41. Shimizu S (2018) Biological roles of alternative autophagy. Mol Cells 41:50–54. https://doi.org/10.14348/molcells.2018.2215
42. Casares-crespo L, Calatayud-baselga I, García-corzo L, Mira H (2018) On the role of basal autophagy in adult neural stem cells and neurogenesis 12:1–9. https://doi.org/10.3389/fncel.2018.00339
43. Lv X, Jiang H, Li B, Liang Q, Wang S, Zhao Q, JJ-S reports (n.d.) Undefined 2014, the crucial role of Atg5 in cortical neurogenesis during early brain development. Nature.Com. https://sci-hub.do/https://www.nature.com/articles/srep06010. Accessed August 11, 2021
44. Morgado AL, Xavier JM, Dionísio PA, Ribeiro MFC, Dias RB, Sebastião AM, Solá S, Rodrigues CMP (2015) MicroRNA-34a modulates neural stem cell differentiation by regulating expression of synaptic and Autophagic Proteins, 1168–1183. https://doi.org/10.1007/s12035-014-8794-6
45. Wu X, Fleming A, Ricketts T, Pavel M, Virgin H, Menzies FM, Rubinsztein DC (2016) Autophagy regulates Notch degradation and modulates stem cell development and neurogenesis. Nat Commun 7. https://doi.org/10.1038/NCOMMS10533
46. Ze-wei TAO, Long-gui LI (2007) Cell therapy in congestive heart failure 8:647–660. https://doi.org/10.1631/jzus.2007.B0647
47. Ro SH, Jang Y, Bae J, Kim IM, Schaecher C, Shomo ZD (2019) Autophagy in adipocyte browning: emerging drug target for intervention in obesity. Front Physiol 10:1–11. https://doi.org/10.3389/fphys.2019.00022
48. Zhang C, He Y, Okutsu M, Ong LC, Jin Y, Zheng L, Chow P, Yu S, Zhang M, Yan Z (2021) Autophagy is involved in adipogenic differentiation by repressesing proteasome-dependent PPARγ2 degradation. Am Physiol Soc 4:530–539. https://doi.org/10.1152/ajpendo.00640.2012
49. Levine B (2011) Autophagy in mammalian development and differentiation 12:823–830. https://doi.org/10.1038/ncb0910-823.Autophagy
50. Pellegrini C, Columbaro M, Schena E, Prencipe S, Andrenacci D, Iozzo P, Guzzardi MA, Capanni C, Mattioli E, Loi M, Araujo- D, Squarzoni S, Cinti S, Morselli P, Giorgetti A, Zanotti L, Gambineri A, Lattanzi G (2019) Altered adipocyte differentiation and unbalanced autophagy in type 2 Familial Partial Lipodystrophy: an in vitro and in vivo study of adipose tissue browning. Exp Mol Med. https://doi.org/10.1038/s12276-019-0289-0
51. Guo L, Huang JX, Liu Y, Li X, Zhou SR, Qian SW, Liu Y, Zhu H, Huang HY, Dang YJ, Tang QQ (2013) Transactivation of Atg4b by C/EBPβ promotes autophagy to facilitate adipogenesis. Mol Cell Biol 33:3180–3190. https://doi.org/10.1128/MCB.00193-13

52. Singh R, Xiang Y, Wang Y, Baikati K, Cuervo AM, Luu YK, Tang Y, Pessin JE, Schwartz GJ, Czaja MJ (2009) Autophagy regulates adipose mass and differentiation in mice 119:3329–3339. https://doi.org/10.1172/JCI39228DS1
53. Yin X, Zhou C, Li J (2019) Autophagy in bone homeostasis and the onset of osteoporosis. Bone Res. https://doi.org/10.1038/s41413-019-0058-7
54. Montaseri A, Giampietri C, Rossi M, Riccioli A, Del Fattore A, Filippini A (2020) Biomolecules The role of autophagy in osteoclast differentiation and bone resorption function, 1–16
55. Tsukamoto S, Yamamoto A, Tsukamoto S, Yamamoto A (2013) The role of autophagy in early mammalian embryonic development the role of autophagy in early mammalian embryonic development 30:86–94
56. Nakashima A, Aoki A, Kusabiraki T, Shima T, Yoshino O, Cheng S, Sharma S, Saito S (2017) Role of autophagy in oocytogenesis, embryogenesis, implantation, and pathophysiology of pre-eclampsia 43:633–643. https://doi.org/10.1111/jog.13292
57. Mizushima N (2007) FIP200, a ULK-interacting protein, is required for autophagosome formation in mammalian cells 9:497–510. https://doi.org/10.1083/jcb.200712064
58. Frudd K, Burgoyne T, Burgoyne JR (n.d.) Oxidation of Atg3 and Atg7 mediates inhibition of autophagy. Nat Commun, 1–15. https://doi.org/10.1038/s41467-017-02352-z
59. Maruyama T, Noda NN (2018) Autophagy-regulating protease Atg4: structure, function, regulation and inhibition. J Antibiot (Tokyo) 71:72–78. https://doi.org/10.1038/ja.2017.104
60. Fujita N, Itoh T, Omori H, Fukuda M, Noda T, Yoshimori T (2008) The Atg16L complex specifies the site of LC3 Lipidation for membrane biogenesis in autophagy 19:2092–2100. https://doi.org/10.1091/mbc.E07
61. Feng Y, Klionsky DJ (2017) Autophagic membrane delivery through ATG9. Cell Res 27:161–162. https://doi.org/10.1038/cr.2017.4
62. Ungermann C, Reggiori F (2018) Atg9 proteins, not so different after all. Autophagy 14:1456–1459. https://doi.org/10.1080/15548627.2018.1477382
63. Alers S, Wesselborg S, Stork B (2014) ATG13: just a companion, or an executor of the autophagic program?. Autophagy 10(6):944–956
64. Popelka H, Klionsky DJ (2017) The molecular mechanism of Atg13 function in autophagy induction: what is hidden behind the data ? Autophagy 13:449–451. https://doi.org/10.1080/15548627.2016.1277312
65. Kang R, Zeh HJ, Lotze MT, Tang D (2011) The Beclin 1 network regulates autophagy and apoptosis. Cell Death Differ 18:571–580. https://doi.org/10.1038/cdd.2010.191
66. Menon MB, Dhamija S (2018) Beclin 1 phosphorylation—at the center of autophagy regulation. Front Cell. Dev Biol 6:1–9. https://doi.org/10.3389/fcell.2018.00137
67. Rostislavleva K, Soler N, Ohashi Y, Zhang L, Pardon E, Burke JE, Masson GR, Johnson C, Steyaert J, Ktistakis NT, Williams RL (2015) Structure and flexibility of the endosomal Vps34 complex reveals the basis of its function on membranes. Science 80(350):1–25. https://doi.org/10.1126/science.aac7365
68. Anding AL, Baehrecke EH (2015) Vps15 is required for stress induced and developmentally triggered autophagy and salivary gland protein secretion in Drosophila. Cell Death Differ 22:457–464. https://doi.org/10.1038/cdd.2014.174
69. Iershov A, Nemazanyy I, Alkhoury C, Girard M, Barth E, Cagnard N, Montagner A, Chretien D, Rugarli EI, Guillou H, Pende M, Panasyuk G (2019) The class 3 PI3K coordinates autophagy and mitochondrial lipid catabolism by controlling nuclear receptor PPARα. Nat Commun 10:1–18. https://doi.org/10.1038/s41467-019-09598-9
70. Grimmel M, Backhaus C, Proikas T (2015) WIPI-mediated autophagy and longevity. Cells 4:202–217. https://doi.org/10.3390/cells4020202
71. Proikas T, Takacs Z, Dönnes P, Kohlbacher O (2015) WIPI proteins: essential PtdIns3P effectors at the nascent autophagosome. J Cell Sci 128:207–217. https://doi.org/10.1242/jcs.146258
72. Liu WJ, Ye L, Huang WF, Guo LJ, Xu ZG, Wu HL, Yang C, Liu HF (2016) P62 links the autophagy pathway and the Ubiqutin-proteasome system upon Ubiquitinated protein degradation. Cell Mol Biol Lett 21:1–14. https://doi.org/10.1186/s11658-016-0031-z

73. Kageyama S, Gudmundsson SR, Sou YS, Ichimura Y, Tamura N, Kazuno S, Ueno T, Miura Y, Noshiro D, Abe M, Mizushima T, Miura N, Okuda S, Motohashi H, Lee JA, Sakimura K, Ohe T, Noda NN, Waguri S, Eskelinen EL, Komatsu M (2021) p62/SQSTM1-droplet serves as a platform for autophagosome formation and anti-oxidative stress response. Nat Commun 12. https://doi.org/10.1038/s41467-020-20185-1
74. Tanida I, Ueno T, Kominami E (2008) LC3 and autophagy. Methods Mol Biol 445:77–88. https://doi.org/10.1007/978-1-59745-157-4_4
75. Runwal G, Stamatakou E, Siddiqi FH, Puri C, Zhu Y, Rubinsztein DC (2019) LC3-positive structures are prominent in autophagy-deficient cells. Sci Rep 9:1–14. https://doi.org/10.1038/s41598-019-46657-z
76. Maria Fimia G, Stoykova A,. Romagnoli A, Giunta L, Di Bartolomeo S, Nardacci R, Corazzari M, Fuoco C, Ucar A, Schwartz P, Gruss P, Piacentini M, Chowdhury K, Cecconi F(2007) Ambra1 regulates autophagy and development of the nervous system. Nature 447:1121–1125. https://doi.org/10.1038/nature05925
77. Sun WL (2016) Ambra1 in autophagy and apoptosis: Implications for cell survival and chemotherapy resistance. Oncol Lett 12:367–374. https://doi.org/10.3892/ol.2016.4644
78. Shen X, Kan S, Liu Z, Lu G, Zhang X, Chen Y, Bai Y (2017) EVA1A inhibits GBM cell proliferation by inducing autophagy and apoptosis. Exp Cell Res 352:130–138. https://doi.org/10.1016/j.yexcr.2017.02.003
79. Hu J, Li G, Qu L, Li N, Liu W, Xia D, Hongdu B, Lin X, Xu C, Lou Y, He Q, Ma D, Chen Y (2016) TMEM166/EVA1A interacts with ATG16L1 and induces autophagosome formation and cell death. Cell Death Dis 7:1–13. https://doi.org/10.1038/cddis.2016.230

Chapter 3
Autophagy in Embryonic Stem Cells and Neural Stem Cells

Deepika Puri, Shalmali Bivalkar-Mehla, and Deepa Subramanyam

Abstract Autophagy is a conserved cytoprotective catabolic pathway that plays a crucial role in cellular turnover and homeostasis in eukaryotic cells. A large body of work has implicated autophagy in normal development and differentiation of mammalian cells. In particular, precise regulation of autophagy is important for the maintenance of stemness and differentiation of embryonic stem cells. Extensive reports also indicate a critical role of autophagy in neural stem cells and during embryonic and adult neurogenesis. One of the critical stages of autophagy regulation occurs at the level of expression of autophagy genes. Transcription factors and chromatin modulators govern the accessibility of autophagy genes to the transcription machinery and regulate their expression. Understanding autophagy regulation in stem cells becomes critical as aberrant autophagy leads to numerous degenerative and neuropsychological diseases. This chapter aims to provide an overview of autophagy in embryonic stem cells and neural stem cells with a focus on regulation of autophagy genes in these cellular states.

Keywords Autophagy · Embryonic stem cells · Neural stem cells · Neurogenesis · Transcription

---

D. Puri and S. Bivalkar-Mehla—These authors contributed equally.

D. Puri · S. Bivalkar-Mehla · D. Subramanyam (✉)
National Centre for Cell Science, SP Pune University Campus, Ganeshkhind, Pune 411007, India
e-mail: deepa@nccs.res.in

D. Puri
Institute for Stem Cell Biology, Helmholtz Institute for Biomedical Engineering, Pauwelsstraße 20, 52074 Aachen, Germany

B. V. Shravage and K. Turksen (eds.), *Autophagy in Stem Cell Maintenance and Differentiation*, Stem Cell Biology and Regenerative Medicine 73,
https://doi.org/10.1007/978-3-031-17362-2_3

Abbreviations

AD	Alzheimers disease
ATG	Autophagy related genes
ESC	Embryonic stem cell
mTOR	Mammalian target of rapamycin
NCS	Neural stem cell
PTSD	Post Traumatic Stress Disorder
SGZ	Subgranular Zone
SVZ	Subventricular zone

Introduction

Autophagy is a well-conserved lysosomal degradation pathway in eukaryotic cells, serving as a quality control checkpoint for cellular proteins and organelles. In addition to the cellular proteasomal pathway, intracellular bulk degradation can occur through the process of autophagy. Autophagy can be categorized into three types based on the mode of lysosomal delivery of the substrate to be degraded: macroautophagy—where cellular components are encapsulated into an autophagosome and delivered to the lysosome; microautophagy—in which the protein substrate or organelle is directly engulfed by the lysosome; and chaperone mediated autophagy—where specific cargo proteins are delivered to the lysosome via chaperones [9].

Macroautophagy, simply referred to as autophagy, was first described in yeast and found to also be functional in higher eukaryotes [53, 54, 78]. It is responsible for the degradation of redundant and nonfunctional organelles, long-lived protein aggregates and misfolded damaged proteins [53, 68, 70, 80]. It works in conjunction with the ubiquitin proteosome system (UPS), and shows both selective and non-selective modes for substrate recognition [46]. In addition to a central conserved catabolic role in general, autophagy plays cell-type specific roles in a context-dependent manner [44, 57, 94, 118, 123, 129]. At the tissue level, it has been implicated in regulating the cell population by causing autophagy-mediated cell death. In addition to being a catabolic process, autophagy also works as a machinery for recycling and fueling, by providing a source of energy and building blocks for the synthesis of macromolecules.

The autophagy pathway has been studied in multiple cell types and in numerous model systems. Autophagy initiation is accompanied by the inactivation of mammalian target of Rapamycin (mTOR) kinase and the activation of ULK1/2 kinases. These form a complex with ATG13 and FIP200 to recruit the VPS15 (BECLIN-ATG14-VPS34-VPS15) complex and initiate phagophore formation. This leads to a cascade of activation and recruitment of ATG family proteins such as ATG5, ATG12, and ATG16L1, which result in the elongation and maturation of the phagophore, and selection and sequestration of cargo. Other ATG family proteins such as ATG7 and ATG3 help in the conversion of LC3I to LC3II by phosphatidylethanolamine conjugation. This facilitates closure of the autophagosome and its fusion with the lysosome. The cargo is then broken down by lysosomal hydrolases [19].

Over the years, literature has provided strong evidence showing that autophagy is involved in degradative and recycling roles right from the beginning of mammalian embryogenesis and differentiation into various lineages [70]. Embryonic stem cells (ESCs) are an ideal model system to recapitulate and study developmental processes and to understand the involvement of specific cellular pathways. Numerous studies have implicated a well-regulated autophagy pathway in the maintenance of stemness and differentiation of ESCs. Autophagy has also been studied extensively in the context of neural stem cells (NSCs) which play a crucial role in embryonic as well as adult neurogenesis. Aberrant autophagy in these cells leads to numerous debilitating neurodegenerative and neuropsychological diseases. This chapter aims to discuss the current literature in the context of autophagy, as well as the regulation of autophagy genes in ESCs and NSCs.

Embryonic Stem Cells

Embryonic stem cells (ESCs) are derived from the inner cell mass of the blastocyst, an important pre-implantation embryonic stage around 3.5–5.5 days post fertilization in the mouse embryo, and 4–9 days post fertilization in case of the human embryo [23, 64, 65] (Fig. 3.1). By virtue of their pluripotency, ESCs can differentiate into the three primary germ layers, namely: endoderm, mesoderm, and ectoderm. They are also capable of maintaining their own population through the process of self-renewal. During the course of early development, ESCs undergo multiple rounds of divisions and differentiation to generate an array of cell types in the embryo. ESCs can be cultured in vitro under conditions that permit the retention of pluripotency. Further, these cells can be utilized in various assays to unravel the events of early embryogenesis. ESCs were first isolated from early mouse embryos and were cultured in vitro in 1981 by Martin Evans and Gail Martin in separate experiments. Numerous studies over the last few decades have revealed an intricate interaction of various cellular pathways that regulate the stemness of ESCs. For example, in vitro studies on mouse ESCs (mESCs) have shown that the presence of a cytokine, leukemia inhibitory factor (LIF) along with bone morphogenetic protein (BMP) a member of the transforming growth factor beta (TGFβ) family (available in the serum added to culture medium), is critical for the maintenance of mESC self-renewal and pluripotency [76]. LIF/GP130 interaction activates multiple pathways including JAK/STAT3, PI3K/AKT, and MAPK/ERK resulting in STAT3 activation and nuclear translocation. Additionally, transcription of the core pluripotency factors OCT4, SOX2, and NANOG is directly under the control of STAT3 [30], Romito and

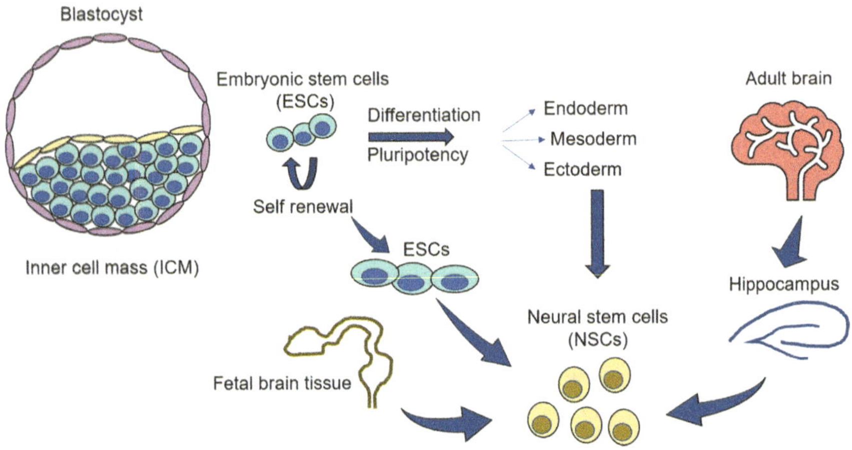

Fig. 3.1 ESCs and NSCs: ESCs are derived from ICM of the pre-implantation stage embryo, the blastocyst. ESCs are pluripotent, which means they can differentiate into any of the three germ cell lineages (endoderm, mesoderm and ectoderm) that give rise to all the cell types of an embryo. ESCs maintain their own population by self-renewal. ESCs can be differentiated into NSCs with appropriate cues. Neural stem cells can be derived from fetal brain tissue as well as from the hippocampus of adult brains

Cobellis [91]. Similarly, in addition to basic signalling pathways, processes such as endocytosis, autophagy, the ubiquitin–proteasome system (UPS), cytoskeletal dynamics, integrin signalling, and cell–cell interactions participate in controlling and maintaining ESC properties [18, 73–75, 111]. During embryogenesis, upon receipt of appropriate stimuli, ESCs exit pluripotency and commit to differentiation down specific lineages. Autophagy is a catabolic, nutrient-sensing stress response pathway that has been reported to assist in the maintenance of ESC pluripotency as well as promote differentiation in a context-dependent manner. Cellular macromolecular turnover along with organelle quality control and clearance are mediated by autophagy to promote ESC homeostasis and makeover during differentiation. In addition to its role in uncommitted, undifferentiated ESCs, autophagy plays homeostatic and quality control roles later in embryogenesis, and in stem cells that form specific tissues in the embryo such as neural stem cells (NSCs). In this chapter, we will attempt to review the significance of autophagy in ESCs and NSCs. Table 3.1 lists the effects of depletion of key autophagy proteins in ESCs and NSCs.

Autophagy in ESCs and During Embryogenesis: Evidence and Importance

Autophagy was found to be induced shortly after fertilization to facilitate removal of the paternal (sperm-derived) mitochondria during early embryonic development

Table 3.1 Effect of deletion of autophagy genes in ESCs and NSCs

Gene name	ESC	NSC	References
Beclin1	• Defective EB formation • Embryonic lethality • Smaller embryos • Developmental delay • Defective differentiation	• Smaller neurospheres, reduced differentiation	[8, 85, 125, 127]
Atg5	• Defective EB formation • Post-natal lethality • Low birth weight	• Reduced differentiation and neurogenesis • Increased apoptosis	[47, 62, 85, 108, 119]
Atg9	• Post-natal lethality • Low birth weight		
Atg7	• Post-natal lethality • Low birth weight	• Reduced autophagic cell death	[38]
Atg3	• Compromised self-renewal and differentiation		[59]
Atg16l1	• Post-natal lethality	• Developmental retardation • Impaired differentiation	[93, 118]
Lamp2a	• Delayed differentiation		[124]
LC3	• Disrupted homeostasis of pluripotency factors		[14]
Ambra1		• Defective neurosphere formation	[108, 125]
Eva1a1		• Reduced newly generated neurons	[56]
Vps34		• Decreased neurogenesis, neuronal migration and axon growth	[33]
Wdr45b		• Cerebral atrophy • Accumulation of autophagosomes	[35]
Fip200		• Loss of NSCs, reduction of neurospheres	[113]

[2, 95]. Autophagy, along with ubiquitin–proteasome system (UPS), was involved in the clearance of maternal factors to promote oocyte to embryo transition [104, 107]. Tsukamoto et al. have demonstrated that autophagic activity in 4-cell stage mouse embryos was associated with embryo viability [106].

Questions about the role of autophagy during early embryonic development can be answered to some extent by studying the autophagy pathway in ESCs. Due to their high proliferation rate, ESCs require constant clearance of unwanted and damaged macromolecules and organelles, which is largely taken care of by autophagy. Similarly, during differentiation, drastic changes in the transcriptome and proteome require clearance and recycling of existing components. In addition to its catabolic role, autophagy serves as a recycling machinery providing building blocks and an energy generation plant for the dividing and differentiating cells. To understand the involvement of autophagy in various physiological roles during development and in later life, multiple conventional mouse autophagy gene knockout models were created and studied [70]. In order to dissect basic steps of the autophagy pathway in mammalian cells, Mizushima et al. generated mESCs that lacked expression of certain key autophagy genes such as *Beclin1* and *Atg5* [71]. These, and similarly developed cell lines were further used to understand the role of autophagy in early mammalian development.

mESC clones deficient in *Beclin1* expression, and thus defective in autophagy, were generated by Yue et al. and assayed for embryoid body (EB) formation, as a proxy for early embryogenesis [127]. ESCs when cultured in nonadherent, pluripotency-suppressive conditions, form three-dimensional solid spheres of cells termed EBs. Upon culture for longer periods, EBs begin to cavitate and mimic differentiation events similar to those observed in the actual embryo. *Beclin1*$^{-/-}$ mESCs failed to form expanded cystic EBs. The visceral endoderm cells in these EBs were also reported to be abnormally large and less well-organized. *Beclin1*$^{-/-}$ mESCs were further used to generate *Beclin1* null embryos. Early embryonic lethality was reported at E7.5 stage with significant developmental delay, evident from the smaller *Beclin1*$^{-/-}$ embryos compared to wild type embryos at the same stage. Embryos lacking *Beclin1* expression showed enhanced cell death in contrast to controlled and patterned cell death in wild type embryos. The phenotype was attributed to loss of *Beclin1* expression and the resultant failure of autophagy [127]. We have also generated *Beclin1* knockout mESCs through CRISPR-Cas9 mediated genome editing. EBs derived from these cells showed defective differentiation into endodermal and mesodermal lineages. Similarly, teratomas generated by injecting *Beclin1* knockout mESCs into NOD/SCID mice showed fewer foci of endoderm and mesoderm lineages compared to teratomas generated from wild type ESCs [8]. Interestingly, *Beclin1* knockout mESCs showed active autophagy hinting towards an autophagy-independent role of BECLIN1 protein in mESC lineage differentiation.

In another study, differentiation of *Atg5-/-* or *Beclin1-/-* ESCs showed accumulation of cell corpses, arising from programmed cell death (PCD) resulting in compromised EB cavitation [85]. Clearance of dead cells by phagocytosis depends on soluble mediators ('Find me' signals) promoting phagocyte chemotaxis, and surface molecules ('Eat me' signals) exposed on apoptotic cells allowing identification by phagocytes [87]. Qu et al. showed that the underlying reason for failure of EB cavitation was failure of the dead cells at the EB core to express the surface phosphatidylserine acting as an 'Eat me' signal. Similarly, 'Come get me' or 'Find me' signals through secreted lysophosphatidylcholine (LPC) which serve as a chemoattractant for phagocyte recruitment were also suppressed resulting in persistence of dead cells in the EBs. These studies clearly demonstrated the involvement of autophagy in early stages of embryogenesis.

Kuma et al. confirmed that mice lacking *Atg5* expression, which is essential for the lipidation of LC3 protein and its recruitment in autophagosome membrane, were born normal at birth. These mice, however, failed to survive beyond a day after birth.

It was observed that *Atg5* deficient neonatal mice had lower amino acid levels in their plasma and tissues accompanied by an energy-depleted state [47]. It is possible that the defects seen in later stages could arise from abnormalities occurring due to autophagy failure in ESCs at earlier stages. One of the functions of autophagy is to provide amino acids and fatty acids as substrates for tricarboxylic acid cycle (TCA) to generate ATP [61]. This could also explain the phenotype observed in the *Atg5*$^{-/-}$ mouse model with defective autophagy.

Similar phenotype of neonatal lethality was presented in *Atg7*- [45], *Atg3*- [99], *Atg9a*- [92] and *Atg16l1*-[93] deficient mice models. In particular, in *Atg3*$^{-/-}$ mESCs, mitochondrial turnover was affected due to the inability to clear unwanted mitochondria, resulting in elevated reactive oxygen species (ROS) levels compared to wild type cells. mESCs lacking expression of *Atg3* presented compromised self-renewal and pluripotency with delayed differentiation potential [59, 3]. Collectively, it is evident from these reports that autophagy is an important pathway in mESCs and is required for maintenance of their properties and function during embryogenesis.

Systemic knockouts of some autophagy genes (*Atg3, Atg5, Atg7, Atg9, Atg16L1, Beclin1, Fip200,* and *Ambra1*) resulted in either embryonic or neonatal lethality, whereas knockouts of other autophagy genes did not present any phenotype (*Lc3b, Gabarap*) (reviewed in [70]). These studies emphasize the importance of autophagy proteins and pathway during mammalian development, and hint at a temporal role of these proteins. However, they are limited to describing the phenotype at a physiological level and further investigation is required to uncover the underlying mechanism at a cellular level. It is important to note that mice lacking expression of *Atg3, Atg5, Atg7, Atg9,* and *Atg16l1* had low birth weight and failed to survive due to suckling defects and starvation. Most of these proteins are involved in autophagosome formation following induction of autophagy. However, mice null for *Beclin1, Fip200,* and *Ambra1*, whose protein products play a regulatory role in induction of autophagy, were lethal at early embryonic stages. The range of phenotypic presentation resulting from the loss of proteins involved in the same pathway could possibly result from the autophagy-independent functions of some of these proteins. No discernible defects were reported in *Lc3b* and *Gabarap* knockout mice which may be due to alternative mechanisms compensating for the loss of these proteins [12, 121].

In an attempt to visualize autophagic activity in human ESCs (hESCs), Tra et al. established recombinant hESC lines transduced with a lentiviral vector containing a GFP-tagged LC3 (LC3-GFP) protein expression cassette. When autophagosomes form, the tagged LC3 protein gets recruited onto the autophagosomes allowing their visualization by fluorescence from LC3-GFP. These cell lines served as an informative tool to visualize the dynamics of autophagy in the pluripotent and differentiating state of hESCs. Autophagy, evident from increased LC3-GFP puncta, was induced in these cells following differentiation by use of the TGFβ receptor II inhibitor. Treatment with Rapamycin, an mTOR inhibitor, also showed increased number of GFP positive puncta, indicative of induced autophagy [105]. This study clearly showed for the first time that autophagy is active in hESCs.

Specific Roles of Autophagy in ESCs

Autophagy, although an alternative catabolic pathway, plays specialized roles in specific cell types and situations. In ESCs, autophagy plays a crucial role in maintaining the stemness of these cells until differentiation stimuli are received. Once ESCs are ready to lose pluripotency and acquire a new fate, autophagy brings about the makeover to ensure proper differentiation. Specific roles of autophagy in ESCs include both macromolecule and organelle clearance.

Autophagy and Proteostasis in ESCs

To maintain their pluripotency and immortality through self-renewal, ESCs have high-fidelity mechanisms ensuring the quality of their proteome. ESCs have an enhanced translation rate, along with elevated levels of ribosomal and chaperone complex and proteasomal components to meet the needs of a highly proliferative state. Multiple pathways including autophagy converge and interact to form a network working towards attaining protein homeostasis (proteostasis) [89]. Proteostasis includes multiple cellular processes throughout the life cycle of the protein, starting from protein synthesis, folding, trafficking, interaction, and finally degradation. Any irregularity in proteostasis leads to accumulation of damaged proteins, compromising ESC properties and functions [51, 110]. Apart from being a degradative pathway, autophagy also works as a recycling plant for providing building blocks for the synthesis of new macromolecules required for cellular infrastructure. Autophagy is thus a critical pathway that functions in both the synthesis and degradative steps of proteostasis, and eventually regulates ESCs stemness.

Autophagy and Pluripotency Factor Turnover

The role of autophagy was first demonstrated in hESCs showing elevated levels of GFP-LC3 puncta upon acute induction of differentiation upon treatment with TGFβ receptor II inhibitor [105]. Pluripotency-associated (PA) transcription factors OCT4, SOX2 and NANOG are key players in establishing and maintaining ESC pluripotency. Human ESCs were reported to depend upon autophagy for regulating the levels of PA factors (OCT4, SOX2, NANOG), suggesting a link between autophagy and ESC pluripotency [14]. Autophagy inhibition either by Rapamycin, a mTOR inhibitor, or by siRNA mediated knockdown of LC3 in hESCs resulted in accumulation of PA factors. It was also demonstrated by immunofluorescence that OCT4 colocalized with sequestosome 1 (SQSTM1/p62), an adaptor protein present in autophagosomes. This finding was corroborated by immunogold electron microscopy where OCT4 was found localized in the autophagosomes. Homeostasis of PA factors was primarily modulated by autophagy in coordination with the UPS to maintain the stemness of hESCs [14].

A recent report from Xu et al. revealed an interesting link between chaperone mediated autophagy (CMA) and epigenetic regulation of pluripotency factor expression. CMA is a specialized form of autophagy that selectively targets individual proteins bearing a signal sequence for lysosomal degradation with the help of chaperones. LAMP2A, a lysosomal outer membrane protein acts as a receptor that allows translocation of the substrate protein into the lysosome. Xu et al. reported that levels of LAMP2A increased when ESCs were induced to differentiate either by withdrawal of LIF, or by suppression of pluripotency factors. On the other hand, overexpression of LAMP2A and a resultant increase in CMA promoted ESC differentiation, whereas knockout or knockdown of LAMP2A supported maintenance of stemness and resulted in delayed differentiation [124]. CMA was shown to degrade isocitrate dehydrogenase 1 and 2 (IDH1 and IDH2), enzymes in the tricarboxylic acid (TCA) cycle that produce α-ketoglutarate (α-KG). α-KG is required by the histone and DNA demethylases for epigenetic changes in the genome that regulate the expression of pluripotency factors. Upon differentiation, induced CMA reduces IDH1 and IDH2 levels resulting in reduction of α-KG, thus inhibiting the activity of demethylases and suppressing the expression of *Oct4* and *Sox2* [124].

Along with protein substrates, autophagy is known to be involved in the clearance of damaged and unwanted organelles. In somatic cells, autophagy is implicated in the turnover of almost all organelles such as mitochondria, peroxisomes, lysosomes, and endoplasmic reticulum (ER) [3]. In ESCs, autophagy has been implicated in the clearance of mitochondria and the midbody ring which directly modulate ESC properties.

Mitophagy in ESCs

Mitophagy is a selective form of autophagy where mitochondria are encapsulated in vesicles and delivered to lysosomes for degradation [20]. ESCs are highly proliferative cells and hence they need to maintain the number and quality of mitochondria to avoid oxidative stress. ESCs possess immature mitochondria that are small, globular, and show underdeveloped cristae [15]. ATG3 deficient mESCs were found to show impaired self-renewal with accumulation of mitochondrial mass, increased ROS production, and decreased ATP synthesis. ATG3-dependent mitophagy was shown to be involved in regulating homeostasis of the mitochondrial pool in ESCs [59]. Mitophagy was elevated during the early events of hESC differentiation, and similarly following induction of mitochondrial damage by chemicals such as ethidium bromide (causes mitochondrial and genomic DNA damage) and CCCP (causes depolarization of mitochondrial membrane) to ensure the timely removal of damaged and harmful mitochondria [39]. This study demonstrated the occurrence of mitophagy in hESCs and its importance during ESC differentiation.

During differentiation, several structural, transcriptional, and metabolic changes occur allowing the ESCs to transform into the differentiated cell. Mitochondria in ESCs also become more mature and elongated with well-developed cristae. It has been reported that enhanced mitophagy due to Resveratrol (a polyphenolic compound) treatment promotes ESC pluripotency by rejuvenation of active mitochondria through elimination of the damaged ones [102].

Midbody Ring, Autophagy, and ESCs

Autophagy is responsible for bringing out some of the remodelling changes in organelles that are required for differentiation of ESCs, e.g., mitochondrial changes, midbody degradation. Midbody ring is a circular structure that forms a bridge between two daughter cells during cytokinesis and assists separation of daughter cells [97]. Midbodies are proteinaceous organelles inherited by daughter cells, and were shown to accumulate in hESCs to promote self-renewal [7, 48]. Autophagy has been implicated in removal of the midbody ring by recruiting NBR1 autophagy receptor that interacts with CEP55 midbody protein [34, 48, 83]. Midbodies were reported to be eliminated by autophagy when ESCs underwent differentiation. ESCs were able to accumulate midbodies by evading the engulfment and degradation by autophagosome. When cells are ready, autophagy comes into play to eliminate the midbodies to promote cellular differentiation, thus acting as a regulator of cellular fate changes [48].

Regulation of Autophagy in ESCs

Being highly proliferative cells, ESCs need to ensure that various cellular processes run at optimum speed. Autophagy flux was shown to be maintained at a level that promoted mESC pluripotency. In 2017, Liu et al. showed that to achieve optimum autophagy flux (a measure of autophagic degradation activity represented by formation of new autophagosome and their lysosomal degradation), mESCs were dependent on a forkhead transcription factor FOXO1, that possesses a winged helix DNA binding domain. FOXO1 was found to directly bind to promoters of core autophagy machinery genes including *Ulk1, Vps34, Beclin1, Atg5, Atg7, Atg12,* and *Lc3b* as shown by chromatin immunoprecipitation (ChIP) assay. mESCs deficient in *Foxo1* expression displayed suppressed autophagy gene expression that directly affected ESC pluripotency marker expression [60].

In somatic cells, autophagy is induced by several intrinsic and extrinsic stimuli such as ER stress due to unfolded proteins, oxidative stress due to ROS accumulation, hypoxia, low ATP levels, bacterial, viral infections, nutrient starvation, and so on. In somatic cells, mTOR is a well-studied kinase responsible for sensing the metabolic state of the cells and growth factor availability [49]. mTOR negatively regulates autophagy by phosphorylating and inactivating autophagosome initiation complex components such as ULK1, ATG13, etc. [21]. In an interesting study, blastocysts treated with Rapamycin were found to gain a state of reversible developmental pause, with a global suppression of transcription. These blastocysts were found to retain their ability to give rise to live and fertile mice with pluripotent ESCs for in vitro culture. Autophagy was active in these blastocysts and was found to be an important factor for retaining viability and properties of the cells [10].

The AMP-activated protein kinase (AMPK) plays a role when cellular ATP levels are low and thereby AMP levels are raised. In low energy states, the autophagy pathway is activated by AMPK through phosphorylation of ULK1, another kinase that initiates autophagy. It was demonstrated that Resveratrol, a natural polyphenolic compound, activated AMPK in mESCs leading to ULK1 activation and autophagy induction [102], thus revealing another mechanism of autophagy regulation in ESCs. Both mTOR and AMPK are nutrient and energy-sensing molecules which are implicated in autophagy regulation in somatic cells and were found to be functioning in ESCs in a similar manner.

Autophagy has been reported to be induced when ESCs receive differentiation cues. hESCs were shown to upregulate autophagy when treated with TGFβ receptor II inhibitor [105]. Similarly, when mESCs form EBs, autophagy was induced during early phases of differentiation [85]. During ESC differentiation, autophagy was induced to allow clearance of midbody rings [48]. These reports highlight the remodelling role of autophagy in addition to its catabolic role during ESC differentiation events.

Autophagy has been found to play important roles in macromolecular breakdown and clearance during differentiation into various lineages as a rapid response to differentiation cues. Autophagy has been shown to help in maintaining differentiation across numerous lineages such as haematopoietic, mesenchymal, muscle, hepatic, and many others (reviewed in Hu et al. [32]). A wealth of literature emphasizes the role of autophagy in neural stem cells and neurogenesis, and aberrant autophagy in these cells leads to a myriad of neurodegenerative and neuropsychiatric disorders. The following section summarizes the role of autophagy in neural stem cells.

Autophagy and Neural Stem Cells

Autophagy plays a fundamental role in the nervous system, especially in neurons that are post mitotic and require precise maintenance of homeostasis and macromolecular turnover. Aberrant autophagy leads to the development of aggregates and neurodegeneration with numerous neurodegenerative diseases linked to defects in autophagy. However, recent studies have implicated autophagy and autophagy genes in neural stem cells (NSCs) and the normal progression of neurogenesis. Aberrant neurogenesis has been found to be a primary proximal risk factor for a variety of neuropsychological disorders such as major depressive disorder, PTSD, anxiety disorders, and many more. Thus, understanding the role of autophagy in neural stem cells becomes imperative as a possible therapeutic approach to combat these disorders (reviewed in Puri and Subramanyam [84]).

Neural Stem Cells

The formation of the central nervous system in most mammals takes place at the prenatal stage and involves neurogenesis (formation of neurons) and gliogenesis (formation of glial cells). It was believed that neurogenesis in mammals only happens during embryonic development. However, studies have now unequivocally proven that neurogenesis takes place in the adult brain and is dependent on the presence of neural stem cells (NSCs), which have the ability for long-term self-renewal and ability to form neurons and glia [6, 26]. NSCs are primarily present in the ventricular-subventricular zone (V-SVZ) of the lateral ventricle and the subgranular zone (SGZ) of the dentate gyrus (DG) of the hippocampus [41]. These quiescent cells give rise to progenitor cells that further differentiate into neuroblasts and finally neurons (Fig. 3.1). Accumulating evidence has indicated that autophagy in NSCs carries out the crucial function of macromolecular turnover, homeostasis, and repair, helping stem cell function.

Evidence for Autophagy in Neural Stem Cells

A growing body of evidence implicates the presence of constitutive autophagy in embryonic as well as adult NSCs.

Autophagy in Embryonic NSCs

Autophagy genes such as *Atg5, Atg7, Becn1, LC3II, Ambra1, and Eva1a1* are expressed in NSCs, and their expression increases during neurogenesis [56, 62, 5, 108, 5, 125]. In 2012 [108, 5], Vazquez et al. used NSCs of the mouse olfactory bulb (OB) as a model system to show that AMBRA1 deficiency leads to defective neurosphere formation and reduced differentiation. Moreover, *Atg5* deficient NSCs also displayed reduced neural differentiation. Pharmacological inhibition of autophagy using Wortmannin, a PI3K inhibitor, impaired embryonic neuronal differentiation in OB NSCs as well as those of the foetal forebrain [26]. ATG5, required for autophagosome formation was also found to be expressed in mouse embryonic cortical neural progenitor cells (NPCs). *Atg5* knockdown led to decreased cortical differentiation and impaired morphology of cortical neurons [62, 5]. EVA1A1, a lysosomal transmembrane autophagy regulator, facilitates self-renewal of NSCs. Li et al., used the NSC-specific Nestin promoter to generate NSC-specific *Eva1a1* knockdown. Nestin-cre driven *Eva1a1* knockdown led to the reduced number of newly generated neurons accompanied by impaired autophagy [56]. Silencing of *Vps34,* an early autophagy regulator, by in utero electroporation in mice led to the disruption of neurogenesis, decreased neuron migration, and axonal growth [33]. In NPCs derived from the mouse forebrain, autophagy, measured by the levels of LC3II/LC3I, increased during neurogenesis, accompanied by an increase in ATG9A levels [72]. A seminal study by Wu et al. [118] elucidated the mechanism of ATG16L1-mediated regulation of embryonic neurogenesis. *Atg16l1* hypomorphs exhibited developmental retention and impaired differentiation of NSCs. Primary cortical cultures of these mice showed an increase in NSCs as measured by Nestin levels. This study suggests that complete inhibition and partial perturbation of autophagy may result in distinct effects in NSCs.

Autophagy in Adult NSCs

Most studies in adult NSCs involve the perturbation of autophagy genes at the embryonic stage while investigating their effect on adult NSCs. However, few recent studies have shown the effect of autophagy disruption directly in adult NSCs. Quiescent hippocampal NSCs contain more, and larger lysosomes than dividing NSCs and activation of autophagy by starvation, implicating autophagy in the maintenance and function of adult NSCs [52]. One key study implicated the autophagy gene *Fip200* in post-natal neurogenesis. *Fip200* was conditionally deleted in mouse NSCs using *Gfap-cre.* Loss of *Fip200* led to an aberrant increase in reactive oxygen species, loss of NSCs, and reduction of neurospheres from SVZ cultures, implicating FIP200 and, in turn, autophagy in post-natal NSCs and neurogenesis [113]. GFAP-dependent deletion of *Fip200* also led to a reduction in neural differentiation, accompanied by microglia activation and niche infiltration. However, this effect was not observed upon depletion of *Atg5, Atg7 or Atg16l1* [112]. In addition to *Fip200,* a recent study reported the relevance of autophagy genes *Beclin1* and *Ambra1* in post-natal NSCs of the SVZ [125]. *Ambra1* and *Beclin1* were both expressed in SVZ NSCs as well as NPCs. NSCs isolated from *Beclin1*$^{-/+}$ heterozygous mice exhibited smaller neurospheres and defective differentiation. This was accompanied by reduced autophagic flux. The reduction of NSCs was accompanied by an increase in apoptosis hinting at autophagy and apoptosis being balanced in normal NSCs. Similar results were seen in *Ambra1* knockout mice. ATG5 was also found to be important for NPC neurogenic potential. Using the fluorescently tagged autophagy protein LC3 to visualize autophagosomes, *Atg5* was knocked out in dividing NPCs in the adult brain. In addition to decreased autophagic flux, these cells exhibited reduction in survival, deficit in neurogenesis, and concomitant increase in apoptosis, which could be rescued by inactivation of the pro-apoptotic gene *Bax* [119, 5]. A recent study has implicated the mouse homolog of *Atg18, Wdr45b* in neural homeostasis [35]. *Wdr45b* deficient mice exhibited accumulation of autophagosomes, swollen axons, and cerebral atrophy accompanied by aggregates of autophagy substrates such as p62. This is of interest because *Wdr45b* deficiency in humans has been linked to intellectual disability [101, 45], and the link with autophagy may shed light on the possible mechanism of this defect.

Regulation of Autophagy in NSCs

Numerous stimuli have led to the activation or repression of autophagy in NSCs. An interesting connection between autophagy and neuronal cell death was identified, where neuronal cells exhibited massively upregulated autophagy, which, in turn, push the cells into a cell death cascade, distinct from apoptosis [17]. Many reports on NSCs implicate aberrant activation of autophagy in autophagic neural cell death. Withdrawal of insulin in adult hippocampal NSCs led to the activation of autophagy accompanied by an upregulation of autophagy genes such as *Beclin1* and *Lc3* [126], with GSK3β functioning as a critical upstream regulator of autophagy in these cells [29]. Additionally, phosphorylation of the autophagy protein, p62, by AMPK in insulin-withdrawn hippocampal NSCs seems to direct cell death pathways from apoptosis to autophagic cell death [28]. Autophagic cell death was also induced upon oxygen and/or glucose deprivation of hippocampal NSCs [16]. Conversely, mouse NSCs are protected from autophagic cell death upon depletion of ATG7 [38]. Thus, autophagy seems to regulate cell death pathways in a context-dependent manner. A recent study showed that autophagy played an important role in the death of NSCs in juvenile mouse brains in response to irradiation [115]. NSC-specific depletion of the autophagy gene *Atg7* reduces radiation-induced cell death and microglia activation, indicating that autophagy reduction may be an avenue to reduce cell death in response to radiation in cancer treatment. Interestingly, chronic restraint stress reduced hippocampal neurogenesis by inducing autophagic cell death of NSCs, which could be reversed by conditional *Atg7* knockdown [37]. Simvastatin, a statin family drug, protects NSCs from hydrogen peroxide mediated cell death by activation of autophagy and reduction in ROS [86]. Metabolites such as homocysteine also regulate autophagy in NSCs, and elevated levels of homocysteine have been linked with an increased risk of ischaemic stroke accompanied by high levels of autophagy which ultimately led to autophagic cell death [114]. This points to the importance of tightly regulated autophagy levels in NSC homeostasis. Gestational diabetes is associated with foetal neuropathy which is characterized by dysfunction of NSCs. Melatonin plays a neuroprotective function in this scenario. Studies have shown that in response to hyperglycaemia, autophagy increases in NSCs along with an increase in the transcription of autophagy genes. Melatonin activates mTOR and thereby represses downstream autophagy genes, preventing excessive autophagy, thus helping maintain NSC homeostasis during hyperglycemia [55]. Interestingly, studies have also implicated aberrant autophagy in human foetal NSCs upon Zika virus infection, which leads to defective neurogenesis and cellular dysregulation [57]. These studies indicate that the regulation of precise levels of autophagy is critical for maintaining the neural stem cell state and further neural differentiation.

Transcriptional Regulation of Autophagy Genes

Recent studies have helped shed light on the regulation of autophagy genes at the transcriptional level. Numerous transcription factors have been identified that regulate the expression of autophagy genes. Forkhead box transcription factors such as FOXO1 and FOXO3 are highly expressed in mouse embryonic SVZ as well as in multipotent adult neural progenitors [79], with *Foxo* depleted NSCs exhibiting decreased self-renewal and increased levels of ROS. Autophagy was impaired in NPC cultures derived from *Foxo 1/3/4* knockout mice, which was accompanied by a decrease in transcription of autophagy genes such as *Atg5, Atg7, and Atg9.* Chromatin immunoprecipitation experiments revealed that FOXO transcription factors were enriched over the enhancers of autophagy genes indicating their role in direct regulation [96]. The defect in autophagy upon *Foxo* knockout could be rescued by treatment with activators such as Rapamycin or trehalose. A recent study also identified FOXO3 as a core transcription factor activating numerous autophagy genes in NSCs [4], with its depletion leading to the accumulation of aggregates in these cells. These results indicated that this transcription factor was crucial in maintaining normal levels of autophagy in NSCs. Transcription factor EB (TFEB) has been identified as a master regulator of autophagy, driving the expression of autophagic and lysosomal genes [98]. TFEB plays a role in maintaining NSC quiescence by maintaining lysosomal activity. Conditional knockdown of *Tfeb* in NSCs delayed quiescence and increased NSC proliferation in the murine dentate gyrus [43]. The zinc finger transcription factor MIZ was found to target numerous genes playing a role in vesicular transport, endocytosis, and autophagy. The autophagy activator AMBRA1 was found to be a target for MIZ1 in neural progenitor cells. Depletion of MIZ1 led to a blockage in autophagic flux and degeneration of NPCs [117]. In addition to transcription factors, miRNAs as well as chromatin modifiers also regulate the expression of autophagy genes. Overexpression of miR34a impaired the differentiation of NSCs by targeting *Atg9a*, accompanied by the reduction of LC3II and, in turn, autophagy [72]. The let-7 family of miRNAs was also found to positively regulate autophagy in new-born neurons and NSCs, thereby inducing radial migration and neurogenesis [81]. The histone deacetylase inhibitor, SAHA, was found to induce autophagy in glioblastoma stem cells by downregulating mTOR signalling. This was accompanied by an increase in acidic vesicles, LC3II, and BECLIN1 levels, and a reduction in p62 levels [13]. Valproic acid, another HDAC inhibitor activated neuronal differentiation of NSCs, via the mTOR signalling pathway, results in the decrease in levels of DNA methyl transferases and activation of differentiation gene *Ngn1* [128]. Using a small molecule inhibitor, the histone methyl transferase G9a was found to supress the expression of autophagy genes such as *Lc3* and Atg family genes in neuroblastoma cells. G9a inhibition led to activation of autophagy and decreased tumorigenicity of neuroblastoma cells [40]. Treatment of mouse primary cortical NSCs with inhibitors for DOT1L (an H3K79 methyl transferase) led to impaired proliferation, survival, and neuronal differentiation. This was accompanied by impaired activation of genes of the *Atf3/Atf4/Ddit3* cascade, which in addition to playing a role in ER stress, also act as activators of the autophagy pathway. While the directed effects of DOT1L on autophagy have not been studied, the effect on autophagy activator genes and the subsequent increase in apoptosis in NSCs point to a possible autophagy-mediated disruption of NSC homeostasis [90].

Implications of Autophagy Defects in Development and Disease

In addition to being a housekeeping catabolic pathway, autophagy also serves as an adaptive pathway involved in cellular homeostasis and stress management. Naturally, mutations in autophagy genes and their consequential disturbances in the functioning of the pathway manifest as pathologies (either morbidity or mortality) in the organism. In recent years, as the importance of autophagy in embryonic development and cellular homeostasis is established, multiple reports have emerged unravelling the mutations in autophagy genes culminating in pathologies [36, 54]. Mutations in core autophagy genes reported to be associated with diseases in humans are discussed here.

A homozygous missense mutation in the *Atg5* gene resulting in E122D substitution is responsible for impaired interaction between ATG5 and ATG12 proteins. This leads to defective autophagy and is reported to be associated with congenital ataxia with impaired coordination of movement and balance during voluntary activity [42]. A homozygous *Atg5* variant with low expression was associated with childhood cerebral palsy [122]. Cerebral palsy was characterized by permanent motor disorders and disturbances of sensation, perception, cognition, and epilepsy.

ATG16L1 is an autophagy component that non-covalently associates with the ATG5-ATG12 complex involved in the lipidation of LC3 during autophagosome membrane elongation [69]. A missense mutation, T300A, in the *Atg16l1* allele was found to be a susceptibility allele for the development of Crohn's disease, a type of inflammatory bowel disease [50]. The presence of this mutation was found to be the cause of abnormal Paneth cells with reduced selective autophagy levels, increased cytokine secretion, and hampered intracellular bacterial clearance. BECLIN1 is another component of the PI3K complex that synthesizes PI3P which is required for autophagosome initiation and elongation step. Monoallelic deletion of *Beclin1* gene has been associated with 40–75% of breast, ovarian, and prostate cancers. Reduced BECLIN1 protein expression and poorer prognostic outcomes have been reported in multiple types of tumors including breast, ovarian, oral, lung, and renal [109]. VPS15 is a membrane targeting component of the PI3K complex [5]. A missense mutation L1224A was found to be associated with severe cortical and optic nerve atrophy, localized cortical dysplasia, epilepsy, intellectual impairment, ataxia, muscle wasting, spasticity, and other related symptoms [27]. It is important to note that mutations in different autophagy genes show pathologies in specific tissues or organ systems. It is highly likely that these mutations result in the hampered development of that particular organ or tissue right from the embryonic stage. It is thus intriguing to uncover the specific roles of these autophagy components and pathways as a whole, chronologically during development. Gaining more insight into specific roles of autophagy during development and organogenesis may allow the development of more targeted therapeutic interventions.

Numerous neuronal, neurodegenerative, and neuropsychological disorders have been linked to aberrant autophagy. Diseases such as Alzheimer's disease (AD), Parkinson's disease (PD), Amyotrophic lateral sclerosis (ALS), and polyglutamine accumulation diseases (Poly Q) are characterized by the accumulation of aggregates containing misfolded proteins. Studies have now implicated aberrant autophagy in the aggregation, defective degradation, and clearance of misfolded proteins in these neurons.

AD is characterized by the accumulation of neurotoxic beta-amyloid plaques, and impaired autophagy is observed in the neurites of human AD patients [77]. Studies have also shown a reduction in BECLIN1 levels in AD brain samples [82]. Activators of autophagy such as lithium, trehalose, and berberine have shown to exhibit neuro-protective effects, improved neurogenesis, and reduced AB aggregates in mouse models of AD [22, 24]. SQSTM1, a receptor for selective autophagy, is implicated in ALS. Insoluble SQSTM1 aggregates were found to accumulate in the brains of ALS patients and resulted in motor neuron degradation [103]. This suggests that aberrant clearance of this substrate by autophagy may lead to aggregate formation. PolyQ diseases such as Huntington's disease (HD) are characterized by the aberrant expansion of CAG repeats, translating into Poly Q tracts in certain genes such as Huntingtin (HTT). In neurons of HD mice, defective autophagy, caused by ineffective cargo recognition and binding to p62, leads to the accumulation of mutant HTT. Furthermore, this mutant HTT was found to impair the function of the autophagy protein BECLIN1, thus perpetuating the cycle of dysfunctional autophagy and aggregate formation [25, 63]. PD is caused by a mutant alpha synuclein that encodes for PARK1. This causes dysfunction and death of motor neurons in the substantia nigra which results in PD-specific motor symptoms. Studies have implicated that alpha synuclein is a target, as well as a regulator of autophagy, and that increased alpha synuclein led to reduced autophagy by mislocalization of ATG9 [11, 116].

In addition to neurodegenerative disorders, increasing evidence suggests that neuropsychological disorders such as major depressive disorder (MDD), schizophrenia, and stress disorders are also consistent with dysfunctional autophagy. Gene expression analysis of the superior dentate gyrus of schizophrenia patients indicated reduced expression of autophagy genes *Beclin1* and *Atg3* [31]. This reduction of autophagy genes was also seen in mouse models of schizophrenia [66], indicating that autophagy plays a prominent role in the pathophysiology of the disorder. Stress response has been shown to be highly correlated with depressive behaviour, and debilitating disorders such as MDD [88]. According to the neurogenic theory of depression, defective neurogenesis is one of the causes of depressive disorders [67] and aberrant autophagy plays a crucial role in maintaining normal neurogenesis, implicating this pathway in stress response. Antidepressants are found to reverse behavioural effects of chronic stress in an mTOR-dependent manner [1]. In addition to antidepressants, nicotine was found to alleviate stress-induced behavioural effects by upregulating autophagy genes such as *Beclin1*, *Lc3*, and *p62* [120]. Interestingly, some stressors exhibit a context-dependent change in autophagy in different regions of the brain. Maternal separation led to decreased autophagy in the hippocampus, while increasing autophagy in the prefrontal cortex of rat brains [58]. These reports, and more, emphasize the importance of maintaining precise autophagy levels in different regions of the brain in normal conditions as well as in response to different stressors.

Conclusions and Outlook

Autophagy is a conserved catabolic pathway that plays a homeostatic role during starvation, growth factor deprivation, infection, remodelling of cellular infrastructure, and protein and organellar turnover. It is proving to be one of the crucial cellular processes that modulate early mammalian embryonic development. In the context of early mammalian development, autophagy seems to be involved in the regulation of cell division, cell death, and clearance of cell debris during cell fate decisions, as well as in controlling metabolic homeostasis during early neonatal stages. The association of autophagy gene mutations with certain diseases emphasizes the role of autophagy in various cellular processes. The generation of animal models with specific autophagy gene mutations can provide valuable tools to understand the underlying mechanisms leading to pathological conditions. CRISPR-Cas9-based genome editing has great potential in generating such animal models. It is intriguing to note that in mouse knockout models, the loss of different autophagy genes lead to a range of phenotypic presentations, pointing towards the unique functions of these components in addition to their autophagic role. In somatic cells, some of the autophagy proteins have already been implicated in functions that are independent of autophagy, e.g., BECLIN1 and FIP200 (reviewed in Subramani and Malhotra [100]). These proteins may play characteristic roles in ESCs during embryonic development resulting in varied phenotypes. Another possibility is that the autophagy pathway performs context-dependent functions at different stages and in different cell types. These hypotheses need to be tested to further understand the importance of autophagy in mammalian development.

The rate of autophagic degradation of various substrates is crucial in various situations. For example, during ESC replication, cells need rapid clearance of damaged mitochondria. During differentiation, ESCs are required to clear macromolecules that promote pluripotency. In such situations, autophagy is upregulated by mechanisms involving enhanced autophagy gene expression or activation of autophagy regulators. Tools are available to visualize autophagosomes either through the use of dyes that accumulate within the autophagosome, or fluorescently tagged LC3 protein which gets incorporated into the autophagosome membrane. While these tools are suitable for cells with high cytoplasm to nuclear ratio, in the case of ESCs which have low cytoplasm to nuclear ratio, visualization of autophagosomes using these tools is challenging. Better visualization methods with high resolution are an absolute necessity to study autophagosome turnover in ESCs. Similarly, reporter systems that will allow tracking of autophagy induction and inhibition in response to various stimuli experienced by ESCs would be useful. This will expand our understanding of the array of stimuli and signals that change the autophagic status in ESCs in different situations.

Neurons are post mitotic cells and require precise regulation of cellular pathways for normal homeostasis. Autophagy plays a role not only during embryonic and adult neurogenesis, but also during the lifetime of the neuron to maintain a steady turnover of macromolecules, and also in response to various positive and negative stimuli. Understanding the exact role of autophagy in NSCs and NPCs becomes essential given the fact that a delicate balance of autophagy levels is required for normal neural function. Decreased autophagy leads to a reduction in the neurogenic potential of NSCs and aggregate formation in mature neurons that further leads to neurodegenerative diseases; while aberrant activation of autophagy leads to autophagic cell death which is detrimental to neuronal functions. In this background, molecules that regulate autophagy are promising putative therapeutic agents to either increase or decrease autophagy in a context-dependent manner with an aim to combat autophagy-mediated neural disorders.

Autophagy and its components are becoming attractive targets for the treatment of various disorders linked to the pathway and mutations in autophagy genes. For this purpose, understanding the regulation and function of autophagy and its interaction with other cellular processes is critical. Putting the pieces of this puzzle together will allow looking at the bigger picture of how this conserved catabolic pathway can be exploited to design better therapeutic strategies for targeting autophagy-related diseases and disorders.

Acknowledgements We thank members of the Subramanyam lab for discussions.

Compliance with ethical standards

Disclosure of Interests All the authors declare that they have no conflict of interest.

Ethical Approval This article does not contain any studies with human participants or animals performed by any of the authors.

References

1. Abelaira HM, Réus GZ, Neotti MV, Quevedo J (2014) The role of mTOR in depression and antidepressant responses. Life Sci 101:10–14
2. Al Rawi S, Louvet-Vallée S, Djeddi A, Sachse M, Culetto E, Hajjar C, Boyd L, Legouis R, Galy V (2011) Postfertilization autophagy of sperm organelles prevents paternal mitochondrial DNA transmission. Science 334:1144–1147
3. Anding AL, Baehrecke EH (2017) Cleaning house: selective autophagy of organelles. Dev Cell 41:10–22
4. Audesse AJ, Dhakal S, Hassell L-A, Gardell Z, Nemtsova Y, Webb AE (2019) FOXO3 directly regulates an autophagy network to functionally regulate proteostasis in adult neural stem cells. PLoS Genet 15:e1008097
5. Backer JM (2016) The intricate regulation and complex functions of the Class III phosphoinositide 3-kinase Vps34. Biochem J 473:2251–2271

6. Baptista P, Andrade JP (2018) Adult hippocampal neurogenesis: regulation and possible functional and clinical correlates. Front Neuroanat 12
7. Baumann K (2011) Inheritance for pluripotency. Nat Rev Mol Cell Biol 12:691–691
8. Bivalkar-Mehla S, Puri D, Singh SB, Subramanyam D (2021) Understanding the role of Beclin1 in mouse embryonic stem cell differentiation through CRISPR-Cas9-mediated gene editing. J Biosci 46:18
9. Boya P, Reggiori F, Codogno P (2013) Emerging regulation and functions of autophagy. Nat Cell Biol 15:713–720
10. Bulut-Karslioglu A, Biechele S, Jin H, Macrae TA, Hejna M, Gertsenstein M, Song JS, Ramalho-Santos M (2016) Inhibition of mTor induces a paused pluripotent state. Nature 540:119–123
11. Button RW, Roberts SL, Willis TL, Hanemann CO, Luo S (2017) Accumulation of autophagosomes confers cytotoxicity. J Biol Chem 292:13599–13614
12. Cann GM, Guignabert C, Ying L, Deshpande N, Bekker JM, Wang L, Zhou B, Rabinovitch M (2008) Developmental expression of LC3α and β: absence of fibronectin or autophagy phenotype in LC3β knockout mice. Dev Dyn 237:187–195
13. Chiao M-T, Cheng W-Y, Yang Y-C, Shen C-C, Ko J-L (2013) Suberoylanilide hydroxamic acid (SAHA) causes tumor growth slowdown and triggers autophagy in glioblastoma stem cells. Autophagy 9:1509–1526
14. Cho Y-H, Han K-M, Kim D, Lee J, Lee S-H, Choi K-W, Kim J, Han Y-M (2014) Autophagy regulates homeostasis of pluripotency-associated proteins in hESCs. STEM CELLS 32:424–435
15. Cho YM, Kwon S, Pak YK, Seol HW, Choi YM, Park DJ, Park KS, Lee HK (2006) Dynamic changes in mitochondrial biogenesis and antioxidant enzymes during the spontaneous differentiation of human embryonic stem cells. Biochem Biophys Res Commun 348:1472–1478
16. Chung Y, Lee J, Jung S, Lee Y, Cho JW, Oh YJ (2018) Dysregulated autophagy contributes to caspase-dependent neuronal apoptosis. Cell Death Dis 9:1–19
17. Clarke PG (1990) Developmental cell death: morphological diversity and multiple mechanisms. Anat Embryol (Berl) 181:195–213
18. David BG, Fujita H, Yasuda K, Okamoto K, Panina Y, Ichinose J, Sato O, Horie M, Ichimura T, Okada Y et al (2019) Linking substrate and nucleus via actin cytoskeleton in pluripotency maintenance of mouse embryonic stem cells. Stem Cell Res 41:101614
19. Dikic I, Elazar Z (2018) Mechanism and medical implications of mammalian autophagy. Nat Rev Mol Cell Biol 19:349–364
20. Ding W-X, Yin X-M (2012) Mitophagy: mechanisms, pathophysiological roles, and analysis. Biol Chem 393:547–564
21. Dossou AS, Basu A (2019) The emerging roles of mTORC1 in macromanaging autophagy. Cancers 11:1422
22. Durairajan SSK, Liu L-F, Lu J-H, Chen L-L, Yuan Q, Chung SK, Huang L, Li X-S, Huang J-D, Li M (2012) Berberine ameliorates β-amyloid pathology, gliosis, and cognitive impairment in an Alzheimer's disease transgenic mouse model. Neurobiol Aging 33:2903–2919
23. Evans MJ, Kaufman MH (1981) Establishment in culture of pluripotential cells from mouse embryos. Nature 292:154–156
24. Fiorentini A, Rosi MC, Grossi C, Luccarini I, Casamenti F (2010) Lithium improves hippocampal neurogenesis, neuropathology and cognitive functions in APP mutant mice. PLoS One 5
25. Fu Y, Wu P, Pan Y, Sun X, Yang H, Difiglia M, Lu B (2017) A toxic mutant huntingtin species is resistant to selective autophagy. Nat Chem Biol 13:1152–1154
26. Götz M, Huttner WB (2005) The cell biology of neurogenesis. Nat Rev Mol Cell Biol 6:777–788
27. Gstrein T, Edwards A, Přistoupilová A, Leca I, Breuss M, Pilat-Carotta S, Hansen AH, Tripathy R, Traunbauer AK, Hochstoeger T et al (2018) Mutations in Vps15 perturb neuronal migration in mice and are associated with neurodevelopmental disease in humans. Nat Neurosci 21:207–217

28. Ha S, Jeong S-H, Yi K, Chung KM, Hong CJ, Kim SW, Kim E-K, Yu S-W (2017) Phosphorylation of p62 by AMP-activated protein kinase mediates autophagic cell death in adult hippocampal neural stem cells. J Biol Chem 292:13795–13808
29. Ha S, Ryu HY, Chung KM, Baek S-H, Kim E-K, Yu S-W (2015) Regulation of autophagic cell death by glycogen synthase kinase-3β in adult hippocampal neural stem cells following insulin withdrawal. Mol Brain 8:30
30. Hirai H, Karian P, Kikyo N (2011) Regulation of embryonic stem cell self-renewal and pluripotency by leukaemia inhibitory factor. Biochem J 438:11–23
31. Horesh Y, Katsel P, Haroutunian V, Domany E (2011) Gene expression signature is shared by patients with Alzheimer's disease and schizophrenia at the superior temporal gyrus. Eur J Neurol 18:410–424
32. Hu Y-X, Han X-S, Jing Q (2019) Autophagy in development and differentiation. In: Qin Z-H (ed) Autophagy: biology and diseases: basic science. Springer, Singapore, pp 469–487
33. Inaguma Y, Matsumoto A, Noda M, Tabata H, Maeda A, Goto M, Usui D, Jimbo EF, Kikkawa K, Ohtsuki M et al (2016) Role of class III phosphoinositide 3-kinase in the brain development: possible involvement in specific learning disorders. J Neurochem 139:245–255
34. Isakson P, Lystad AH, Breen K, Koster G, Stenmark H, Simonsen A (2013) TRAF6 mediates ubiquitination of KIF23/MKLP1 and is required for midbody ring degradation by selective autophagy. Autophagy 9:1955–1964
35. Ji C, Zhao H, Li D, Sun H, Hao J, Chen R, Wang X, Zhang H, Zhao YG (2020) Role of Wdr45b in maintaining neural autophagy and cognitive function. Autophagy 16:615–625
36. Jiang P, Mizushima N (2014) Autophagy and human diseases. Cell Res 24:69–79
37. Jung S, Choe S, Woo H, Jeong H, An H-K, Moon H, Ryu HY, Yeo BK, Lee YW, Choi H et al (2020) Autophagic death of neural stem cells mediates chronic stress-induced decline of adult hippocampal neurogenesis and cognitive deficits. Autophagy 16:512–530
38. Jung S, Jeong H, Yu S-W (2020) Autophagy as a decisive process for cell death. Exp Mol Med 52:921–930
39. Kao L-P, Wolvetang E (2018) Mitophagy during differentiation of human embryonic stem cells. FASEB J 32:653.1
40. Ke X-X, Zhang D, Zhu S, Xia Q, Xiang Z, Cui H (2014) Inhibition of H3K9 methyltransferase G9a repressed cell proliferation and induced autophagy in neuroblastoma cells. PLoS One 9:e106962
41. Kempermann G, Song H, Gage FH (2015) Neurogenesis in the adult hippocampus. Cold Spring Harb Perspect Biol 7:a018812
42. Kim M, Sandford E, Gatica D, Qiu Y, Liu X, Zheng Y, Schulman BA, Xu J, Semple I, Ro S-H, et al (2016) Mutation in ATG5 reduces autophagy and leads to ataxia with developmental delay. eLife 5
43. Kobayashi T, Piao W, Takamura T, Kori H, Miyachi H, Kitano S, Iwamoto Y, Yamada M, Imayoshi I, Shioda S et al (2019) Enhanced lysosomal degradation maintains the quiescent state of neural stem cells. Nat Commun 10:5446
44. Komatsu M, Waguri S, Chiba T, Murata S, Iwata J, Tanida I, Ueno T, Koike M, Uchiyama Y, Kominami E et al (2006) Loss of autophagy in the central nervous system causes neurodegeneration in mice. Nature 441:880–884
45. Komatsu M, Waguri S, Ueno T, Iwata J, Murata S, Tanida I, Ezaki J, Mizushima N, Ohsumi Y, Uchiyama Y et al (2005) Impairment of starvation-induced and constitutive autophagy in Atg7-deficient mice. J Cell Biol 169:425–434
46. Korolchuk VI, Menzies FM, Rubinsztein DC (2010) Mechanisms of cross-talk between the ubiquitin-proteasome and autophagy-lysosome systems. FEBS Lett 584:1393–1398
47. Kuma A, Hatano M, Matsui M, Yamamoto A, Nakaya H, Yoshimori T, Ohsumi Y, Tokuhisa T, Mizushima N (2004) The role of autophagy during the early neonatal starvation period. Nature 432:1032–1036
48. Kuo T-C, Chen C-T, Baron D, Onder TT, Loewer S, Almeida S, Weismann CM, Xu P, Houghton J-M, Gao F-B et al (2011) Midbody accumulation through evasion of autophagy contributes to cellular reprogramming and tumorigenicity. Nat Cell Biol 13:1214–1223

49. Laplante M, Sabatini DM (2012) mTOR signaling in growth control and disease. Cell 149:274–293
50. Lavoie S, Conway KL, Lassen KG, Jijon HB, Pan H, Chun E, Michaud M, Lang JK, Gallini Comeau CA, Dreyfuss JM et al (2019) The Crohn's disease polymorphism, ATG16L1 T300A, alters the gut microbiota and enhances the local Th1/Th17 response. eLife 8
51. Lee HJ, Gutierrez-Garcia R, Vilchez D (2017) Embryonic stem cells: a novel paradigm to study proteostasis? FEBS J 284:391–398
52. Leeman DS, Hebestreit K, Ruetz T, Webb AE, McKay A, Pollina EA, Dulken BW, Zhao X, Yeo RW, Ho TT et al (2018) Lysosome activation clears aggregates and enhances quiescent neural stem cell activation during aging. Science 359:1277–1283
53. Levine B, Kroemer G (2008) Autophagy in the pathogenesis of disease. Cell 132:27–42
54. Levine B, Kroemer G (2019) Biological functions of autophagy genes: a disease perspective. Cell 176:11–42
55. Li H, Zhang Y, Liu S, Li F, Wang B, Wang J, Cao L, Xia T, Yao Q, Chen H et al (2019) Melatonin enhances proliferation and modulates differentiation of neural stem cells via autophagy in hyperglycemia. Stem Cells 37:504–515
56. Li M, Lu G, Hu J, Shen X, Ju J, Gao Y, Qu L, Xia Y, Chen Y, Bai Y (2016) EVA1A/TMEM166 regulates embryonic neurogenesis by autophagy. Stem Cell Rep 6:396–410
57. Liang Q, Luo Z, Zeng J, Chen W, Foo S-S, Lee S-A, Ge J, Wang S, Goldman SA, Zlokovic BV et al (2016) Zika virus NS4A and NS4B proteins deregulate Akt-mTOR signaling in human fetal neural stem cells to inhibit neurogenesis and induce autophagy. Cell Stem Cell 19:663–671
58. Liu C, Hao S, Zhu M, Wang Y, Zhang T, Yang Z (2018) Maternal separation induces different autophagic responses in the hippocampus and prefrontal cortex of adult rats. Neuroscience 374:287–294
59. Liu K, Zhao Q, Liu P, Cao J, Gong J, Wang C, Wang W, Li X, Sun H, Zhang C et al (2016) ATG3-dependent autophagy mediates mitochondrial homeostasis in pluripotency acquirement and maintenance. Autophagy 12:2000–2008
60. Liu P, Liu K, Gu H, Wang W, Gong J, Zhu Y, Zhao Q, Cao J, Han C, Gao F et al (2017) High autophagic flux guards ESC identity through coordinating autophagy machinery gene program by FOXO1. Cell Death Differ 24:1672–1680
61. Lum JJ, DeBerardinis RJ, Thompson CB (2005) Autophagy in metazoans: cell survival in the land of plenty. Nat Rev Mol Cell Biol 6:439–448
62. Lv X, Jiang H, Li B, Liang Q, Wang S, Zhao Q, Jiao J (2014) The crucial role of Atg5 in cortical neurogenesis during early brain development. Sci Rep 4
63. Shibata M, Lu T, Furuya T, Degterev A, Mizushima N, Yoshimori T, MacDonald M, Yankner B, Yuan J (2006) Regulation of intracellular accumulation of mutant Huntingtin by Beclin 1. J Biol Chem 281:14474–14485
64. Martello G, Smith A (2014) The nature of embryonic stem cells. Ann Rev Cell Dev Biol 30:647–675
65. Martin GR (1981) Isolation of a pluripotent cell line from early mouse embryos cultured in medium conditioned by teratocarcinoma stem cells. Proc Natl Acad Sci USA 78:7634–7638
66. Merenlender-Wagner A, Shemer Z, Touloumi O, Lagoudaki R, Giladi E, Andrieux A, Grigoriadis NC, Gozes I (2015) New horizons in schizophrenia treatment: autophagy protection is coupled with behavioral improvements in a mouse model of schizophrenia. Autophagy 10:2324–2332
67. Miller BR, Hen R (2015) The current state of the neurogenic theory of depression and anxiety. Curr Opin Neurobiol 30:51–58
68. Mizushima N (2007) Autophagy: process and function. Genes Dev 21:2861–2873
69. Mizushima N, Kuma A, Kobayashi Y, Yamamoto A, Matsubae M, Takao T, Natsume T, Ohsumi Y, Yoshimori T (2003) Mouse Apg16L, a novel WD-repeat protein, targets to the autophagic isolation membrane with the Apg12-Apg5 conjugate. J Cell Sci 116:1679–1688
70. Mizushima N, Levine B (2010) Autophagy in mammalian development and differentiation. Nat Cell Biol 12:823–830

71. Mizushima N, Yamamoto A, Hatano M, Kobayashi Y, Kabeya Y, Suzuki K, Tokuhisa T, Ohsumi Y, Yoshimori T (2001) Dissection of autophagosome formation using Apg5-deficient mouse embryonic stem cells. J Cell Biol 152:657–668
72. Morgado AL, Xavier JM, Dionísio PA, Ribeiro MFC, Dias RB, Sebastião AM, Solá S, Rodrigues CMP (2015) MicroRNA-34a modulates neural stem cell differentiation by regulating expression of synaptic and autophagic proteins. Mol Neurobiol 51:1168–1183
73. Mote RD, Mahajan G, Padmanabhan A, Ambati R, Subramanyam D (2017) Dual repression of endocytic players by ESCC microRNAs and the Polycomb complex regulates mouse embryonic stem cell pluripotency. Sci Rep 7:17572
74. Mote RD, Yadav J, Singh SB, Tiwari M, V LV, Patil S, Subramanyam D (2020) Pluripotency of embryonic stem cells lacking clathrin mediated endocytosis cannot be rescued by restoring cellular stiffness. J Biol Chem 295:16888
75. Narayana YV, Gadgil C, Mote RD, Rajan R, Subramanyam D (2019) Clathrin-mediated endocytosis regulates a balance between opposing signals to maintain the pluripotent state of embryonic stem cells. Stem Cell Rep 12:152–164
76. Niwa H, Burdon T, Chambers I, Smith A (1998) Self-renewal of pluripotent embryonic stem cells is mediated via activation of STAT3. Genes Dev 12:2048–2060
77. Nixon RA, Wegiel J, Kumar A, Yu WH, Peterhoff C, Cataldo A, Cuervo AM (2005) Extensive involvement of autophagy in Alzheimer disease: an immuno-electron microscopy study. J Neuropathol Exp Neurol 64:113–122
78. Ohsumi Y (1999) Molecular mechanism of autophagy in yeast, Saccharomyces cerevisiae. Philos Trans R Soc Lond B Biol Sci 354:1577–1580; discussion 1580–1581
79. Paik J, Ding Z, Narurkar R, Ramkissoon S, Muller F, Kamoun WS, Chae S-S, Zheng H, Ying H, Mahoney J et al (2009) FoxOs cooperatively regulate diverse pathways governing neural stem cell homeostasis. Cell Stem Cell 5:540–553
80. Pankiv S, Clausen TH, Lamark T, Brech A, Bruun J-A, Outzen H, Øvervatn A, Bjørkøy G, Johansen T (2007) p62/SQSTM1 binds directly to Atg8/LC3 to facilitate degradation of ubiquitinated protein aggregates by autophagy. J Biol Chem 282:24131–24145
81. Petri R, Pircs K, Jönsson ME, Åkerblom M, Brattås PL, Klussendorf T, Jakobsson J (2017) let-7 regulates radial migration of new-born neurons through positive regulation of autophagy. EMBO J 36:1379–1391
82. Pickford F, Masliah E, Britschgi M, Lucin K, Narasimhan R, Jaeger PA, Small S, Spencer B, Rockenstein E, Levine B et al (2008) The autophagy-related protein beclin 1 shows reduced expression in early Alzheimer disease and regulates amyloid beta accumulation in mice. J Clin Invest 118:2190–2199
83. Pohl C, Jentsch S (2009) Midbody ring disposal by autophagy is a post-abscission event of cytokinesis. Nat Cell Biol 11:65–70
84. Puri D, Subramanyam D (2019) Stress-(self) eating: epigenetic regulation of autophagy in response to psychological stress. FEBS J 286:447–2460
85. Qu X, Zou Z, Sun Q, Luby-Phelps K, Cheng P, Hogan RN, Gilpin C, Levine B (2007) Autophagy gene-dependent clearance of apoptotic cells during embryonic development. Cell 128:931–946
86. Varmazyar R, Noori-Zadeh A, Abbaszadeh HA, Ghasemi Hamidabadi H, Rajaei F, Darabi S, Rezaie MJ, Abdollahifar MA, Zafari F, Bakhtiyari S (2019) Neural stem cells neuroprotection by simvastatin via autophagy induction and apoptosis inhibition. Bratisl Lek Listy 120:744–751
87. Ravichandran KS (2010) Find-me and eat-me signals in apoptotic cell clearance: progress and conundrums. J Exp Med 207:1807–1817
88. Richter-Levin G, Xu L (2018) How could stress lead to major depressive disorder? IBRO Rep 4:38–43
89. Rodriguez-Fernandez IA, Qi Y, Jasper H (2019) Loss of a proteostatic checkpoint in intestinal stem cells contributes to age-related epithelial dysfunction. Nat Commun 10:1050
90. Roidl D, Hellbach N, Bovio PP, Villarreal A, Heidrich S, Nestel S, Grüning BA, Boenisch U, Vogel T (2016) DOT1L activity promotes proliferation and protects cortical neural stem cells from activation of ATF4-DDIT3-mediated ER stress in vitro. Stem Cells 34:233–245

91. Romito A, Cobellis G (2016) Pluripotent stem cells: current understanding and future directions. Stem Cells Int 2016:9451492
92. Saitoh T, Fujita N, Hayashi T, Takahara K, Satoh T, Lee H, Matsunaga K, Kageyama S, Omori H, Noda T et al (2009) Atg9a controls dsDNA-driven dynamic translocation of STING and the innate immune response. Proc Natl Acad Sci 106:20842–20846
93. Saitoh T, Fujita N, Jang MH, Uematsu S, Yang B-G, Satoh T, Omori H, Noda T, Yamamoto N, Komatsu M et al (2008) Loss of the autophagy protein Atg16L1 enhances endotoxin-induced IL-1beta production. Nature 456:264–268
94. Salemi S, Yousefi S, Constantinescu MA, Fey MF, Simon H-U (2012) Autophagy is required for self-renewal and differentiation of adult human stem cells. Cell Res 22:432–435
95. Sato M, Sato K (2012) Maternal inheritance of mitochondrial DNA: degradation of paternal mitochondria by allogeneic organelle autophagy, allophagy. Autophagy 8:424–425
96. Schäffner I, Minakaki G, Khan MA, Balta E-A, Schlötzer-Schrehardt U, Schwarz TJ, Beckervordersandforth R, Winner B, Webb AE, DePinho RA et al (2018) FoxO function is essential for maintenance of autophagic flux and neuronal morphogenesis in adult neurogenesis. Neuron 99:1188-1203.e6
97. Schink KO, Stenmark H (2011) Cell differentiation: midbody remnants—junk or fate factors? Curr Biol CB 21:R958-960
98. Settembre C, Di Malta C, Polito VA, Garcia Arencibia M, Vetrini F, Erdin S, Erdin SU, Huynh T, Medina D, Colella P et al (2011) TFEB links autophagy to lysosomal biogenesis. Science 332:1429–1433
99. Sou Y, Waguri S, Iwata J, Ueno T, Fujimura T, Hara T, Sawada N, Yamada A, Mizushima N, Uchiyama Y et al (2008) The Atg8 conjugation system is indispensable for proper development of autophagic isolation membranes in mice. Mol Biol Cell 19:4762–4775
100. Subramani S, Malhotra V (2013) Non-autophagic roles of autophagy-related proteins. EMBO Rep 14:143–151
101. Suleiman J, Allingham-Hawkins D, Hashem M, Shamseldin HE, Alkuraya FS, El-Hattab AW (2018) WDR45B-related intellectual disability, spastic quadriplegia, epilepsy, and cerebral hypoplasia: a consistent neurodevelopmental syndrome. Clin Genet 93:360–364
102. Suvorova II, Pospelov VA (2019) AMPK/Ulk1-dependent autophagy as a key mTOR regulator in the context of cell pluripotency. Cell Death Dis 10
103. Teyssou E, Takeda T, Lebon V, Boillée S, Doukouré B, Bataillon G, Sazdovitch V, Cazeneuve C, Meininger V, LeGuern E et al (2013) Mutations in SQSTM1 encoding p62 in amyotrophic lateral sclerosis: genetics and neuropathology. Acta Neuropathol (Berl) 125:511–522
104. Toralova T, Kinterova V, Chmelikova E, Kanka J (2020) The neglected part of early embryonic development: maternal protein degradation. Cell Mol Life Sci 77:3177–3194
105. Tra T, Gong L, Kao L-P, Li X-L, Grandela C, Devenish RJ, Wolvetang E, Prescott M (2011) Autophagy in human embryonic stem cells. PLoS One 6:e27485
106. Tsukamoto S, Hara T, Yamamoto A, Kito S, Minami N, Kubota T, Sato K, Kokubo T (2014) Fluorescence-based visualization of autophagic activity predicts mouse embryo viability. Sci Rep 4:4533
107. Tsukamoto S, Tatsumi T (2018) Degradation of maternal factors during preimplantation embryonic development. J Reprod Dev 64:217–222
108. Vázquez P, Arroba AI, Cecconi F, de la Rosa EJ, Boya P, Pablo FD (2012) Atg5 and Ambra1 differentially modulate neurogenesis in neural stem cells. Autophagy 8:187–199
109. Vega-Rubín-de-Celis S (2019) The role of Beclin 1-dependent autophagy in cancer. Biology 9
110. Vilchez D, Simic MS, Dillin A (2014) Proteostasis and aging of stem cells. Trends Cell Biol 24:161–170
111. Vitillo L, Kimber SJ (2017) Integrin and FAK regulation of human pluripotent stem cells. Curr Stem Cell Rep 3:358–365
112. Wang C, Chen S, Yeo S, Karsli-Uzunbas G, White E, Mizushima N, Virgin HW, Guan J-L (2016) Elevated p62/SQSTM1 determines the fate of autophagy-deficient neural stem cells by increasing superoxide. J Cell Biol 212:545–560

113. Wang C, Liang C-C, Bian ZC, Zhu Y, Guan J-L (2013) FIP200 is required for maintenance and differentiation of postnatal neural stem cells. Nat Neurosci 16:532–542
114. Wang M, Liang X, Cheng M, Yang L, Liu H, Wang X, Sai N, Zhang X (2019) Homocysteine enhances neural stem cell autophagy in in vivo and in vitro model of ischemic stroke. Cell Death Dis 10:1–14
115. Wang Y, Zhou K, Li T, Xu Y, Xie C, Sun Y, Zhang Y, Rodriguez J, Blomgren K, Zhu C (2017) Inhibition of autophagy prevents irradiation-induced neural stem and progenitor cell death in the juvenile mouse brain. Cell Death Dis 8:e2694
116. Winslow AR, Chen C-W, Corrochano S, Acevedo-Arozena A, Gordon DE, Peden AA, Lichtenberg M, Menzies FM, Ravikumar B, Imarisio S et al (2010) α-Synuclein impairs macroautophagy: implications for Parkinson's disease. J Cell Biol 190:1023–1037
117. Wolf E, Gebhardt A, Kawauchi D, Walz S, von Eyss B, Wagner N, Renninger C, Krohne G, Asan E, Roussel MF et al (2013) Miz1 is required to maintain autophagic flux. Nat Commun 4:2535
118. Wu X, Fleming A, Ricketts T, Pavel M, Virgin H, Menzies FM, Rubinsztein DC (2016) Autophagy regulates Notch degradation and modulates stem cell development and neurogenesis. Nat Commun 7:10533
119. Xi Y, Dhaliwal JS, Ceizar M, Vaculik M, Kumar KL, Lagace DC (2016) Knockout of Atg5 delays the maturation and reduces the survival of adult-generated neurons in the hippocampus. Cell Death Dis 7:e2127
120. Xiao X, Shang X, Zhai B, Zhang H, Zhang T (2018) Nicotine alleviates chronic stress-induced anxiety and depressive-like behavior and hippocampal neuropathology via regulating autophagy signaling. Neurochem Int 114:58–70
121. Xin Y, Yu L, Chen Z, Zheng L, Fu Q, Jiang J, Zhang P, Gong R, Zhao S (2001) Cloning, expression patterns, and chromosome localization of three human and two mouse homologues of GABA(A) receptor-associated protein. Genomics 74:408–413
122. Xu J, Xia L, Shang Q, Du J, Zhu D, Wang Y, Bi D, Song J, Ma C, Gao C et al (2017) A variant of the autophagy-related 5 gene is associated with child cerebral palsy. Front Cell Neurosci 11
123. Xu X, Araki K, Li S, Han J-H, Ye L, Tan WG, Konieczny BT, Bruinsma MW, Martinez J, Pearce EL et al (2014) Autophagy is essential for effector CD8(+) T cell survival and memory formation. Nat Immunol 15:1152–1161
124. Xu Y, Zhang Y, García-Cañaveras JC, Guo L, Kan M, Yu S, Blair IA, Rabinowitz JD, Yang X (2020) Chaperone-mediated autophagy regulates the pluripotency of embryonic stem cells. Science 369:397–403
125. Yazdankhah M, Farioli-Vecchioli S, Tonchev AB, Stoykova A, Cecconi F (2014) The autophagy regulators Ambra1 and Beclin 1 are required for adult neurogenesis in the brain subventricular zone. Cell Death Dis 5:e1403–e1403
126. Yu S-W, Baek S-H, Brennan RT, Bradley CJ, Park SK, Lee YS, Jun EJ, Lookingland KJ, Kim E-K, Lee H et al (2008) Autophagic death of adult hippocampal neural stem cells following insulin withdrawal. Stem Cells Dayt Ohio 26:2602–2610
127. Yue Z, Jin S, Yang C, Levine AJ, Heintz N (2003) Beclin 1, an autophagy gene essential for early embryonic development, is a haploinsufficient tumor suppressor. Proc Natl Acad Sci USA 100:15077–15082
128. Zhang X, He X, Li Q, Kong X, Ou Z, Zhang L, Gong Z, Long D, Li J, Zhang M et al (2017) PI3K/AKT/mTOR signaling mediates valproic acid-induced neuronal differentiation of neural stem cells through epigenetic modifications. Stem Cell Rep 8:1256–1269
129. Zhang Y, Morgan MJ, Chen K, Choksi S, Liu Z (2012) Induction of autophagy is essential for monocyte-macrophage differentiation. Blood 119:2895–2905

Chapter 4
Autophagy in Germline Stem Cells

Kaitlin E. Kosinski and Alicia Meléndez

Abstract Macroautophagy (autophagy) is a catabolic process that delivers intracellular constituents to the lysosome for recycling. Reproduction is an energetically expensive process for living organisms, and investing in progeny that has little chance of survival is not an effective use of resources. The decision for an organism to invest in its offspring or survival thus requires communication between the conditions in the environment and the germ line. Autophagy provides a potential mechanism for signaling to the germline that conditions require diverting resources for individual survival rather than reproduction. Although typically upregulated in response to cellular stress conditions, basal levels of autophagy can also function as a quality control mechanism that serves to maintain cellular homeostasis in response to developmental cues. Our current understanding of the role of autophagy and the fundamental mechanisms that control autophagy have been elucidated by experiments that focus on animal development. In this review, we focus on the role of autophagy in germline stem cell proliferation, differentiation and homeostasis in *C. elegans*, Drosophila and mouse development.

Keywords Autophagy · Macroautophagy · Microautophagy · Chaperone mediated autophagy · Germline stem cells · P granules · *C. elegans* · *Drosophila* · TOR · IIS

Abbreviations

CMA	Chaperone-mediated autophagy
GSC	Germline stem cell
PGC	Primordial germline cells
DTC	Distal tip cell

K. E. Kosinski · A. Meléndez (✉)
Biology Department, Queens College, CUNY, Flushing, NY 11367, USA
e-mail: alicia.melendez@qc.cuny.edu

The Graduate Center, City University of New York (CUNY), New York, NY 10016, USA

B. V. Shravage and K. Turksen (eds.), *Autophagy in Stem Cell Maintenance and Differentiation*, Stem Cell Biology and Regenerative Medicine 73,
https://doi.org/10.1007/978-3-031-17362-2_4

POI	Primary ovarian insufficiency
IIS	Insulin-like IGF-1 signaling
AMPK	AMP-activated kinase
TOR	Target of Rapamycin
DILPs	Drosophila insulin-like peptides
CySCs	Cyst stem cells
CC	Cyst Cells
EGFR	Epidermal growth factor receptor
mtDNA	Mitochondrial DNA
CPEB	Cytoplasmic polyadenylation element binding protein

Overview of Autophagy

Autophagy is an evolutionarily conserved cellular process that delivers intracellular constituents, such as long-lived proteins, damaged proteins and/or organelles, to the lysosome to promote catabolism. The turnover of cellular components can be controlled in response to intracellular and extracellular signals. In higher eukaryotes, there are several different types of autophagy, including chaperone-mediated autophagy (CMA), microautophagy, and macroautophagy. In microautophagy, the lysosomal membrane invaginates to internalize cargo in vesicles that pinch off. Microautophagy has been mostly studied in yeast, where selective forms that target specific cellular components, and non-selective forms of microautophagy have been described, such as microautophagy of cytosolic components [13, 60, 71], mitochondria [8] peroxisomes [89, 105], endoplasmic reticulum (ER) [91], vacuolar membrane proteins [114], and lipid droplets [107]. In mammals, a form of endosomal microautophagy (eMI) has been reported where late endosomal membranes generate invaginating vesicles that internalize ubiquitinated membrane proteins [87, 90]. In Drosophila, a form of eMI has also been documented [68]. In Chaperone-Mediated Autophagy (CMA), cytosolic proteins are selected on the basis of a specific pentapeptide motif (KFERQ) as they are recognized by hsc70 (heat-shock cognate protein of 70 kDa), unfolded and translocated to the lysosome for degradation [12, 14]. Transport through the lysosomal membrane requires the multimerization of the lysosome-associated membrane protein type 2A (LAMP2A) [4], however, orthologs based on sequence homology to LAMP2A have not been found in *C. elegans* nor Drosophila. Macroautophagy (hereafter autophagy) is a multi-step process that includes the formation of a de novo double-membrane bound organelle (the autophagosome), which engulfs bulk cytoplasmic material and fuses with the lysosome to form the autophagolysosome. The contents of the autophagosome undergo lysosomal degradation, and macromolecular components are then recycled. Autophagy occurs at low or basal levels under normal conditions in the cell, but can be upregulated under conditions of stress, such as nutrient starvation, heat stress, oxidative stress, pathogen infection or DNA damage [65, 67].

The step-wise autophagy process initiates with induction, nucleation, and formation of an isolation membrane or phagophore (a vesicle containing a continuous membrane) [67, 111] (Fig. 4.1). The phagophore then transforms into a cup-shaped structure that engulfs a portion of the cytoplasm, and then it expands to become an autophagosome, which consists of two separate inner and outer membranes (Fig. 4.1). Following fusion with the lysosome and degradation of contents by lysosomal hydrolases, autolysosome reformation occurs by tubulation and scission (essentially pinching off) of small portions of autolysosome membrane. These small tubules become vesicles known as proto-lysosomes that mature will contribute to the lysosomal pool. A number of proteins referred to as ATG proteins, and their regulators, control the different steps of the autophagic process. Briefly, autophagosome formation requires the UNC-51/Atg1/ULK kinase complex, the class III phosphatidylinositol 3-kinase (PtdIns3K) VPS-34/Vps34/PI3KC3 complex, two ubiquitin-like protein (LGG-3/Atg12 and LGG-1,2/Atg8a,b/LC3) conjugation systems, and the transmembrane protein ATG-9/Atg9 and VMP1 [67, 115].

Perturbations of autophagic flux (the formation and degradation of autophagosomes) during development can lead to a wide range of phenotypes, including effects on germ cell development (Table 4.1). Autophagy is exquisitely sensitive to environmental changes as cells adapt themselves to external stressors. Germline stem cells respond to external signals during development, and autophagy appears to play a role in how these signals are transmitted to the stem cell population. In this review, we

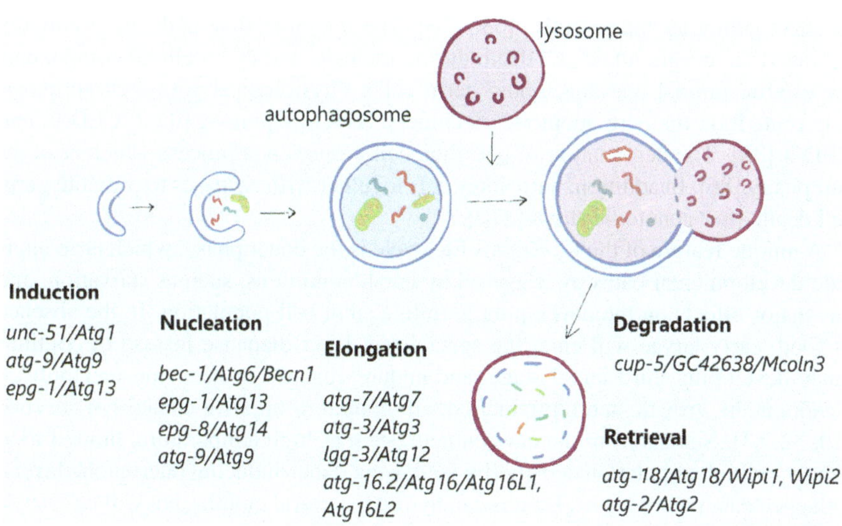

Fig. 4.1 Genetic regulation of macroautophagy. A step-wise schematic of autophagy listing the genes acting at each step of induction, nucleation, elongation, retrieval and degradation. Only genes which have a demonstrated role in GSC development are listed. For a complete list, please refer to [50, 77]

evaluate the current knowledge on the role of autophagy in regulating the germline stem cell niche and the germline stem cell population. Autophagy contributes to the maintenance and homeostasis of germline stem cells, as well as playing roles in differentiation and programmed cell death of germ cells.

Overview of GSCs in Model Organisms

The field of *C. elegans* genetics and its utility in studying germline development has provided a wealth of information on germline stem cell biology. This topic has been reviewed in-depth [43] but briefly, the model nematode *C. elegans* provides a genetically tractable and rapidly developing germline stem cell system that shares the features of stem cell systems in other more complex organisms. Similarities include their ability for renewal, differentiation and multipotency, the specific requirements of a local stem cell niche, Notch signaling, cycles of quiescence and active division, and the cells' regeneration and depletion with age [43]. *C. elegans* is a free-living, translucent nematode. It has two sexes, males and self-fertile hermaphrodites. The hermaphrodite animal is essentially female, but in the last larval stage (L4) produces a limited supply of sperm, which migrates to the spermatheca and is available to fertilize the oocytes produced in adulthood. Subsequently, germ cell development switches from spermatogenesis to oogenesis, with any germ cells that enter meiosis after this point in development becoming oocytes. Not all the differentiating cells become gametes, some will act to support the development of neighboring oocytes, in effect acting as "nurse cells" [33, 116]. These "nurse"-like cells are eventually degraded via programmed cell death during meiosis, and their cellular components are used to nourish the developing germ cells. Physiological germ cell apoptosis is executed via the core apoptotic machinery, which requires CED-3, CED-9, and CED-4 [33]. The degradation of germline cell corpses is a process which requires autophagy [86]. In addition, autophagy acts together with caspases to promote germ cell death after genotoxic stress [110].

A unique feature of the *C. elegans* life cycle is the dauer phase, which is an alternate developmental pathway triggered by harsh conditions, such as starvation, and has major effects on the developing germline stem cell population. In the absence of food, early larvae will enter the specialized dauer diapause instead of continuously developing third larval stage, and endure changes in behavior, metabolism, development, growth, and reproduction, all thought to improve chances of survival [10, 32, 84]. Signals from the environment, such as high temperature, limited food supply, or high population density (by increasing dauer-inducing pheromone levels) influence dauer entry levels. Dauers can live up to several months, but will recover, if conditions improve. In dauer animals, the developing germline is reversibly arrested [42]. As this alternative dauer developmental arrest is induced by similar triggers as that of the autophagy process in cells, that is stress and lack of nutrients, it is not unsurprising that dauer morphogenesis is dependent upon the cellular remodeling

Table 4.1 Autophagy gene activity in germline stem cells

Gene Name (Black=*C. elegans*; Blue=*D. melanogaster*; Green=*M. Musculus)*	Function (Black=*C. elegans*; Blue=*D. melanogaster*; Green=*M. Musculus)*	Reference (Black=*C. elegans*; Blue=*D. melanogaster*; Green=*M. Musculus)*
unc-51/Atg1/Ulk1, Ulk2	Required for engulfment of cell corpses in germ line; Required for GSC development; Required for programmed cell death of germ cells	Huang et al, 2013; Demarco, et al 2020; Nezis et al 2010
atg-3/Atg3/Atg3	Required for GSC proliferation	Ames et al 2017
Atg5/Atg5/Atg5	Sertoli cell-germ cell communication; testosterone synthesis; GSC development	Demarco et al 2020; Liu et al 2016; Gao et al 2018;
bec-1/Atg6/Becn1	Required for GSC proliferation; Required for engulfment of cell corpses in germ line; Required for GSC development; Progesterone synthesis	Ames et al 2017; Huang et al, 2013; Demarco et al 2020 Gawriluk et al 2014;
atg-7/Atg7/Atg7	Required for GSC proliferation; Required for engulfment of cell corpses in germ line; Required for GSC development; Spermatogenesis; testosterone synthesis	Ames et al, 2017; Huang et al, 2013; Demarco et al 2020; Wang et al, 2014; Gawriluk et al 2011; Song et al, 2015; Gao et al 2018
lgg-2/Atg8/Lc3	Required for GSC development	Demarco et al, 2020
atg-9/Atg9/ Atg9a, Atg9b	Required for GSC proliferation; Required for GSC development	Ames et al 2017; Demarco et al 2020
lgg-3/Atg12/Atg12	Required for GSC proliferation; Required for GSC development;	Ames et al, 2017; Demarco et al, 2020
atg-16.2/Atg16/ Atg16l1, Atg16l2	Required for GSC proliferation; Required for mitophagy during spermatogenesis	Ames et al, 2017; Zhang et al, 2020
atg-18/Atg18/ Wipi1, Wipi2	Required for GSC proliferation; Required for engulfment of cell corpses in germ line	Ames et al 2017; Huang et al, 2013
cup-5/CG42638/Mcoln3	Required for GSC proliferation	Ames et al 2017
epg-1/ Atg13/Atg13	Required for GSC proliferation; Required for GSC development	Ames et al 2017; Demarco et al, 2020
epg-8/Atg14/ Atg14L, Barkor	Required for GSC proliferation	Ames et al 2017
*vps-34/*Pi3K59F/Pik3c3	Required for GSC development; Required for programmed cell death of germ cells	Demarco et al, 2020; Nezis et al 2010

resulting from autophagy [64]. We will discuss the dauer pathway of development and its effects on the germline later in this chapter.

The germline is the sole stem cell population in *C. elegans*, and the only actively dividing cell population in the adult animal. The germline lineage in the worm arises during the first four cell divisions of embryogenesis, when the sole germline blastomere (P4) divides to form the primordial germ cells Z2 and Z3. The somatic gonad precursor cells, Z1 and Z4, then migrate to join the other two "Z cells" ultimately forming the gonad primordium during late embryogenesis [48, 101].

As the embryos hatch and begin the larval stages of development, the primordial germline cells (PGCs) first undergo a period of quiescence. This occurs during the first half of the L1 phase, but if sufficient food is detected in the environment, and the somatic gonad precursor cells are present, the PGCs begin to divide, forming the initial pool of germline stem cells, referred to as the proliferative zone. During the middle of the L3 phase, the progenitor pool accumulates rapidly to give rise to more progenitors, and germ cells that will enter the meiotic pathway [34, 47, 80]. This results in the germline adopting its characteristic distal to proximal axis. The germline is often described as an "assembly line," or "factory," along which germ cells progress from the progenitor population at the distal tip of the gonad to the proximal end, where meiotic oocytes arrest in the diakinesis stage of Prophase I.

The distal end of the hermaphrodite germ line is overlaid with a somatic cell known as the Distal Tip Cell (DTC) (Fig. 4.2). This specialized cell has an elaborate network of projections (a "plexus") which extend as finger-like projections between the germ cells and reach into the first several cell diameters of the distal germline [7]. This DTC niche expresses the GLP-1/Notch receptor ligands which maintain the progenitor cell population at the distal end by promoting the stem cell fate [3, 16]. As the PGCs divide, they progress away from the DTC and the Notch signal, and thus begin the meiotic program [25]. In animals that carry a loss of function mutation in *glp-1* (encoding the Notch Receptor), *e2141*, PGCs enter prematurely the meiotic fate and differentiate [3, 17, 79], whereas animals that carry a gain of function mutation in *glp-1*, *ar202*, display a Tumorous phenotype, where all germ cells continuously proliferate and do not differentiate.

The genetic model fruit fly *Drosophila melanogaster* has also been instrumental in the study of germline stem cell (GSC) biology. An in-depth discussion of germline development in this animal is beyond the scope of this review, but it has been reviewed in detail elsewhere [18]. In Drosophila, GSCs are derived from embryonic pole cells, which migrate to meet somatic gonadal precursor cells to form the embryonic gonad [18]. This structure consists of roughly ten primordial germ cells which divide and remain undifferentiated until the development of the stem cell niche in the anterior gonad. This niche, similarly to the distal tip cell in *C. elegans*, acts as a signaling source to maintain the primordial germ cells in their undifferentiated state while they remain spatially close to the niche. Adherens junctions develop between the GSCs and the niche, which orients cell division by anchoring the germline stem cell to the niche. Dividing daughter cells move away from the signals from the niche and they begin to differentiate [96].

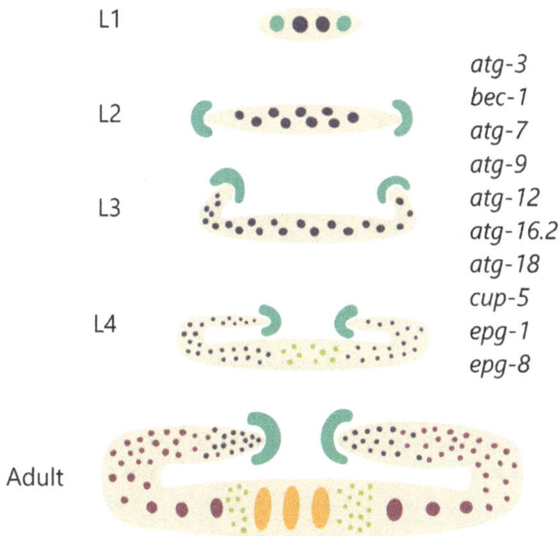

Fig. 4.2 Development of the *C. elegans* germ line. The distal gonad and stem cell niche in the *C. elegans* hermaphrodite. The distal tip cell (DTC) [teal] develops a network of projections which promote the stem cell niche. The germline stem cells (GSCs) [purple] divide and move away from the DTC and differentiate. In the L4 stage, a small amount of GSCs differentiate into sperm (green), whereafter all differentiating GSCs become oocytes (pink). Oocytes are fertilized as they pass through the spermatheca and become embryos (yellow). Autophagy genes known to have an effect on *C. elegans* GSCs are listed

In the female fly ovary, structures known as the ovarioles are the functional subunit of the ovary. The ovariole resembles pearls on a string, where egg chambers are arranged in a linear fashion, with the most mature egg chamber at the posterior. The stem cell niche is at the anterior end, within a structure named the germarium. The germarium consists of germline stem cells, escort stem cells and somatic stem cells, which will eventually develop into follicle cells (Fig. 4.3). The germline stem cells can be distinguished by their contact with the cap cells and possession of an organelle known as the spectrosome attached to the site of contact with the niche [18]. During mitosis, the spectrosome becomes elongated, while connected to the mother and daughter cells, before it is unequally divided between the two cells, with the mother cell (the stem cell) regaining the majority portion [20].

In the fly testis, the two stem cells that organize spermatogenesis are the GSCs and the cyst progenitor cells (Fig. 4.3). Each GSC is encapsulated by two somatic cyst cells, which in turn maintain a connection with the hub—the niche. Male GSCs also contain the spectrosome structures seen in the ovary, which in both cell types eventually become the structure known as the fusome. This germ cell specific structure works to coordinate cell divisions in the germline [49].

In both Drosophila and *C. elegans*, the determination of germline stem cell fate is regulated by the localization of maternally deposited pole plasm. Pole plasm, or

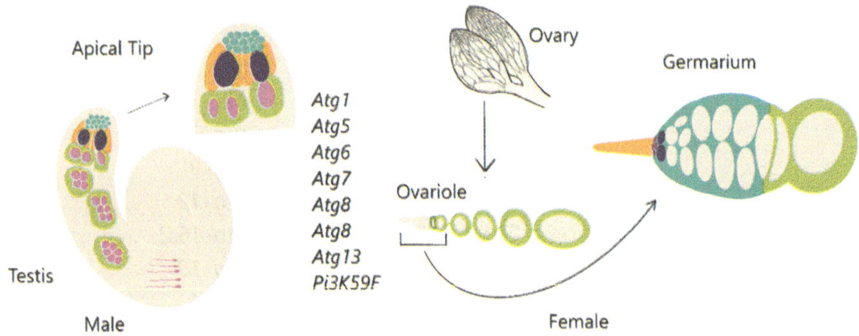

Fig. 4.3 The Drosophila GSC niches (ovary and testis). In the testis (left), the cap cells [teal] along with the cyst stem cells [yellow] promote the GSC [purple] niche. As the stem cells divide, the daughters that move away from the cap cells differentiate [pink]. In the ovary (left) the cap cells [magenta] and escort stem cells [teal] maintain the niche for GSCs [purple] until they differentiate [tan]. Autophagy genes known to have an effect on Drosophila GSCs are indicated

germ plasm, is cytoplasm containing aggregates of RNA and protein, known as polar granules, and the presence of these components in daughter cells is sufficient for the determination of germ cell fate [62]. In Drosophila, several components of germ plasm have been identified including Oskar (Osk), which initiates the polar granule assembly through recruitment of Vasa and Tudor [39, 62]. The RNAs *gcl* and *pgc* and Piwi are also present in germ plasm and play a role in maintaining GSC fate and division through the microRNA pathway [36, 63].

In *C. elegans*, autophagy plays a critical role in the degradation of P granules in the early embryo [103, 118, 120, 121]. P granules are a type of protein-RNA aggregate [38, 92, 100], (Brangwynne et al. 2009), which contain RNA silencing complexes that monitor germline gene expression, essential for germ cell differentiation [46, 98, 106] The daughter cells destined for the somatic lineage require autophagy for the complete degradation of the P granules during embryonic development [123]. P granules begin dispersed throughout the cytoplasm during early embryogenesis, but eventually localize to the blastomeres that will give rise to the germline via asymmetric divisions and their selective degradation in the somatic lineage [123]. The formation of the P granules in *C. elegans* requires the protein SEPA-1, which is also required for autophagic degradation of the P granule components PGL-1 and PGL-3 [123]. SEPA-1 interacts with both PGL-3 and the autophagy protein LGG-1 (LC3 in mammals) [123]. This process of autophagic degradation of P granules in daughter cells not destined for germline cell fate is regulated by mTORC1 signaling, which modulates liquid–liquid phase separation in order to separate and degrade P granules [120]. Whether similar requirements exist in higher eukaryotes for the autophagic degradation of protein aggregates in cells destined for somatic lineages remains to be determined, however, RNA condensates analogous to P granules have been found in the Drosophila germ line [78]. Like P granules, nucleoli, stress granules and the polar granules of Drosophila all contain distinct phases or compartments [24, 44, 75, 104], (Little et al. 2015).

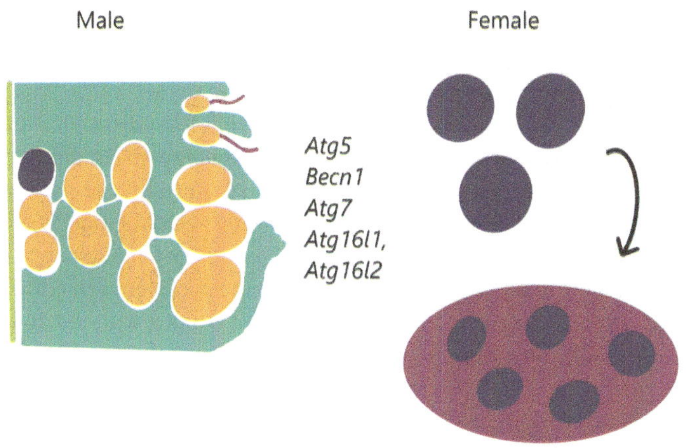

Fig. 4.4 Proposed model of GSC niche in mouse. Left: In XY animals, Sertoli cells [teal] promote the GSC niche for the GSCs [purple] while they differentiate into spermatocytes [yellow]. Right: XX GSCs in fetal mouse. The current model states female mammalian GSCs (purple) colonize the early ovary prior to birth, and then arrest during meiosis as oocytes. Oocyte development continues after birth. Autophagy genes known to have an effect on mouse GSCs are listed

In mammals such as *Mus musculus*, the germline lineage is also separated from somatic lineage early in development (Fig. 4.4). However, whereas the germ cell lineage in the prior two systems is through the inheritance of germ plasm, mammals specify the germ cell lineage through induction of pluripotent embryonic stem cells, known as epigenesis [23]. In mammals, the PGCs are the first germline stem cell population, which colonize the developing gonads by active migration [83, 88]. As opposed to *C. elegans* and *Drosophila*, female (XX) GSCs are set aside early in development in mammals and arrested in the diplotene stage of prophase I of meiosis until fertilization allows entry into the second meiotic division and embryogenesis (Fig. 4.4) [37, 97]. PGCs in males (XY) enter a period of quiescence in G0/G1 during the final phases of embryogenesis, and begin proliferation in the first few postnatal days, with some cells being set aside to become spermatogonial stem cells (Fig. 4.4) [99]. For an in-depth review of primordial germ cells in mice, see Saitou and Yamaji [88].

Perturbed Autophagy Affects Fertility

Mutations in autophagy genes often result in severe developmental defects or embryonic lethality. Through the use of genetic studies with hypomorphic mutant alleles and tissue specific knock outs, several essential autophagy genes have been shown to regulate different steps of germ cell development (Table 4.1). The autophagy related gene Atg7 was shown to affect both oogenesis and spermatogenesis in mice. In male mice, Atg7 was shown to regulate the formation of the acrosome in spermatozoa

[110]. Compromised autophagy through the loss of Becn1 (or even a dose reduction in heterozygote animals), or germ cell specific loss of Atg7 in female mouse ovaries, led to a dramatic reduction in the germ cell population [30, 95]. Germ cell specific knockout of Atg7 in female mice resulted in a loss of oocytes during post-natal starvation conditions [95]. This effect mimics the human condition of primary ovarian insufficiency (POI), in which there is a premature loss of ovarian follicles in women, and is often a cause of infertility. In fact, recent exome sequencing of patients affected by POI suggested links to *Atg7* and *Atg9* insufficiency [19]. In the mouse testis, the Sertoli cells play a similar role to the cyst stem cells in *Drosophila*, providing support and nourishment through physical contact with the dividing germ cells in the niche. The junction between the Sertoli cells and the developing spermatid is a highly dynamic process that requires precise control of the cytoskeleton [5]. The specific disruption of autophagy through knockout of *Atg5* and *Atg7* in the Sertoli cells leads to reduced fertility due to aberrant spermatocyte morphology resulting from impaired Sertoli cell-germ cell communication [61]. The authors found that autophagy was required for the degradation of PDLIM1, a negative regulator of cytoskeleton organization [61].

BEC-1 in *C. elegans* is a critical regulator of autophagy and is evolutionarily conserved from yeast (Atg6/Vps30) to mammals (Beclin1/Becn1) [64]. It is required both for the formation of the autophagosome, and fusion with the lysosome [9]. Loss of function mutations of *bec-1/Beclin1* result in sterility, with no surviving offspring and a 50% reduction in the number of stem cells in the proliferative zone of the distal gonad [1]. In addition to the reduced proliferative zone, loss of *bec-1/Beclin1* also resulted in disorganized membranes within the developing germline [1]. Other autophagy related genes in this study similarly displayed a reduction in the germline stem cell pool, including *atg-7*, *atg-3*, *atg-16.2*, and *atg-18*. RNAi depletion of the lysosomal biogenesis gene *cup-5* also results in a decrease in the stem cell pool, suggesting that degradation is required to establish the stem cell population in adult hermaphrodites.

Autophagy and Signaling in GSCs

Autophagy is important for the cellular response to nutrient starvation. Germline proliferation responds to nutritional signals in the *Drosophila* ovary; female flies supplied with a nutritionally poor source of food showed a 60-fold reduction in the rate of oogenesis [22]. This reduction in fertility was not due to a reduced number of the stem cell population, but a reduced rate of division of germline stem cells and an increased programmed cell death in later stages of oogenesis [22].

Several signaling systems involved in responding to environmental nutrient conditions are highly evolutionarily conserved. Insulin-like IGF-1 signaling (IIS), AMP-activated Kinase (AMPK), and Target-of-Rapamycin (TOR) kinase are major responders to nutrient conditions and all act to modulate autophagy. Insulin signaling has been shown to regulate the proliferation of germline stem cells via the niche in

Drosophila. The niche is located in the anterior germarium of each ovariole in the fly ovary and is composed of cap cells, terminal filament cells and escort cells [49]. As in the distal gonad of *C. elegans*, the niche provides physical contact and signals with GSCs to maintain the stem cell microenvironment. Germline stem cell division was shown to be regulated by Drosophila insulin-like peptides (DILPs) produced by neural cells [22, 52]. GSC maintenance likewise requires insulin signaling, as loss of the *Drosophila insulin receptor* (*dinr*) in female flies resulted in ovaries which contained fewer germline stem cells and underwent rapid stem cell loss through differentiation [40]. Mosaic analysis revealed that insulin signaling acts non-cell autonomously to maintain the germline stem cell niche [40]. It was later shown that insulin signaling controls the competence of the niche to respond to Notch ligands in cap cells, in turn promoting the germline stem cell niche environment [40].

Poor nutritional conditions also lead to quiescence of the germline in *C. elegans* hermaphrodites [42]. During dauer diapause, widespread cell cycle arrest halts the expansion of the germline stem cell pool (Narbonne and Roy 2006). Insulin signaling is required for this stem cell expansion in the L3 and L4 stages of development, as reduction of insulin-like IGF-1 receptor DAF-2(INR/IIR) function leads to entry into the alternative dauer pathway even in the presence of food [58]. The DAF-2/IIR pathway initiates an intracellular signaling cascade upon the activation of the receptor by insulin-like peptides, leading to recruitment of the phosphoinositide protein-kinases (PI3K), and activation of serine/threonine kinases (PDK-1, AKT-2, AKT-2) which phosphorylate transcription factors, including DAF-16/FoxO, HSF-1 and SKN-1/Nrf [69, 70] DAF-2/IIR signaling regulates longevity via repression of the pro-longevity transcription factor DAF-16/FOXO, the sole member of the FOXO family of transcription factors in *C. elegans* [58]. DAF-2/IIR signaling also promotes the proliferation of germline stem cells during development, and it acts independently of GLP-1/Notch [66].

In *C. elegans*, the interaction between the role of autophagy in germline development and DAF-2/IIR signaling is complex and remains to be fully investigated. The lifespan extension seen in *daf-2/IIR* mutants requires autophagy genes [64]. In the germline, BEC-1/Beclin 1 required DAF-18/PTEN and SKN-1/Nrf for the promotion of germline stem cell proliferation but acts independently of DAF-16/FOXO [1]. This interaction is similar to the starvation-induced germline cell cycle arrest in G2 phase seen in early larvae (L1), which requires DAF-18/PTEN but not DAF-16/FOXO [26]. Loss of *bec-1, atg-18*, and *atg-16.2* in *C. elegans* resulted in a delay in the cell cycle with a prolonged G2 phase [1]. However, the effects ATG-16.2/ATG16L and ATG-18/WIPI1/2 on the germline required both DAF-16/FOXO and DAF-18/PTEN [1]. DAF-18/PTEN has been shown to antagonize germline development in *C. elegans* and is required for the regulation of germline proliferation in response to nutrient replete conditions [72]. Interestingly, ATG-7 was found to act with the TGF-β homologue DAF-7 to promote germline development [1]. These results suggested that autophagy genes potentially regulate germline proliferation through autophagy-dependent and autophagy-independent mechanisms [2].

In *Drosophila,* TOR signaling has also been shown to act on the germline. Loss of *Tor* leads to decreased proliferation of germline stem cells resulting from slower progression of GSCs through G2 [53]. This regulation of G2 by *Tor* was independent of insulin signaling [53]. This study also illustrated that *Tor* was required for the maintenance of GSCs in aging female flies, with *Tor* mutant females losing GSCs more rapidly. Tor is a known regulator of autophagy, but inhibition of autophagy via *Atg7* mutation did not ameliorate this effect, suggesting that the loss is not due to Tor-mediated autophagic death [53].

Tor signaling in *C. elegans* development has also been thoroughly reviewed in *C. elegans*, [6], but briefly, Tor is proposed to play several roles during germline development. During starvation-induced diapause in L1 larvae, it is thought that AMP-activated kinase (AMPK) negatively regulates TOR leading to germline developmental arrest, although parallel pathways independent of TOR signaling may also be active in this process [27]. TOR also acts to positively regulate cell cycle progression in larval germline development, although this activity is independent of DAF-2/IIR function [51, 66]. It is unknown so far how Tor-mediated autophagy functions in these roles.

The loss of AMPK has major effects on germline development. The quiescence of GSCs in the dauer diapause in *C. elegans* is dependent upon AMPK, and its absence results in the failure of the GSCs to arrest. This leads to overproliferation and failure to regain fertility upon the resumption of normal growth in post-dauer development, due to oocytes failing to progress to diakinesis during meiosis [45]. This disruption of normal GSC development was linked to abnormal chromatin modifications in dauer germ cells, which were then preserved as the germ cells developed post-dauer. Interestingly, these effects were regulated through the endogenous small RNA pathway, and found to act in a partially non-cell-autonomous manner [45]. In *Drosophila*, AMPK plays diverse roles in oogenesis in that its activity has diet-dependent and diet-independent functions as well as cell-autonomous and cell non-autonomous functions [56]. AMPK was required cell-autonomously for the reduction in the rate of GSC and follicle cell proliferation resulting from poor nutritional conditions in the ovary, but surprisingly was not required in the germline, but instead only in the follicle cells for the regulation of growth [56]. Additionally, this work showed that AMPK regulated follicle cell encapsulation of cysts in the germarium, which was previously shown to be diet-independent [22, 56].

Autophagy, Lipid Metabolism, and Germline Stem Cells

Reproduction is an energy-intensive process that is affected by the overall metabolic functioning of the organism. Fat metabolism has been linked to germline stem cell proliferation in many model organisms, including *C. elegans*. Temperature sensitive GLP-1/Notch mutants, such as *e2141* mutant, experience a loss of proliferating stem cells due to premature differentiation (when shifted to the non-permissive temperature). They also display altered lipid storage levels, suggesting that germ cells themselves modulate lipid storage and mobilization during development and adulthood [109]. As previously mentioned, autophagy genes are required for the lifespan extension seen in DAF-2/IIR mutants, and are required for many of the metabolic and physiologic changes associated with the dauer phenotype [54, 64]. Indeed, BEC-1/Beclin 1/Becn1 and other autophagy genes were required for the normal storage of lipids during development, as well as the increase in lipid storage seen in *glp-1* and *daf-2* loss of function mutants [55]. However, it is not clear if these changes in lipid storage directly affect germline development in autophagy mutants.

In the *Drosophila* testis, two stem cell populations are located at the apical tip, the germline stem cells and the somatic cyst stem cells (CySCs), which surround somatic support cells, referred to as the hub. The germline stem cells divide asymmetrically to self-renew and to give rise to daughter gonialblast, which then undergoes four rounds of mitotic, transit amplification divisions with incomplete cytokinesis to generate a cyst of 16 interconnected spermatogonia. The spermatogonia will then mature into spermatocytes and differentiate through meiosis to produce 64 haploid spermatids and then mature sperm [28, 35]. Division of the somatic cyst stem cells (CySCs) maintains the somatic stem cell population giving rise to somatic cyst cells (CCs) that differentiate in close contact with the germ line and during spermatogenesis will encapsulate the developing germ cells and provide signals for self-renewal as well as differentiation [113]. Interestingly, basal levels of autophagy were found to be active in somatic cyst stem cells (CySCs) and in the early somatic cyst cells (CCs) to support stem cell maintenance and regulate progenitor cell differentiation [102]. A reduction of autophagy resulted in loss of CySCs, and aberrant spermatogonial transit amplification divisions, which mimicked the effect of reduced EGFR signaling. Disruption of autophagy in this population of cells in the *Drosophila* testis leads to aberrant lipid metabolism and mobilization, visualized by significant accumulation of lipid droplets [102]. Since the reduction in the germline stem cell population phenotypically resembled the effects of reduced EGFR signaling, further investigation revealed that autophagy was activated in the somatic CySCs, and CCs, in response to EGFR signaling through AP-1/Fox [102]. The authors also found that TOR signaling is required for the suppression of autophagy, which allows the CySC cells to differentiate into cyst Cells [102]. Thus, CySCs and their daughter CC cells, which initiate differentiation, may have distinct metabolic needs, and the metabolic remodeling provided by autophagy may be required for this transition. The lipid accumulation may be due to defects in lipophagy, a selective form of autophagy that is involved in lipid catabolism [93] or a consequence from the lack of fatty acid oxidation due to damaged mitochondria [59, 74, 117, 120].

Building evidence suggests that autophagy plays an important role at the juncture of metabolic and reproductive functions. Autophagy appears to regulate lipid metabolism to maintain the germline stem cell niche, which in turn is required for the promotion of correct germline stem cell development and differentiation. In mice, autophagy is active in the Leydig cells and is required for proper cholesterol metabolism to maintain sufficient testosterone levels to support reproduction [29]. Conditional knockout of *Atg7* or *Atg5* in the Leydig cells of mice leads to the down-regulation of scavenger receptor class B, type I (SR-BI) due to accumulation of its negative regulator NHERF2, which is usually degraded via the autophagy-lysosome pathway to facilitate proper cholesterol uptake [29]. Additionally, Becn1 deficiency in the murine ovary leads to insufficient progesterone production, resulting in pregnancy loss, further implicating autophagy in the regulation of steroidal metabolism, which largely depends on lipid homeostasis [31]. Autophagy's role in lipid metabolism is currently an exciting area of investigation. Further exploration into additional mechanisms of action should provide insights into autophagy's role in modulating lipid metabolism to sustain germline stem cell function.

Autophagy and Organelle Homeostasis in GSCs

Autophagy also plays a critical role in the management of organelle homeostasis in the cell, selectively degrading damaged or aging organelles, such as mitochondria. Mitophagy is the targeted degradation of damaged or aged mitochondria through the autophagic pathway. The proper function of mitochondria in the germline stem cell environment is required for proper germ cell development in *C. elegans* [11]. Autophagy has been shown to be important for the degradation of paternal mitochondria to ensure proper inheritance of maternal mitochondria in the embryos of worms, flies and mice [94]. In *C. elegans* and others, the autophagy receptor SQSTM1/p62 accumulates near sperm mitochondria and binds to LC3/LGG-1, targeting paternal mitochondria for degradation [21]. Degradation of both paternal cellular structures and mitochondrial DNA immediately after fertilization in the embryo require LC3-dependent autophagy [82].

In Drosophila, Lieber et al. has shown that a form of mitophagy is involved in the selection against deleterious mitochondria DNA mutations [57]. This selection appears to be germline specific and only in female flies. The mtDNA selection occurs early in oogenesis, during germline cyst differentiation. Interestingly, the developmentally regulated fragmentation of cyst mitochondria was found to be needed so that mitochondria possessing mutant mtDNA can be selected through a mitophagy process that involves Atg1 and the NIX/BNIP3L ortholog (CG5059), but not Atg8 or Parkin, a mechanism similar to the one involved in the clearance of mitochondria during red blood cell maturation [108, 119, 121].

Mitophagy has recently been shown to be required during spermatogenesis in mice [122]. A novel gene, *Spata33*, was shown to act as a testis-specific mediator for mitophagy [122]. *Spata33* was required to promote mitophagy in mouse testis cells through its interaction with both the mitochondrial outer membrane protein VDAC2 and the autophagy protein ATG16L1, targeting these mitochondria for degradation through the lysosomal-autophagic pathway [122]. These results suggest that selective autophagy plays an indispensable role during the development and differentiation of germline cells in male mice, with the autophagy mediator SPATA33 being required for the maintenance of mitochondrial homeostasis through mitophagy. Further identification of tissue specific mediators of autophagy will provide more insight into the role it plays in germline development.

Selective autophagy, such as chaperone-mediated autophagy (CMA) has also been shown to be important in the regulation of embryonic stem cells in mammalian systems, with low levels of CMA promoting stemness, and elevated levels of CMA promoting differentiation [112]. Selective autophagy is another field that is currently very actively investigated and should provide additional insight into the role of autophagy in germline stem cell development.

Autophagy in Programmed Cell Death and Germline Development

The physiologic programmed cell death of germline cells is an important process in gametogenesis. In *C. elegans*, a population of the germline stem cells that have entered meiosis will undergo programmed cell death, a process which is dependent on the caspase protein CED-3, along with other core machinery EGL-1 and CED-4 [15]. This pathway leads to the activation of CED-3, which in turn mediates numerous effectors which enact the cell death program, resulting in the degradation of DNA and eventual engulfment and degradation of the dying cell by a neighboring cell [15]. The cell corpse is contained in a phagosome, which is delivered to the lysosome, degraded, and the cellular components recycled to feed the developing germ cells [33]. Autophagy contributes to this process of programmed cell death. Inactivation of numerous autophagy genes acting at different steps of autophagy, including *bec-1, unc-51, atg-7, atg-18*, and others, leads to the accumulation of cell corpses in the gonad [86]. Transmission electron microscopy analysis of the germ cell corpses and their surroundings found several examples of dying germ cells completely engulfed by gonadal sheath cells, but not digested. Corpse clearance defects have also been observed in *bec-1* mutant embryos, and were rescued by BEC-1 expression in the engulfing cells [41]. *Bec-1* mutant enhanced corpse clearance defects with simultaneous mutations in the engulfing genes *ced-1, ced-6*, and *ced-12* suggest that autophagy proteins function in parallel to known pathways involved in corpse removal [41].

Similarly, autophagy plays an important role in the later stages of oogenesis in *Drosophila* by mediating the programmed cell death of nurse cells [73]. The egg chambers which make up the Drosophila ovary each contain the oocyte, nurse cells, and follicle cells, with the latter two normally degraded during the maturation of the oocyte [81]. Nurse cells deficient in autophagy genes atg1 or vps34 failed to degrade the *Drosophila* inhibitor of apoptosis (IAP) dBruce, which results in nurse cell nuclei with fragmented DNA persisting in the egg chambers [73]. Thus, autophagy regulates programmed cell death of germ cells in *Drosophila* via regulation of DNA fragmentation in nurse cells [73].

Autophagy also appears to be tightly regulated during the initial phases of oogenesis in *Drosophila*. The Orb protein, the homolog of human Cytoplasmic polyadenylation element binding protein (CPEB), acts in translational regulation via interaction and alteration of mRNA poly(A) tails [85]. Orb functions to regulate autophagy by interacting with the *Atg12* mRNA, which restricts autophagy induction during the transition to meiosis and oogenesis, with loss of Orb function leading to abnormally high levels of cell death [85]. Orb was also shown to interact with *Atg1/unc-51/Ulk* and *Atg7* in ovarian extracts, which suggests that Orb regulation of autophagy plays a critical role in oogenesis [85].

Conclusions

There is a developing body of evidence that autophagy plays a vital role in the maintenance of the germline stem cell niche, and is required for the proper development of germline cells across metazoans. There remain many questions to answer in determining the specific activities which require autophagy in germline stem cells. For example, we do not know whether stem cell quiescence requires autophagy. It is known that many of the mediators of autophagy, such as TOR and insulin, act during cell quiescence. In *C. elegans,* insulin signaling was found to act during the transition from L1 quiescence to resumption of development through the re-initiation of the cell cycle [76]. It is not yet known what role autophagy may play in this interaction. Additionally, it is still unknown whether selective autophagy plays other roles in germline stem cell homeostasis, in addition to the regulation of mitophagy. The field of lipid metabolism continues to develop and will undoubtedly uncover additional roles of autophagy for the maintenance of the lipid pool in germline stem cell development. This work will continue to provide insight into the pathological mechanisms of infertility, as well as into the carcinogenesis of germline stem cell derived-tumors.

Acknowledgements We thank Drs. Cathy Savage-Dunn and Hannes Bülow for helpful discussions and/or critical reading of the manuscript. This work was supported in part by National Institutes of Health grant number 2R15GM102846-03 and the US-Israel Binational Science Foundation grant number 2017227 to AM. Dr. Meléndez received a New Scholar in Aging Award from the Ellison Medical Foundation.

Compliance with Ethical Standards

Disclosure of Interests All the authors declare they have no conflict of interest.

Ethical Approval This article does not contain any studies with human participants or animals performed by any of the authors.

References

1. Ames K, Da Cunha DS, Gonzalez B, Konta M, Lin F, Shechter G, Starikov L, Wong S, Bülow HE, Meléndez A (2017) A non-cell-autonomous role of BEC-1/BECN1/Beclin1 in coordinating cell-cycle progression and stem cell proliferation during germline development. Curr Biol CB 27:905–913
2. Ames K, Meléndez A (2017) Non-autonomous autophagy in germline stem cell proliferation. Cell CycleGeorget Tex 16:1481–1482
3. Austin J, Kimble J (1987) glp-1 is required in the germ line for regulation of the decision between mitosis and meiosis in C. elegans. Cell 51:589–599
4. Bandyopadhyay U, Kaushik S, Varticovski L, Cuervo AM (2008) The chaperone-mediated autophagy receptor organizes in dynamic protein complexes at the lysosomal membrane. Mol Cell Biol 28:5747–5763. https://doi.org/10.1128/MCB.02070-07
5. Berruti G, Paiardi C (2014) The dynamic of the apical ectoplasmic specialization between spermatids and Sertoli cells: the case of the small GTPase Rap1. BioMed Res Int 2014:635979
6. Blackwell TK, Sewell AK, Wu Z, Han M (2019) TOR signaling in Caenorhabditis Elegans development, metabolism, and aging. Genetics 213:329–360
7. Byrd DT, Knobel K, Affeldt K, Crittenden SL, Kimble J (2014) A DTC niche plexus surrounds the germline stem cell pool in Caenorhabditis Elegans. PLoS ONE 9:e88372
8. Campbell CL, Thorsness PE (1998) Escape of mitochondrial DNA to the nucleus in yme1 yeast is mediated by vacuolar-dependent turnover of abnormal mitochondrial compartments. J Cell Sci 111(Pt 16):2455–2464
9. Cao Y, Klionsky DJ (2007) Physiological functions of Atg6/Beclin 1: a unique autophagy-related protein. Cell Res 17:839–849
10. Cassada RC, Russell RL (1975) The dauerlarva, a post-embryonic developmental variant of the nematode Caenorhabditis elegans. Dev Biol 46:326–342
11. Charmpilas N, Tavernarakis N (2020) Mitochondrial maturation drives germline stem cell differentiation in Caenorhabditis elegans. Cell Death Differ 27:601–617
12. Chiang HL, Dice JF (1988) Peptide sequences that target proteins for enhanced degradation during serum withdrawal. J Biol Chem 263:6797–6805
13. Chiang HL, Schekman R, Hamamoto S (1996) Selective uptake of cytosolic, peroxisomal, and plasma membrane proteins into the yeast lysosome for degradation. J Biol Chem 271:9934–9941
14. Chiang HL, Terlecky SR, Plant CP, Dice JF (1989) A role for a 70-kilodalton heat shock protein in lysosomal degradation of intracellular proteins. Science 246:382–385
15. Conradt B, Xue D (2005) Programmed cell death. WormBook Online Rev. C Elegans Biol. 1–13
16. Crittenden SL, Troemel ER, Evans TC, Kimble J (1994) GLP-1 is localized to the mitotic region of the C. elegans germ line. Dev Camb Engl 120:2901–2911
17. Crittenden SL, Eckmann CR, Wang L, Bernstein DS, Wickens M, Kimble J (2003) Regulation of the mitosis/meiosis decision in the Caenorhabditis elegans germline. Philos Trans R Soc Lond B Biol Sci 358:1359–1362. https://doi.org/10.1098/rstb.2003.1333
18. Dansereau DA, Lasko P (2008) The development of germline stem cells in Drosophila. Methods Mol Biol Clifton NJ 450:3–26
19. Delcour C, Amazit L, Patino LC, Magnin F, Fagart J, Delemer B, Young J, Laissue P, Binart N, Beau I (2019) ATG7 and ATG9A loss-of-function variants trigger autophagy impairment and ovarian failure. Genet Med Off J Am Coll Med Genet 21:930–938
20. Deng W, Lin H (1997) Spectrosomes and fusomes anchor mitotic spindles during asymmetric germ cell divisions and facilitate the formation of a polarized microtubule array for oocyte specification in Drosophila. Dev Biol 189:79–94
21. Djeddi A, Rawi SA, Deuve JL, Perrois C, Liu Y-Y, Russeau M, Sachse M, Galy V (2015) Sperm-inherited organelle clearance in C. elegans relies on LC3-dependent autophagosome targeting to the pericentrosomal area. Development 142:1705–1716

22. Drummond-Barbosa D, Spradling AC (2001) Stem cells and their progeny respond to nutritional changes during Drosophila oogenesis. Dev Biol 231:265–278
23. Extavour CG, Akam M (2003) Mechanisms of germ cell specification across the metazoans: epigenesis and preformation. Dev Camb Engl 130:5869–5884
24. Feric M, Vaidya N, Harmon TS, Mitrea DM, Zhu L, Richardson TM, Kriwacki RW, Pappu RV, Brangwynne CP (2016) Coexisting liquid phases underlie nucleolar subcompartments. Cell 165:1686–1697
25. Fox PM, Schedl T (2015) Analysis of germline stem cell differentiation following loss of GLP-1 notch activity in Caenorhabditis elegans. Genetics 201:167–184
26. Fukuyama M, Rougvie AE, Rothman JH (2006) C. elegans DAF-18/PTEN mediates nutrient-dependent arrest of cell cycle and growth in the germline. Curr Biol CB 16:773–779
27. Fukuyama M, Sakuma K, Park R, Kasuga H, Nagaya R, Atsumi Y, Shimomura Y, Takahashi S, Kajiho H, Rougvie A et al (2012) C. elegans AMPKs promote survival and arrest germline development during nutrient stress. Biol Open 1:929–936
28. Fuller MT (1993) Spermatogenesis. In: Bate M, Martinez-Arias A (eds) The development of Drosophila. Cold Spring Harbor Laboratory Press, Cold Spring Harbor, NY
29. Gao F, Li G, Liu C, Gao H, Wang H, Liu W, Chen M, Shang Y, Wang L, Shi J et al (2018) Autophagy regulates testosterone synthesis by facilitating cholesterol uptake in Leydig cells. J Cell Biol 217:2103–2119
30. Gawriluk TR, Hale AN, Flaws JA, Dillon CP, Green DR, Rucker EB (2011) Autophagy is a cell survival program for female germ cells in the murine ovary. Reprod Camb Engl 141:759–765
31. Gawriluk TR, Ko C, Hong X, Christenson LK, Rucker EB (2014) Beclin-1 deficiency in the murine ovary results in the reduction of progesterone production to promote preterm labor. Proc Natl Acad Sci 111:E4194–E4203
32. Golden JW, Riddle DL (1984) The Caenorhabditis elegans dauer larva: developmental effects of pheromone, food, and temperature. Dev Biol 102:368–378
33. Gumienny TL, Lambie E, Hartwieg E, Horvitz HR, Hengartner MO (1999) Genetic control of programmed cell death in the Caenorhabditis elegans hermaphrodite germline. Dev Camb Engl 126:1011–1022
34. Hansen D, Hubbard EJA, Schedl T (2004) Multi-pathway control of the proliferation versus meiotic development decision in the Caenorhabditis elegans germline. Dev Biol 268:342–357
35. Hardy RW, Tokuyasu KT, Lindsley DL, Garavito M (1979) The germinal proliferation center in the testis of Drosophila melanogaster. J Ultrastruct Res 69:180–190
36. Hatfield SD, Shcherbata HR, Fischer KA, Nakahara K, Carthew RW, Ruohola-Baker H (2005) Stem cell division is regulated by the microRNA pathway. Nature 435:974–978
37. Hilscher B, Hilscher W, Bülthoff-Ohnolz B, Krämer U, Birke A, Pelzer H, Gauss G (1974) Kinetics of gametogenesis. I. Comparative histological and autoradiographic studies of oocytes and transitional prospermatogonia during oogenesis and prespermatogenesis. Cell Tissue Res 154:443–470
38. Hird SN, Paulsen JE, Strome S (1996) Segregation of germ granules in living Caenorhabditis elegans embryos: cell-type-specific mechanisms for cytoplasmic localisation. Dev Camb Engl 122:1303–1312
39. Houston DW, King ML (2000) A critical role for Xdazl, a germ plasm-localized RNA, in the differentiation of primordial germ cells in Xenopus. Dev Camb Engl 127:447–456
40. Hsu H-J, Drummond-Barbosa D (2011) Insulin signals control the competence of the Drosophila female germline stem cell niche to respond to Notch ligands. Dev Biol 350:290–300
41. Huang S, Jia K, Wang Y, Zhou Z, Levine B (2013) Autophagy genes function in apoptotic cell corpse clearance during C. elegans embryonic development. Autophagy 9:138–149
42. Hubbard EJA, Korta DZ, Dalfó D (2013) Physiological control of germline development. Adv Exp Med Biol 757:101–131
43. Hubbard EJA, Schedl T (2019) Biology of the Caenorhabditis elegans Germline stem cell system. Genetics 213:1145–1188

44. Jain S, Wheeler JR, Walters RW, Agrawal A, Barsic A, Parker R (2016) ATPase-modulated stress granules contain a diverse proteome and substructure. Cell 164:487–498
45. Kadekar P, Roy R (2019) AMPK regulates germline stem cell quiescence and integrity through an endogenous small RNA pathway. PLOS Biol 17:e3000309. https://doi.org/10.1371/journal.pbio.3000309
46. Kawasaki I, Shim YH, Kirchner J, Kaminker J, Wood WB, Strome S (1998) PGL-1, a predicted RNA-binding component of germ granules, is essential for fertility in C. elegans. Cell 94:635–645
47. Killian DJ, Hubbard EJA (2005) Caenorhabditis elegans germline patterning requires coordinated development of the somatic gonadal sheath and the germ line. Dev Biol 279:322–335
48. Kimble J, Hirsh D (1979) The postembryonic cell lineages of the hermaphrodite and male gonads in Caenorhabditis elegans. Dev Biol 70:396–417
49. Kirilly D, Xie T (2007) The Drosophila ovary: an active stem cell community. Cell Res 17:15–25
50. Klionsky et al (2021) Guidelines for the use and interpretation of assays for monitoring autophagy. Autophagy 17(1):1–382. https://doi.org/10.1080/15548627.2020.1797280
51. Korta DZ, Tuck S, Hubbard EJA (2012) S6K links cell fate, cell cycle and nutrient response in C. elegans germline stem/progenitor cells. Dev Camb Engl. 139:859–870
52. LaFever L, Drummond-Barbosa D (2005) Direct control of germline stem cell division and cyst growth by neural insulin in Drosophila. Science 309:1071–1073
53. LaFever L, Feoktistov A, Hsu H-J, Drummond-Barbosa D (2010) Specific roles of Target of rapamycin in the control of stem cells and their progeny in the Drosophila ovary. Development 137:2451–2451
54. Lapierre LR, Gelino S, Meléndez A, Hansen M (2011) Autophagy and lipid metabolism coordinately modulate life span in germline-less C. elegans. Curr Biol CB 21:1507–1514
55. Lapierre LR, De Magalhaes Filho CD, McQuary PR, Chu C-C, Visvikis O, Chang JT, Gelino S, Ong B, Davis AE, Irazoqui JE et al (2013) The TFEB orthologue HLH-30 regulates autophagy and modulates longevity in Caenorhabditis elegans. Nat Commun 4:2267
56. Laws KM, Drummond-Barbosa D (2016) AMP-activated protein kinase has diet-dependent and -independent roles in Drosophila oogenesis. Dev Biol 420:90–99. https://doi.org/10.1016/j.ydbio.2016.10.006
57. Lieber T, Jeedigunta SP, Palozzi JM, Lehmann R, Hurd TR (2019) Mitochondrial fragmentation drives selective removal of deleterious mtDNA in the germline. Nature 570:380–384
58. Lin K, Dorman JB, Rodan A, Kenyon C (1997) daf-16: An HNF-3/forkhead Family member that can function to double the life-span of Caenorhabditis elegans. Science 278:1319–1322
59. Liu K, Czaja MJ (2013) Regulation of lipid stores and metabolism by lipophagy. Cell Death Differ 20:3–11
60. Liu X-M, Sun L-L, Hu W, Ding Y-H, Dong M-Q, Du L-L (2015) ESCRTs cooperate with a selective autophagy receptor to mediate vacuolar targeting of soluble cargos. Mol Cell 59:1035–1042
61. Liu C, Wang H, Shang Y, Liu W, Song Z, Zhao H, Wang L, Jia P, Gao F, Xu Z et al (2016) Autophagy is required for ectoplasmic specialization assembly in sertoli cells. Autophagy 12:814–832
62. Mahowald AP (2001) Assembly of the Drosophila germ plasm. Int Rev Cytol 203:187–213
63. Megosh HB, Cox DN, Campbell C, Lin H (2006) The role of PIWI and the miRNA machinery in Drosophila Germline determination. Curr Biol 16:1884–1894
64. Meléndez A, Tallóczy Z, Seaman M, Eskelinen E-L, Hall DH, Levine B (2003) Autophagy genes are essential for dauer development and life-span extension in C. elegans. Science 301:1387–1391. https://doi.org/10.1126/science.1087782
65. Meléndez A, Levine B (2018) Autophagy in C. elegans (WormBook)
66. Michaelson D, Korta DZ, Capua Y, Hubbard EJA (2010) Insulin signaling promotes germline proliferation in C. elegans. Dev. Camb. Engl. 137:671–680

67. Mizushima N, Yoshimori T, Ohsumi Y (2011) The role of Atg proteins in autophagosome formation. Annu Rev Cell and Dev Biol 27:107–132. https://doi.org/10.1146/annurev-cellbio-092910-154005
68. Mukherjee A, Patel B, Koga H, Cuervo AM, Jenny A (2016) Selective endosomal microautophagy is starvation-inducible in Drosophila. Autophagy 12:1984–1999. https://doi.org/10.1080/15548627.2016.1208887
69. Murphy CT (2013) Insulin/insulin-like growth factor signaling in C. elegans. WormBook 1–43
70. Murphy CT, Hu PJ (2018) Insulin/insulin-like growth factor signaling in C. elegans (WormBook)
71. Müller O, Sattler T, Flötenmeyer M, Schwarz H, Plattner H, Mayer A (2000) Autophagic tubes. J Cell Biol 151:519–528
72. Narbonne P, Maddox PS, Labbé J-C (2015) DAF-18/PTEN locally antagonizes insulin signalling to couple germline stem cell proliferation to oocyte needs in C. elegans. Development 142:4230–4241
73. Nezis IP, Shravage BV, Sagona AP, Lamark T, Bjørkøy G, Johansen T, Rusten TE, Brech A, Baehrecke EH, Stenmark H (2010) Autophagic degradation of dBruce controls DNA fragmentation in nurse cells during late Drosophila melanogaster oogenesis. J Cell Biol 190:523–531
74. Nguyen TB, Louie SM, Daniele JR, Tran Q, Dillin A, Zoncu R, Nomura DK, Olzmann JA (2017) DGAT1-dependent lipid droplet biogenesis protects mitochondrial function during starvation-induced autophagy. Dev Cell 42:9-21.e5
75. Niepielko MG, Eagle WVI, Gavis ER (2018) Stochastic Seeding Coupled with mRNA self-recruitment generates heterogeneous Drosophila Germ Granules. Curr Biol CB 28:1872-1881.e3
76. Olmedo M, Mata-Cabana A, Jesús Rodríguez-Palero M, García-Sánchez S, Fernández-Yañez A, Merrow M, Artal-Sanz M (2020) Prolonged quiescence delays somatic stem cell-like divisions in Caenorhabditis elegans and is controlled by insulin signaling. Aging Cell 19:e13085
77. Palmisano NJ, Meléndez A (2019) Autophagy in C elegans development. Dev Biol Mar 1 447(1):103–125
78. Pek JW, Patil VS, Kai T (2012) piRNA pathway and the potential processing site, the nuage, in the Drosophila germline. Dev Growth Differ 54:66–77
79. Pepper AS-R, Killian DJ, Hubbard EJA (2003) Genetic analysis of Caenorhabditis elegans glp-1 mutants suggests receptor interaction or competition. Genetics 163:115–132
80. Pepper AS-R, Lo TW, Killian DJ, Hall DH, Hubbard EJA (2003) The establishment of Caenorhabditis elegans germline pattern is controlled by overlapping proximal and distal somatic gonad signals. Dev Biol 259:336–350
81. Pritchett TL, Tanner EA, McCall K (2009) Cracking open cell death in the Drosophila ovary. Apoptosis Int J Program Cell Death 14:969–979
82. Rawi SA, Louvet-Vallée S, Djeddi A, Sachse M, Culetto E, Hajjar C, Boyd L, Legouis R, Galy V (2011) Postfertilization autophagy of sperm organelles prevents paternal Mitochondrial DNA transmission. Science 334:1144–1147
83. Richardson BE, Lehmann R (2010) Mechanisms guiding primordial germ cell migration: strategies from different organisms. Nat Rev Mol Cell Biol 11:37–49
84. Riddle DL, Albert PS (1997) Genetic and environmental regulation of Dauer larva development. In: Elegans II C, Riddle DL, Blumenthal T, Meyer BJ, Priess JR (eds) (Cold Spring Harbor (NY): Cold Spring Harbor Laboratory Press), p
85. Rojas-Ríos P, Chartier A, Pierson S, Séverac D, Dantec C, Busseau I, Simonelig M (2015) Translational control of autophagy by orb in the Drosophila Germline. Dev Cell 35:622–631
86. Ruck A, Attonito J, Garces KT, Núnez L, Palmisano NJ, Rubel Z, Bai Z, Nguyen KCQ, Sun L, Grant BD et al (2011) The Atg6/Vps30/Beclin 1 ortholog BEC-1 mediates endocytic retrograde transport in addition to autophagy in C elegans. Autophagy. 7:386–400

87. Sahu R, Kaushik S, Clement CC, Cannizzo ES, Scharf B, Follenzi A, Potolicchio I, Nieves E, Cuervo AM, Santambrogio L (2011) Microautophagy of cytosolic proteins by late endosomes. Dev Cell 20:131–139
88. Saitou M, Yamaji M (2010) Germ cell specification in mice: signaling, transcription regulation, and epigenetic consequences. Reprod Camb Engl 139:931–942
89. Sakai Y, Koller A, Rangell LK, Keller GA, Subramani S (1998) Peroxisome degradation by microautophagy in Pichia pastoris: identification of specific steps and morphological intermediates. J Cell Biol 141:625–636
90. Santambrogio L, Cuervo AM (2011) Chasing the elusive mammalian microautophagy. Autophagy 7:652– 654. https://doi.org/10.4161/auto.7.6.15287
91. Schuck S, Gallagher CM, Walter P (2014) ER-phagy mediates selective degradation of endoplasmic reticulum independently of the core autophagy machinery. J Cell Sci 127:4078–4088
92. Seydoux G (2018) The P Granules of C. elegans: a genetic model for the study of RNA-Protein condensates. J Mol Biol 430:4702–4710
93. Singh R, Kaushik S, Wang Y, Xiang Y, Novak I, Komatsu M, Tanaka K, Cuervo AM, Czaja MJ (2009) Autophagy regulates lipid metabolism. Nature 458:1131–1135
94. Song W-H, Yi Y-J, Sutovsky M, Meyers S, Sutovsky P (2016) Autophagy and ubiquitin–proteasome system contribute to sperm mitophagy after mammalian fertilization. Proc Natl Acad Sci 113:E5261–E5270
95. Song Z-H, Yu H-Y, Wang P, Mao G-K, Liu W-X, Li M-N, Wang H-N, Shang Y-L, Liu C, Xu Z-L et al (2015) Germ cell-specific Atg7 knockout results in primary ovarian insufficiency in female mice. Cell Death Dis 6:e1589
96. Song X, Zhu C-H, Doan C, Xie T (2002) Germline stem cells anchored by adherens junctions in the Drosophila ovary niches. Science 296:1855–1857
97. Speed RM (1982) Meiosis in the foetal mouse ovary. I. An analysis at the light microscope level using surface-spreading. Chromosoma 85:427–437
98. Spike C, Meyer N, Racen E, Orsborn A, Kirchner J, Kuznicki K, Yee C, Bennett K, Strome S (2008) Genetic analysis of the Caenorhabditis elegans GLH family of P-granule proteins. Genetics 178:1973–1987
99. Spradling A, Fuller MT, Braun RE, Yoshida S (2011) Germline stem cells. Cold Spring Harb Perspect Biol 3:a002642
100. Strome S, Wood WB (1982) Immunofluorescence visualization of germ-line-specific cytoplasmic granules in embryos, larvae, and adults of Caenorhabditis elegans. Proc Natl Acad Sci 79:1558–1562. https://doi.org/10.1073/pnas.79.5.1558
101. Sulston JE, Schierenberg E, White JG, Thomson JN (1983) The embryonic cell lineage of the nematode Caenorhabditis elegans. Dev Biol 100:64–119
102. Sênos Demarco R, Uyemura BS, Jones DL (2020) EGFR signaling stimulates autophagy to regulate stem cell maintenance and lipid Homeostasis in the Drosophila Testis. Cell Rep 30:1101-1116.e5
103. Tian Y, Li Z, Hu W, Ren H, Tian E, Zhao Y, Lu Q, Huang X, Yang P, Li X et al (2010) C. elegans screen identifies autophagy genes specific to multicellular organisms. Cell 141:1042–1055
104. Trcek T, Grosch M, York A, Shroff H, Lionnet T, Lehmann R (2015) Drosophila germ granules are structured and contain homotypic mRNA clusters. Nat Commun 6:7962
105. Tuttle DL, Lewin AS, Dunn WA (1993) Selective autophagy of peroxisomes in methylotrophic yeasts. Eur J Cell Biol 60:283–290
106. Updike DL, Knutson AK, Egelhofer TA, Campbell AC, Strome S (2014) Germ-granule components prevent somatic development in the C. elegans germline. Curr Biol CB 24:970–975
107. van Zutphen T, Todde V, de Boer R, Kreim M, Hofbauer HF, Wolinski H, Veenhuis M, van der Klei IJ, Kohlwein SD (2014) Lipid droplet autophagy in the yeast Saccharomyces cerevisiae. Mol Biol Cell 25:290–301. https://doi.org/10.1091/mbc.E13-08-0448
108. Villa E, Marchetti S, Ricci J-E (2018) No Parkin Zone: Mitophagy without Parkin. Trends Cell Biol 28:882–895

109. Wang MC, O'Rourke EJ, Ruvkun G (2008) Fat Metabolism links Germline stem cells and Longevity in C. elegans. Science 322:957–960
110. Wang H, Wan H, Li X, Liu W, Chen Q, Wang Y, Yang L, Tang H, Zhang X, Duan E et al (2014) Atg7 is required for acrosome biogenesis during spermatogenesis in mice. Cell Res 24:852–869
111. Wen X, Klionsky DJ (2016) An overview of macroautophagy in yeast. J Mol Biol 428:1681–1699. https://doi.org/10.1016/j.jmb.2016.02.021
112. Xu Y, Zhang Y, García-Cañaveras JC, Guo L, Kan M, Yu S, Blair IA, Rabinowitz JD, Yang X (2020) Chaperone-mediated autophagy regulates the pluripotency of embryonic stem cells. Science 369:397–403
113. Yamashita YM, Jones DL, Fuller MT (2003) Orientation of asymmetric stem cell division by the APC tumor suppressor and centrosome. Science 301:1547–1550
114. Yang X, Zhang W, Wen X, Bulinski PJ, Chomchai DA, Arines FM, Liu Y-Y, Sprenger S, Teis D, Klionsky DJ et al (2020) TORC1 regulates vacuole membrane composition through ubiquitin- and ESCRT-dependent microautophagy. J Cell Biol 219:e201902127
115. Yang Y, Klionsky DJ (2020) TORC1 regulates vacuole membrane composition through ubiquitin- and ESCRT-dependent microautophagy. J Cell Biol 219(3):e201902127. https://doi.org/10.1083/jcb.201902127
116. Zanetti S, Puoti A (2013) Sex determination in the Caenorhabditis elegans germline. Adv Exp Med Biol 757:41–69
117. Zhang H (2018) Lysosomal acid lipase and lipid metabolism: new mechanisms, new questions, and new therapies. Curr Opin Lipidol 29:218–223
118. Zhang L, Gajewski TF, Kline J (2009) PD-1/PD-L1 interactions inhibit antitumor immune responses in a murine acute myeloid leukemia model. Blood 114:1545–1552
119. Zhang J, Loyd MR, Randall MS, Waddell MB, Kriwacki RW, Ney PA (2012) A short linear motif in BNIP3L (NIX) mediates mitochondrial clearance in reticulocytes. Autophagy 8:1325–1332
120. Zhang G, Wang Z, Du Z, Zhang H (2018) mTOR regulates phase separation of PGL granules to modulate their Autophagic degradation. Cell 174:1492-1506.e22
121. Zhang Y, Yan L, Zhou Z, Yang P, Tian E, Zhang K, Zhao Y, Li Z, Song B, Han J et al (2009) SEPA-1 mediates the specific recognition and degradation of P granule components by autophagy in C. elegans. Cell 136:308–321
122. Zhang Y, Xu X, Hu M, Wang X, Cheng H, Zhou R (2020) SPATA33 is an autophagy mediator for cargo selectivity in germline mitophagy. Cell Death Differ, 1–15
123. Zhao Y, Tian E, Zhang H (2009) Selective autophagic degradation of maternally-loaded germline P granule components in somatic cells during C. elegans embryogenesis. Autophagy 5:717–719

Chapter 5
The Role of Autophagy in the Regulation of Hematopoietic Stem Cells

Pallavi Budgude, Prajakta Teli, Anuradha Vaidya, and Vaijayanti Kale

Abstract Hematopoietic stem cells (HSCs) reside in the bone marrow and replenish the blood cells as per the physiological requirement of the body, thereby maintaining homeostasis. The intrinsic mechanisms within the HSCs and the extrinsic mechanisms exerted by the bone marrow (BM) microenvironment play an important role in the critical regulation of the HSC stem cell fates, such as quiescence, self-renewal, and differentiation. An imbalance in the intrinsic and extrinsic regulatory mechanisms of the HSCs contributes towards the development of different types of blood cancers. Autophagy, a conserved catabolic process, is also involved in the intrinsic and extrinsic programming of HSCs throughout the various developmental stages of the hematopoietic system, right from the yolk sac to the BM. We begin this chapter by giving an introduction to different types of autophagy and its mechanisms, followed by discussing the role of autophagy in regulating different types of stem cells. The major part of the chapter is focused on understanding how autophagy plays an important functional role in normal hematopoiesis and the regulation of different HSC fates. We have also given an overview of how autophagy regulates the ontogeny of the hematopoietic system, and how dysregulation of the autophagy mechanism leads to the development of hematological malignancies. In the end, we discuss how the modulation of autophagy could enhance our knowledge of HSC-associated diseases and perhaps augment the development of efficient treatment strategies.

Keywords Autophagy · Adult stem cells · Hematopoietic stem cells · Aging · Hematological disorders

---

P. Budgude · P. Teli · A. Vaidya · V. Kale (✉)
Symbiosis Centre for Stem Cell Research, Symbiosis International (Deemed University), Lavale, Pune 412115, India
e-mail: vaijayanti.kale@ssbs.edu.in; vaijayanti.kale@gmail.com; head_scscr@siu.edu.in

P. Budgude · P. Teli · A. Vaidya
Symbiosis School of Biological Sciences, Symbiosis International (Deemed University), Pune 412115, India

B. V. Shravage and K. Turksen (eds.), *Autophagy in Stem Cell Maintenance and Differentiation*, Stem Cell Biology and Regenerative Medicine 73,
https://doi.org/10.1007/978-3-031-17362-2_5

Introduction

For decades, hematopoietic stem cells (HSC) have sparked interest due to their proven therapeutic potential. The scarcity of HSCs has always been a significant obstacle to the widespread clinical use of HSC-based therapies [42]. Therefore, by comprehending the biology of HSCs, attempts are being made to develop strategies to enrich HSC numbers or boost their repopulating potential in vivo following transplantation [54]. Autophagy has long been known as a fundamental process for cells to generate energy during nutritional deprivation, respond to cellular stress, and eliminate old or damaged organelles as a quality control mechanism [17]. The role of autophagy in the maintenance and differentiation of HSCs has recently become evident. Autophagy is one of the crucial metabolic pathways involved in the progression of leukemias [50]. This chapter, therefore, aims to provide an overview of the types of autophagy and their role in stem cell biology, with a special focus on HSC fate determination and hematopoietic malignancies.

Autophagy

The term autophagy was derived from the Greek meaning "self-eating" and was coined by Christen de Duve in 1963 at the Ciba Foundation Symposium on Lysosomes, where he described the process of self-eating [47]. Autophagy was named so after observing single membrane vesicles of rat hepatic cells containing cytoplasm and organelles such as mitochondria and endoplasmic reticulum [5]. It is a highly conserved process that plays a crucial role in cell survival and maintenance of homeostasis by degrading unwanted or damaged proteins, macromolecules, and cell organelles [74].

Autophagy delivers the unwanted or damaged components of the cells to the lysosomes to degrade and recycles them into a form that can be reused by the cells [20]. Based on the pathways used to deliver the cargos to the lysosomes, autophagy is broadly divided into three types: Microautophagy, Macroautophagy, and Chaperone-mediated autophagy (CMA) [76]. Even though all three types of autophagy ultimately deliver the targeted cargos to the lysosome for degradation, they are mechanistically different from each other [111]. Microautophagy uses lysosomal membrane invagination or protrusions to capture and invade targeted cargos inside the lumen [61]. CMA uses chaperones containing pentapeptide motifs, which identity and bind to unfolded proteins, and translocate them to the lysosomal membrane [44]. Macroautophagy is a unique and extensively studied process, as compared to microautophagy and CMA. It involves sequestration of cargos by forming a double membrane structure called the autophagosome, which fuses with the lysosome for degradation [22]. Microautophagy and macroautophagy can be selective or non-selective, whereas CMA is non-selective. Selective autophagy is used to clear specific targets like damaged organelles, including peroxisomes and mitochondria, ubiquitinated proteins, and invasive microbes. Each process under microautophagy involves a core set of machinery and distinct components, and accordingly is identified with a unique name such as mitophagy for selective removal of mitochondria, pexophagy for removal of peroxisomes, and xenophagy for removal of microbes [28]. Non-selective autophagy is used under starvation conditions for turnover of bulk cytoplasmic components [46].

Types of Autophagy

Microautophagy

In microautophagy, the lysosomal membrane engulfs the cytoplasmic content by forming a tube-like structure. It helps in the formation and budding of vesicles containing cytoplasmic content into the lysosomal lumen [89]. In the early stages of microautophagy, the membrane expands into the surface of the lysosome by excluding transmembrane proteins and segregating lipids. Invagination moves rapidly and extends laterally, which further specializes into a tubular structure called "autophagic tubes." Cytoplasmic content captured by the tube is degraded by the hydrolases present in the lumen. This lateral extension and tube formation are ATP-dependent [71].

In 2011, Sahu et al. studied selective microautophagy and showed the presence of a microautophagy-like process that transfers cytoplasmic content to the vesicles of late endosome/multivesicular bodies (MVBs) (Fig. 5.1a). This process is called endosomal microautophagy, as it uses the molecular components used in endocytic pathways. The internalization of cytosolic cargos into the vesicles relies on endosomal sorting complexes required for the transport (ESCRT) I and III, and protein cargo delivery relies on chaperones like Heat shock cognate 70 (HSC 70) [89]. Even though chaperones mediate this process, it is different from CMA as it delivers cargos to late endosomes.

Chaperone-Mediated Autophagy (CMA)

CMA is a highly specific process, unlike microautophagy and macroautophagy, which can non-specifically engulf bulk cytoplasm. CMA plays an essential role in eliminating damaged proteins and regulating cellular proteostasis [44]. It also balances the cellular energetics by recycling targeted proteins into amino acids, which can be utilized by the cells. Targeted proteins are translocated to the lysosomal lumen, where they are cleaved into amino acids by luminal proteases called cathepsins. CMA is highly specific because of its ability to recognize and bind to the targeted proteins. This binding occurs because the pentapeptide motif biochemically

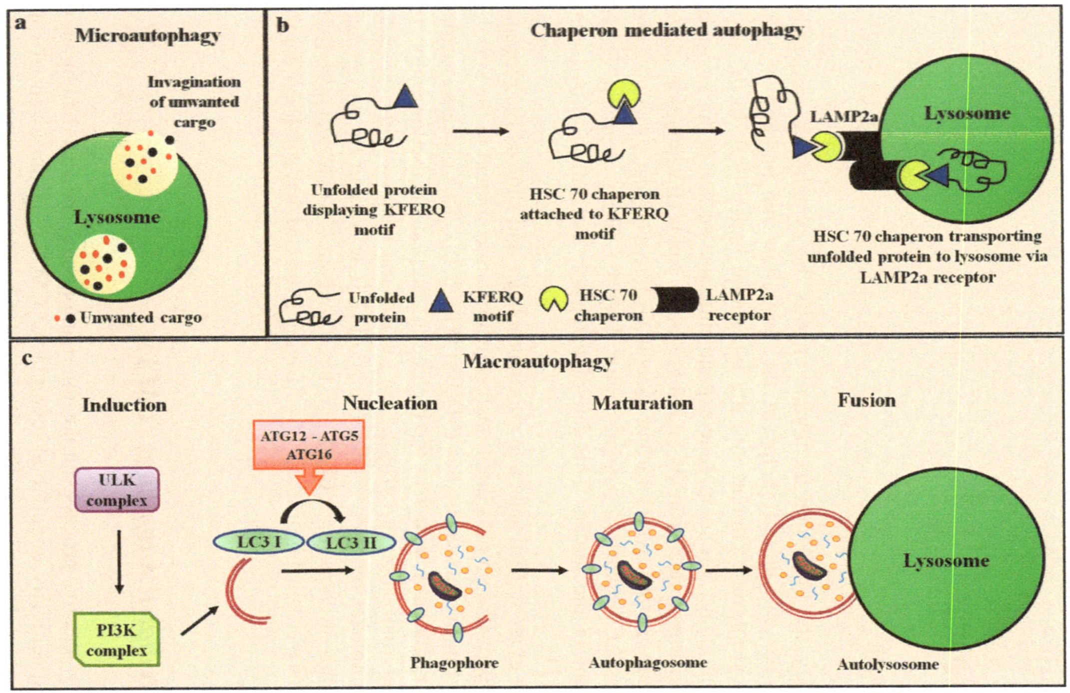

Fig. 5.1 An overview of types of autophagy mechanisms: The illustration depicts mechanisms involved in different types of autophagy such as **a** Microautophagy, which is characterized by the direct absorption of cargo through lysosomal membrane invagination, **b** Chaperon-mediated autophagy, which transports individual unfolded proteins directly across the lysosomal membrane via chaperones, and **c** Macroautophagy, which transfers cargo to the lysosome by the formation of cytosolic double-membrane vesicles known as autophagosomes that fuse with lysosome for degradation of cargos

related to the KFERQ motif, which is present on all CMA substrates [2]. According to analysis, around 30% of cytoplasmic proteins contain this sequence [18].

The constitutively expressed HSC70 recognizes KFERQ motifs on the target proteins. The chaperone HSC70 and co-chaperones like heat shock protein (hsp)A8 bind to the targeted proteins and deliver them to the lysosomal membrane. This protein-chaperone complex binds to a lysosomal monomeric single-span membrane protein called lysosomal-associated membrane protein type-2A (LAMP2a) [32]. After association, LAMP2a initiates multimerization with the help of lysosomal resident HSC70. The target protein is unfolded as per the requirement of translocation machinery and internalized into the lysosomal compartment, where HSC70 regulates the proteolysis of proteins into amino acids with the help of luminal proteases like cathepsins. After degradation, translocation assembly is disassembled by luminal HSC70 returning LAMP2A to its monomeric form to initiate subsequent translocation [15] (Fig. 5.1b). Regulation of translocation process is rate-limiting and is regulated by HSC70 and other factors like hsp90, mechanistic target of rapamycin complex (mTORC) and the kinase AKT [4].

Macroautophagy

Macroautophagy is commonly expressed at a basal level in all cells, but it gets upregulated during various stress conditions. However, too much self-degradation can lead to cell death; and hence, the process needs to be kept in check. Accordingly, multiple inputs regulate the induction of macroautophagy [48]. These include intracellular sensors that respond to extrinsic and intrinsic changes, such as the presence or absence of reactive oxygen species (ROS), growth factors, glucose, amino acids, and nitrogen. Macroautophagy and its molecular mechanisms have been mostly studied in yeast model systems. Over 30 autophagy-related genes (ATG) were identified by screening yeast mutants [60]. Among the ATG protein complexes, an Unc-51-like kinase (ULK) complex is a primary group of components. It is one of the initial ones that dictate the site of autophagosome formation, called pre-autophagosomal structure (PAS). In contrast, other complexes participate in the later phases of autophagosome formation. Interestingly, autophagosome formation is accomplished through sequential events such as induction, nucleation, elongation, and fusion (Fig. 5.1c) [22].

Induction

Induction of autophagosome formation in macroautophagy is regulated by activation of ULK family either by ULK1 or ULK2. Following activation, ULK phosphorylates other autophagy factors like ATG13 and scaffold protein called RB1-inducible coiled-coil 1/Focal adhesion kinase family interacting protein of 200 kD (RB1CC1/FIP200), thus resulting in the assembly of the ULK1-ATG13-RB1CC1-ATG101 complex. This initiation complex is stable and forms regardless of nutrient status. On the other hand, the MTORC1 is influenced by the nutrient status of the cell, which regulates its association/dissociation with the induction complex [82]. When MTORC1 is associated with induction complex, it phosphorylates ULK1/2 and ATG13, thereby inactivating them and inhibiting induction of autophagy. However, upon rapamycin treatment or when the cells are undergoing nutrient starvation, MTORC1 dissociates from the induction complex, resulting in dephosphorylation of the induction complex and leads to induction of macroautophagy. Moreover, the induction complex is also responsible for activating another essential autophagy protein complex, the phosphatidylinositol 3-kinase (PI3K) complex, required for nucleation events of autophagosome formation.

Nucleation

The ATG14-containing class III PI3K complex is the next complex recruited at the site of autophagosome formation. This complex is involved in the nucleation of the phagophore and consists of phosphoinositide 3-kinase catalytic subunit 3 (PIK3C3/VPS34), phosphoinositide 3-kinase regulatory subunit 4 (PIK3R4/p150), and Beclin (BECN1) [110]. The PI3K complex produces PI3P, which is required for macroautophagy, and its regulation occurs primarily through proteins that interact with BECN1. Importantly, BCL2 (anti-apoptotic protein) binds BECN1 and prevents its interaction with PIK3C3, thereby inhibiting macroautophagy [45]. Furthermore, Autophagy and Beclin 1 Regulator 1 (AMBRA1) and Bax-interacting factor 1 (Bif-1) are two positive regulators of the PI3K complex that interact with BECN1 directly to induce macroautophagy.

Elongation

There are two conjugation structures consisting of ubiquitin-like (UBL) proteins responsible for expanding the phagophore. The first system involves forming the ATG12–ATG5-ATG16 complex [30]. Initially, ATG12 binds to ATG5, and this binding is dependent on the E1 and E2-like activating enzymes called ATG7 and ATG10, respectively. Furthermore, ATG16L1 non-covalently binds to ATG5 of the ATG12–ATG5 complex. It further dimerizes and allows its association with the phagophore promoting elongation of the membrane. A microtubule-associated protein 1A/1B-light chain 3 (LC3) conjugation complex makes up the second UBL system involved in phagophore expansion [58]. ATG4 processes LC3 to reveal a C-terminal glycine of ATG8 (LC3-I). Further, ATG7 activates LC3-I and transfers it into ATG3. The C-terminal glycine of ATG8 is covalently attached to the lipid phosphatidylethanolamine (PE). The ATG12–ATG5–ATG16L1 complex functions as an E3 ligase in the conjugation of PE to LC3-I, resulting in LC3-II, which binds to the phagophore [30]. LC3-II can subsequently be cleaved by ATG4 to release LC3, leading to deconjugation, an essential step in macroautophagy because defects in cleavage result in partial autophagic dysfunction.

Fusion

The expanding phagophore eventually matures to form a completed autophagosome, which fuses with the lysosome, becoming an autolysosome. Microtubules are responsible for the trafficking of autophagosomes to lysosomes [108]. Fusion of autophagosomes with lysosomes involves the protein UV-radiation resistance-associated gene (UVRAG), which can associate with the PI3K complex and activate the GTPase RAB7. Upon fusion, lysosomal acidic hydrolases degrade the unwanted cargo carried by autophagosome and recycle them back to the cytoplasm to be reused by the cells [9].

Autophagy and Stem Cells

Stem cells are defined as cells that have the ability to self-renew and differentiate into specific cell types. There are different types of stem cells like embryonic stem cells (ESCs), tissues specific stem cells (hematopoietic, neural, and muscle stem cells), mesenchymal stem cells (MSCs), and induced pluripotent stem cells (iPSCs). Autophagy also plays an important role during embryonic development and in adult stem cells and is important for maintaining cell survival. There are multiple pieces of evidence that demonstrate the essential role of autophagy, primarily macroautophagy (Fig. 5.1c) in the maintenance and function of all types of stem cells [12].

Hematopoietic Stem Cells

Hematopoiesis is defined as the formation of blood cell components, and the stem cells that give rise to blood cells are called as Hematopoietic stem cells (HSCs). Hematopoiesis occurs during embryonic development and throughout adulthood to produce and replenish the blood system. Quiescent HSCs exhibit low oxidative phosphorylation levels, whereas activation of HSCs leads to high oxidative phosphorylation [111]. Studies demonstrate that autophagy plays an essential role in removing activated mitochondria as a mechanism for controlling oxidative metabolism in order to maintain HSCs' quiescence and self-renewal. Therefore, upon deletion of Atg12 in the HSCs, an increased mitochondrial content accompanied by an activated metabolic state was observed, which enhanced myeloid differentiation (aging-like phenotype). The role of autophagy in the regulation of HSCs is explained in the latter part of the chapter.

Neural Stem Cells

Neural stem cells (NSCs) reside in distinct niches of the adult brain, and their progenitor cells differentiate into neurons, astrocytes, and oligodendrocytes. The NSCs in the adult mammalian brain are quiescent, which is necessary for long-term stem cell pool maintenance. Studies have identified lysosome-mediated autophagy as a key pathway for maintaining cellular proteostasis, which retains quiescence and prevents senescence in NSCs [24]. Moreover, to maintain quiescence, rapid degradation of epidermal growth factor receptors by lysosome is required. Therefore, lysosomal activity is more in quiescent NSCs as compared to cycling NSCs. Hence, inhibition of this lysosomal activity leads to the exit of NSCs from a quiescent state [49]. Furthermore, studies have also shown that when quiescent NSCs age, they acquire more protein aggregates in their lysosomes, while a decrease in the abundance of these aggregates causes aged quiescent NSCs to rejuvenate [59]. In addition, autophagy has also shown to promote survival and prevent cell death in the NSCs. Yazdankhah et al. demonstrated that the autophagic proteins AMBRA1 and BECN1 are involved in the early stages of autophagosome formation and are highly expressed in adult NSCs. Hence, their downregulation causes a decrease in NSC proliferation, with an increase in basal apoptosis and DNA-damage-induced death [109]. Additionally, studies have also shown that deletion of Atg5 in cycling neural progenitor cells of the adult brain results in their death, thereby suggesting that autophagy is also crucial for the survival of proliferating neural progenitor cells [107]. Overall, these studies demonstrated the importance of autophagy in survival and clearances of protein aggregate formation, thereby maintaining quiescence in NSCs.

Muscle Stem Cells

Muscle stem cells, also known as satellite cells, mediate muscle homeostasis and regeneration in skeletal muscle tissues. Short-term caloric restriction in mice enhanced satellite cell number, muscle regeneration, and satellite cell transplantation efficiency, suggesting that autophagy induced by nutrient deprivation in satellite cells may promote muscle stem cell activity [11]. Moreover, quiescent satellite cells derived from young mice exhibited robust levels of autophagic flux, which was reduced in aged muscle cells in an age-dependent manner. This demonstrates that aged muscle cells express impaired autophagy levels. Moreover, as determined by transplantation and engraftment assays, inducing autophagy in aged satellite cells improved stem cell function and prevented their aging-induced entry into senescence [25]. Altogether, these studies indicate that induction of autophagy prevents stem cell senescence and may enhance muscle stem cell function with respect to aging.

Induced Pluripotent Stem Cells

The breakthrough discovery of iPSCs has allowed scientists to obtain pluripotent stem cells without the controversial use of embryos, providing a novel and powerful method to "de-differentiate" cells whose developmental fates are traditionally assumed to be determined. In 2006, the pioneering work of Shinya Yamanaka demonstrated that differentiated adult somatic cells can be genetically reprogrammed into undifferentiated cells that resemble embryonic stem cells. These manipulated cells were termed induced pluripotent stem cells (iPSCs). The ectopic expression of pluripotency transcription factors including Oct4, Sox2, Klf4, and c-Myc, triggers somatic cells to revert to an undifferentiated state [95]. Interestingly, autophagy has revealed its importance in maintaining the efficiency of somatic cell reprogramming. Autophagy is enhanced during the initial stages of reprogramming by inhibiting mTOR signaling [103]. Consequently, activation of mTOR impeded the efficiency of the reprogramming process, while mTOR inhibitors such as rapamycin enhanced the reprogramming process [14]. These data suggested that early upregulation of autophagy is an essential, but transient event, and its downregulation is essential to complete the reprogramming process, as inhibition of mTOR at later stages was found to inhibit reprogramming [103]. Moreover, stem cells, in general, have fewer mitochondria as compared to differentiated cells. Hence, numerous autophagy and mitophagy genes are involved in mitochondrial homeostasis and clearance of somatic mitochondria during reprogramming, autophagy facilitates mitochondrial remodeling [36].

Cancer Stem Cells

Cancers are primarily curable by conventional treatments when they are diagnosed at an earlier stage. However, when diagnosed at later stages, they become progressive and metastasize to other organs. Even if cancer is detected and treated early, some residual cells survive and, over time, become a cause of tumor recurrence and metastasis [84]. These residual cells are responsible for causing therapeutic resistance, as they possess stem cell-like properties and functions such as self-renewal and differentiation into multiple cell types known as the cancer stem cells (CSCs) [73]. Interestingly, autophagy plays a controversial role in cancer as evidence suggests that autophagy can both prevent and promote tumorigenesis. On the one hand, autophagy prevents oncogenic stresses such as oxidative stress and DNA damage, thereby impeding chromosomal instability and tumorigenesis. It also triggers inflammatory responses and protects cells from undergoing necrotic cell death. On the other hand, cancer cells upregulate basal autophagy expression when the cells are subjected to hypoxia or nutrient starvation to promote their survival deconvoluting the context-dependent role for autophagy in cancer [106]. Therefore, an increase in levels of autophagy in cancer cells can also prevent the efficacy of anticancer treatment.

Autophagy and Hematopoiesis

Hematopoiesis

HSCs are present in both early embryonic and adult hematopoietic organs. They maintain life-long hematopoiesis by producing all lineages of hematopoietic cells after transplantation. HSCs are characterized as either long-term HSCs (LT-HSCs), which can engraft into an irradiated recipient, or short-term HSCs (ST-HSCs), which have a limited ability for self-renewal and are unable to maintain hematopoiesis for a prolonged period. LT-HSCs differentiate into ST-HSCs and multipotent progenitors (MPPs), capable of producing all lineages but lack the self-renewal capacity needed for long-term engraftment. The MPPs further differentiate into the lymphoid or myeloid lineage. The common myeloid progenitor (CMP) can generate all myeloid cells, either through the granulocyte–macrophage progenitor (GMP) or the megakaryocyte–erythroid progenitor (MEP), while the common lymphoid progenitor (CLP) can generate B-and T-lymphocytes as well as natural killer (NK) cells. Hence, HSC transplantation (HSCT) leverages the use of HSCs as a curative therapy for various hematological disorders due to their potential to repopulate the entire immune system of a host [92].

Role of Autophagy in the Ontogeny of the Hematopoietic System

The development of HSCs is distinct from that of the other types of stem cells because of their migration to different organs during different stages of development. In the vertebrate embryo, hematopoietic development unfolds in waves, with each successive wave producing cohorts of cells with greater blood lineage complexities. The first wave of hematopoiesis in mammals occurs outside the embryo, in the blood islands of the yolk sac [10]. The first wave, also known as the primitive wave, is responsible for the formation of unipotent blood cell types. The primary hematopoietic product of the yolk sac is the large primitive nucleated erythrocytes, with the sporadic presence of primitive macrophages and megakaryocytes. A definitive wave, which generates multipotent HSPCs, follows the first wave. Although HSCs in the yolk sac may play a role in adult hematopoiesis, definitive hematopoiesis in mice occurs around E10.5 in the aorta-gonad mesonephros (AGM) [10]. In the AGM, definitive HSCs develop alongside non-self-renewing hematopoietic progenitor cells. Definitive HSCs can be serially transplanted and have long-term engraftment ability. HSCs then travel to the fetal liver and spleen before establishing in the BM [100]. Autophagy has pleiotropic effects on HSC features throughout the formation and progression of the hematopoietic system (Fig. 5.2), but its involvement in HSC self-renewal and differentiation at different stages of development is unknown. Adult HSCs seldom divide, but embryonic and neonatal HSCs divide rapidly to replenish

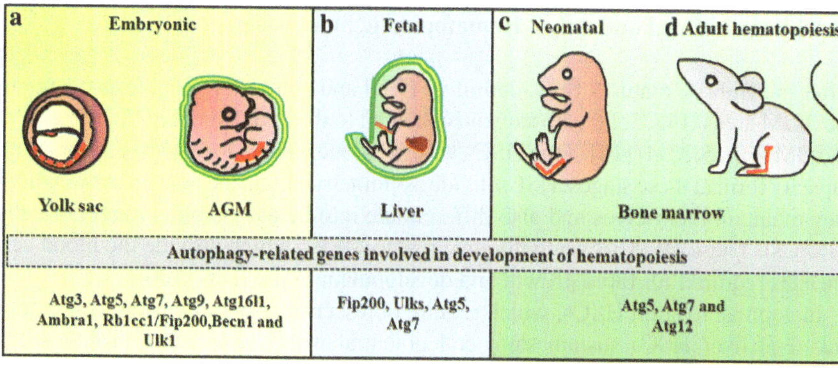

Fig. 5.2 Autophagy-mediated regulation during the development of the hematopoietic system: The illustration depicts the role of autophagy-related genes in regulating hematopoiesis during the **a** Embryonic, **b** Fetal, **c** Neonatal, and **d** Adult stages

the maturing hematopoietic system. Transitioning from embryonic to neonatal to adult hematopoiesis necessitates significant metabolic changes, which are partially controlled by autophagy [31]. Determining the stage-specific role of autophagy in the development of the hematopoietic system is therefore vital for reproducing and tailoring the process of HSC generation in vitro for translational applications.

Autophagy in Embryonic Hematopoietic Stem Cells

Autophagy has been linked to embryonic development and survival in animal models. Early embryonic or perinatal mortality is observed in mouse models lacking essential autophagy genes such as *Atg3, Atg5, Atg7, Atg9, Atg16l1, Ambra1, Rb1cc1/Fip200*, and *Becn1* [67]. In embryonic HSCs, autophagy is required to maintain the balance between quiescence, self-renewal, and differentiation. However, the role of autophagy during embryonic hematopoiesis is still uncertain. Embryonic hematopoiesis is initially observed in the blood islands of the yolk sac during mouse embryogenesis [10] (Fig. 5.2a). Hematopoietic precursors called hemangioblasts have a restricted potential for self-renewal and can only differentiate into specific cell lineages, such as endothelial cells, nucleated red blood cells (RBCs), and macrophages. The removal of organelles such as mitochondria is, required, for the differentiation of yolk sac-derived embryonic erythroid cells. Mitophagy is, therefore, enhanced during this period, and degraded mitochondrion can be seen in the lysosomes [96]. Blood is produced in the fetal spleen and liver as the embryos develop further. ULK1-dependent ATG5-independent autophagy engulfs and digests mitochondria in fetal definitive reticulocytes in the liver [34]. These findings suggest that autophagy is necessary for reticulocyte formation during the early phases of embryonic growth.

Autophagy in Fetal and Adult Hematopoietic Stem Cells

After the embryo matures fully, definitive (fetal and adult) hematopoiesis occurs in the AGM areas (Fig. 5.2a), subsequently moving to the fetal liver (Fig. 5.2b), spleen, and BM (Fig. 5.2c, d) [10]. True HSCs with self-renewal and long-term repopulating capacity form at these stages. HSCs divide asymmetrically in the bone marrow, where they maintain themselves and also differentiate into hematopoietic progenitor cells (HSPCs). These HSPCs eventually give rise to MPPs, which provide the blood cell lineages required for rapid growth and development in the fetus [100].

In contrast to adult HSCs, which require quiescence to preserve stem cell potential, fetal liver HSCs sustain stem cell potential while undergoing proliferation. Fetal HSCs have more mitochondria with increased bioenergetic function and are rapid cycling than adult HSCs [72]. Mitochondrial metabolism is necessary for fetal HSC expansion, increasing mitochondrial membrane potential ($\Delta\Psi$mt) and oxygen consumption [31]. Autophagy helps HSCs to maintain their metabolic state by maintaining a sufficient number of healthy and functional mitochondria. The FIP200 is an autophagy factor that influences cellular activities such as cell growth, proliferation, and migration in conjunction with ULKs. FIP200 is crucial for maintaining fetal HSCs [63]. FIP200 is involved in the initiation of autophagosome nucleation. FIP200 provides fetal HSCs with long-term multilineage reconstitution capabilities, and its loss impairs the maintenance and stemness of fetal HSCs by increasing mitochondrial mass and generating reactive oxygen species (ROS). HSC-specific deletion of Atg5 and Atg7 during the fetal stage causes a change in mitochondrial status, weight loss, severe anemia, and a reduction in the HSC population in mice, leading to mortality. On the contrary, fetal liver Atg7-deficient HSCs can save lethally irradiated recipients, implying that Atg7 is dispensable for fetal HSC function [93, 97, 101].

Late in gestation, HSCs shift to the BM and reside there permanently [10]. Within 4 weeks after birth, HSCs attain cell cycle quiescence [56]. Environmental factors such as oxidative stress, nutritional stress, and microbial infection significantly impact fetal and neonatal development during the perinatal period. Although fetal life develops in hypoxic environments, it offers a rich environment for organ development. The transplacental nutrition supply ceases upon birth, putting neonates in a stressful environment of starvation until milk feeding. Additionally, the fetal-to-neonatal transition might result in hyperoxia. Oxidative stress occurs in neonatal tissues in both of these conditions. Autophagy protects HSCs from harsh conditions in the early neonatal stage and is necessary for effective long-term hematopoiesis in adults [53, 75].

The cell cycle and metabolic status of neonatal HSCs differ from those of adult HSCs. HSCs from neonates multiply rapidly, have a high mitochondrial metabolism, and a reduced efflux capacity. In adult HSCs, excessive proliferation is inversely linked with their long-term reconstitution (LTR) potential, yet in spite of being highly proliferative, neonatal HSCs exhibit an intense LTR activity. Hashimoto et al. found that neonatal HSCs have more autophagy activity than adult HSCs, implying that the autophagy mechanism in neonatal HSCs is more active compared to the adult ones. They also determined that at the neonatal stage, Atg7 deficiency had a minor impact on hematopoiesis and metabolic status of actively dividing HSCs. The LTR activity in Atg7-deficient neonatal HSCs was comparable to wild-type neonatal HSCs. During the transition from neonatal to the adult stage, Atg7-deficient mice show an excess of cell divisions and are unable to maintain a quiescent state. The increased mitochondrial metabolism in Atg7-deficient mice further results in BM failure at the adult stage [31]. In neonatal HSCs, autophagy is thus dispensable for stem cell activity and hematopoietic homeostasis, however, it affects the adult HSCs.

Adult HSCs are distinguished by their phenotypic diversity, multipotency, and self-renewal abilities, as well as their capacity to direct migration to hematopoietic tissues. As a result, adult HSCs have a different metabolic state from those in the fetal and neonatal stages. Autophagy is, therefore, perceived as a crucial mechanism in adult HSCs for metabolic regulation. For example, autophagy is initiated quickly in adult HSCs during starvation via the Forkhead box O 3a (FoxO3a)-driven gene expression mechanism, which keeps HSCs functioning [104]. Autophagy also degrades active healthy mitochondria, allowing HSCs to retain a low metabolic state and quiescence. Mitophagy helps adult HSCs sustain their stemness by reducing their mitochondrial potential and lowering ROS levels, which prevents them from entering the cell cycle [13, 33, 70]. When Atg5 is knocked out in HSCs, the clearing of damaged mitochondria becomes aberrant, and the ability of HSCs to regenerate is reduced [41]. Atg12 plays a role in the expulsion of mitochondria from HSCs, and deletion of Atg12 in HSCs results in an increase in active mitochondria with a high $\Delta\Psi$mt [33]. Likewise, loss of Atg7 causes reduced HSC function and exhaustion and an increase in mitochondrial counts and levels of ROS [70, 26]. Furthermore, conditional knock-out of Atg12 reduces HSC reconstitution capacity and causes premature aging of HSCs [33]. Autophagy deficiency in adults reduces HSC quiescence by inducing stress-like hematopoiesis. In adult HSCs, autophagy is, therefore, crucial for the maintenance of LTR activity.

Role of Autophagy in Hematopoietic Stem Cell Fates

Adult HSCs have self-renewal capacity and are either multipotent or unipotent, and, thus, are capable of producing differentiated progeny. They can exist in a quiescent state or enter the cell cycle as and when required. Adult HSCs divide either symmetrically to generate two identical stem cells, or asymmetrically, to produce two daughter cells—one destined for self-renewal and the other for differentiation. Moreover, autophagy has emerged as an important mechanism that regulates the fate of HSCs at an extrinsic and intrinsic level (Fig. 5.3) [50].

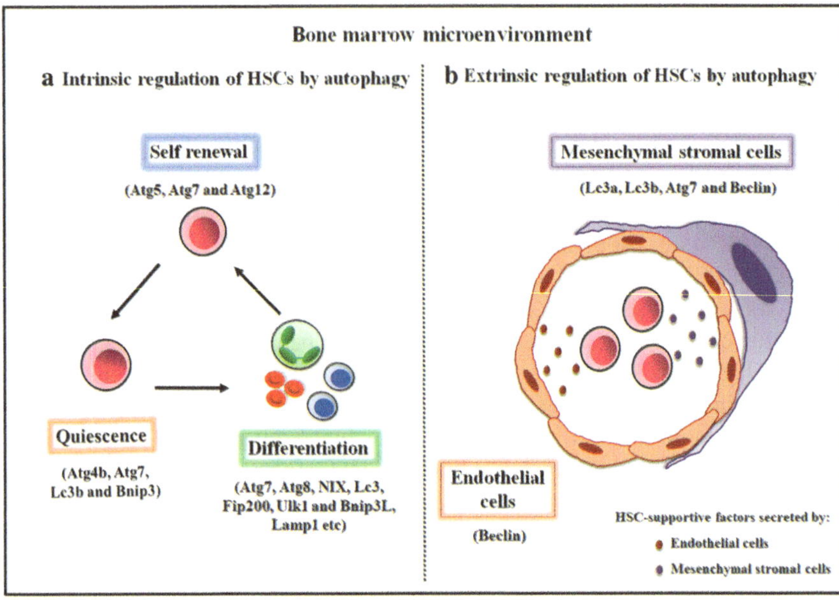

Fig. 5.3 Autophagy-mediated regulation of HSC fate: The illustration depicts the role of autophagy-related genes in regulation of HSC fates such as quiescence, self-renewal and differentiation **a** intrinsically and **b** extrinsically

Autophagy in HSC Quiescence

Quiescence is a critical characteristic that preserves HSC potency throughout life, despite their incredible in vivo repopulation potential. It maintains the flow of mature blood cells while also ensuring that HSCs are not depleted during an individual's lifespan. Under a steady state condition, most HSCs remain quiescent, and only a small number enters the cell cycle. LT-HSCs are predominantly found in the BM in a hypoxic niche distant from the blood circulation. The majority of these LT-HSCs are in a quiescent G_0/G_1 reversible phase of the cell cycle. The low oxygen concentration in the BM niche maintains HSCs in a quiescent state [21]. Quiescence is interconnected to cellular metabolism, which is significantly influenced by HSC commitment. The maintenance or exit of HSCs from a quiescent state is often linked to the metabolic changes that occur in response to their surroundings. Quiescent HSCs suppress mitochondrial respiration and rely primarily on glycolysis for their maintenance. HSCs switch to mitochondrial metabolism and become metabolically active as they migrate to a more oxygen-rich environment in the bloodstream. As a result, mitochondria are engaged in maintaining homeostatic HSC function, whereas lysosomal breakdown and clearance of mitochondria via mitophagy, a kind of selective autophagy, is essential for the maintenance of the HSC quiescence. Mitochondrial defects during oxidative phosphorylation (OXPHOS) increase ROS levels, causing HSCs to differentiate. Because of its role in mitophagy and selective mitochondrial breakdown, autophagy is crucial at this stage. Mitophagy regulates ROS levels, which govern the transition of HSCs from a dormant to an active state. Mitophagy dysfunction in HSCs leads to increased ROS levels, rapid HSC cell division, loss of quiescence, proliferation and eventually loss of stemness. Therefore, myeloproliferative diseases develop in mice lacking autophagy [27].

Mitochondrial function and autophagy are both regulated by factors including FOXO3 and TSC1. By restricting cellular mitochondrial content and activity, the transcription factor FOXO3 maintains HSCs in a state of quiescence. FOXO3 regulates HSCs by directing the autophagy proteins Atg4b, Lc3b, and BCL2 Interacting Protein 3 (Bnip3) (Fig. 5.3a). It also controls stress-induced autophagy in HSCs, which is required for HSC survival during aging and starvation. TSC1-mediated regulation of the mTOR pathway, which promotes autophagy by suppressing mitochondrial biogenesis and ROS generation, also contributes to HSC quiescence [86]. Autophagy also contributes to the maintenance and regulation of the cell cycle of HSPCs in a nutrient-dependent manner. Autophagy activation promotes cell cycle entry under physiologic or nutrient-rich environments, loss of Atg7 thus results in HSPC cell cycle elimination. Furthermore, inhibition of early, but not late, autophagy-signaling processes accelerated the G_1/S transition under low nutritional circumstances. Therefore, autophagy has a dual function in nutrient-dependent regulation of cell cycle activation and the G_1/S transition of HSPCs. By controlling cyclin D3, autophagy maintains a healthy cell cycle in HSPCs and normal hematopoiesis in adults [35]. Hence, autophagy in HSCs is required to maintain the cells in a quiescent state.

Autophagy in HSC Self-renewal

HSCs can be maintained throughout an organism's lifespan by striking the right balance between self-renewal and differentiation. Autophagy is indispensable during the self-renewal of HSCs (Fig. 5.3a). Numerous signaling pathways, including the Bmi1 and Wnt pathways, have been discovered to control HSC self-renewal, which also involves autophagy activation. Autophagy inhibition by pharmacological compounds or targeted knockdown of ATG5 causes human adult HSCs to lose their colonogenic capacity [78]. Reduced LT-HSC self-renewal and function, as well as a bias toward differentiation into the myeloid lineage, are associated with loss of Atg12 within the hematopoietic system [75]. Mice lacking ataxia-telangiectasia mutated kinase (ATM), a protein that is indirectly implicated in autophagy, demonstrate a dysfunction in HSC self-renewal. ATM is found in the cytoplasm, where it works as a ROS sensor and can communicate via LKB1, AMPK, and TSC2 to deactivate mTORC1, thereby inducing autophagy [1].

Tie2$^+$ HSCs have a remarkable capacity for self-renewal and reconstitution. The induction of mitophagosome formation and up-regulation of *Parkin* and *Pink1* plays a vital role in maintaining the stemness of *Tie2*$^+$ HSCs. Additionally, defective mitophagy impairs the self-renewal of *Tie2*$^+$ HSCs [37]. Loss of Atg7 in the HSCs causes mitochondrial accumulation which leads to increased ROS, enhanced proliferation, and DNA damage in them. Furthermore, the absence of the critical autophagy genes Atg7 or Atg5 affects HSC function, resulting in lethal anemia accompanied by lymphopenia, severe myeloproliferation, and BM failure. HSCs lacking Atg7 and Atg5 lose their LTR ability following transplantation, confirming the importance of autophagy in self-renewal and maintenance [37, 104].

Autophagy in HSC Differentiation

Autophagy is critical for the differentiation function of HSCs (Fig. 5.3a) because of the enormous rearrangement of intracellular organelles and macromolecular protein complexes that must be coordinated during HSC maturation. Additionally, the amount of ROS produced by HSCs as a result of their metabolic status can also influence their fate [50].

Primitive HSCs have low ROS levels in their quiescent state, whereas progenitor cells with short-term repopulation potential and a propensity for myeloid differentiation have higher ROS levels. In the cell, mitochondria are the chief generator of ROS. Severe mitochondrial malfunction results in excessive ROS production, which triggers the induction of mitophagy. ROS production is associated with a paucity of mitophagy-related ATG proteins. ATG7 deficiency, for example, leads to an increase in ROS levels in HSCs due to mitochondrial generation of superoxide anions [70].

The elimination of mitochondria in developing RBCs is one of the most well-studied examples of autophagy's function in HSC differentiation. RBCs degrade organelles and proteins as they mature, leaving only hemoglobin behind, which allows them to pass through even the finest capillaries. Interestingly, autophagy is responsible for the elimination of mitochondria during the maturation and enucleation stages of RBCs. During erythroid development, programmed mitophagy targets normal functioning mitochondria with intact membrane polarization. NIX(BNIP3L), a BH3-only outer mitochondrial membrane (OMM) stimulates mitophagy by recruiting LC3/GABARAP proteins during terminal erythroid lineage differentiation a step critical during programmed mitophagy [19, 87, 90]. Moreover, deletion of autophagy genes including Ulk1 [55], Bnip3L [90], Fip200 [63], or Atg7[69], the molecules required for autophagosome elongation, results in erythroid differentiation defects and anemia. Additionally, GATA-1 is a transcription factor that regulates mitophagy during erythropoiesis and is essential for normal erythropoiesis. GATA-1 upregulates autophagy genes such as ATG8 homologs, ATG4B, and ATG12, Atp6v0e, Clcn7, Ctsb, Neu1, and Lamp1 as well as BNIP3L in early human erythroblasts by direct transcriptional regulation. GATA-1 is, therefore, regarded as a master regulator of programmed mitophagy in the differentiation of erythroid cells [43]. Hence, autophagy facilitates RBC differentiation, which necessitates extensive remodeling to meet specialized cellular functions.

Autophagy reduces apoptosis during HSC differentiation by limiting ROS production, endoplasmic reticulum (ER) stress, and DNA damage. HSCs differentiate into monocytes, which further differentiate into macrophages or dendritic cells. Monocytes, on the other hand, are programmed to perish in the absence of stimuli. The onset of monocyte-macrophage differentiation not only promotes cellular changes but also inhibits monocyte apoptosis. When monocytes are stimulated to differentiate, autophagy is activated. A differentiation signal activates JNK, which releases Becn1 from Bcl-2 and prevents Atg5 cleavage, inducing autophagy. The induction of autophagy is, therefore, essential for monocyte survival and differentiation [112].

Neutrophils, a type of granulocytes, are the most ubiquitous and short-lived immune cells in the body, and deficits in their quantity or function have been related to severe immunological disorders. The transition from glycolysis to OXPHOS during normal granulocyte differentiation necessitates lysosomal lipid breakdown for mitochondrial fatty acid oxidation. This transition to OXPHOS is followed by an autophagy-dependent reduction in lipid droplets in mature granulocytes. Lipophagy dysfunction in Atg7-deficient granulocytes hinders the metabolic transition to OXPHOS essential for neutrophil maturation, leading to lipid droplet deposition [85].

Autophagy activity is thus required not only for HSC self-renewal but also for controlling the terminal differentiation of various blood cell types in order to maintain hemostasis.

Role of Autophagy in HSC Aging

In animals, aging is a complex process that causes changes in tissue structure as well as a decrease in numerous functions and activities. Proteostasis is required for the majority of biological activities, including genetic replication, catalysis of metabolic reaction, and immunological response. Proteostasis failure can result in aggregation of toxic undesired proteins, causing cellular dysfunction. Protein homeostasis impairment and stem cell exhaustion are two main mechanisms involved in the loss of regenerative capacity as a result of age-related damage accumulation [91]. HSCs lose their regenerative abilities as they age, and they also exhibit an autophagy deficit. Autophagy deficit leads to the deposition of macroautophagy vesicles, increased intracellular p62 protein levels, increased LC3II expression, and ubiquitin-positive inclusions [64]. Furthermore, nearly 30% of aged HSCs showed significant levels of basal autophagy, retaining a low metabolic state and great long-term regeneration potential, similar to young HSCs [83]. The basal levels of autophagy ensure HSC function during aging and under conditions of intense regenerative stress. The residual population of aged HSCs, on the other hand, exhibits autophagy loss, resulting in an activated metabolic state, rapid myeloid differentiation, and reduced HSC self-renewal activity and regenerative potential. As a result, transplantation of HSCs defective in autophagy genes including Atg12 and Atg5 from adult and aged mice results in a substantially accelerated age-related decrease in donor chimerism.

These recipients exhibit premature blood aging in the adult stage, including increased cellularity, a skewed ratio of circulating myeloid vs lymphoid cells, and a phenotype similar to the myeloid bias HSCs as seen in aged mice [64].

In adult HSCs, altered mitophagy causes an accumulation of ROS, which causes premature aging and senescence. Young HSCs with the Atg12 mutation have metabolically active mitochondria, increased OXPHOS, higher protein synthesis rates, and enhanced cell cycle activity, all of which are linked with aged HSCs and result in the loss of HSC quiescence and acceleration of myeloid differentiation. In addition, autophagy-deficient HSCs facilitate irregular changes in fate decisions as a result of epigenetic reprogramming that affects gene expression in HSCs [64].

Role of Autophagy in Niche-Mediated Regulation of Hematopoietic Stem Cells

HSCs reside within a heterogeneous milieu in the BM. The primary components of the HSC niche are MSCs and MSC-derived progenies such osteoblasts and adipocytes, which are known to regulate HSC quiescence, self-renewal, and differentiation in the BM. HSC functioning is affected by extrinsic changes in the composition and function of different HSC niche cells (Fig. 5.3b). For example, as MSCs age, they gain increased AKT signaling, which results in a decrease in autophagy-related genes such as Lc3a, Lc3b, Becn1 and Atg7 in them. MSCs regulate HSCs in a paracrine manner by secretion of microvesicles (MVs) and exosomes. Reduction of autophagy in MSCs also reduces the autophagy-related genes in the MSC-derived MVs. Additionally, partitioning of miRNAs like miR-17 and 32b, which are negative regulators of autophagy-related mRNAs, into their exosomes increases [51]. This leads to the aging of HSCs, as they not only receive the MVs which are deficient in autophagy-inducing mRNAs but additionally, also receive exosomes containing miRNAs that degrade autophagy-inducing mRNAs. This niche-mediated aging of HSCs involving the autophagy process underscores the importance of autophagy in HSC functionality.

Endothelial cells (ECs) are another type of accessory cells that aid hematopoiesis in the BM niche by delivering essential signals such as colony-stimulating factor-1 (CSF-1) that regulates HSCs. The autophagy status of ECs influences their capacity to support hematopoiesis via modulation of the Beclin-1 pathway. The capacity of ECs to promote hematopoiesis is significantly reduced when autophagy is suppressed by the deletion of BECN1. Furthermore, rapamycin-treated ECs activate autophagy via BECN1 overexpression, thus restoring their ability to support HSCs [65].

Hence, it is crucial to examine the niche-mediated (extrinsic) function of autophagy in the regulation of HSCs to develop effective strategies for HSC-based interventions.

Role of Autophagy in Hematological Disorders

Dysregulation of the fine balance between quiescence, self-renewal, and differentiation can lead to the progression of a variety of blood disorders and malignancies. The significance of autophagy in cancer cell death and survival is currently debated. Autophagy may appear to be a dichotomy at first look because of its tumor-suppressive and tumor-promoting qualities. The deletion of genes involved in the autophagy process could either increase or decrease the mortality rate of cancer cells. Basal autophagy is thought to act as a tumor suppressive mechanism by preventing cancer from developing. Moreover, autophagy assures the removal of damaged organelles, such as mitochondria, which can produce increased levels of ROS, as well as protecting cells from genomic instability and inflammation, which can lead to cancer. Increased DNA damage, elevated ROS levels, aneuploidy, and abnormal accumulation of p62/SQSTM1 and ER chaperones have all been linked to an altered autophagic process, highlighting the critical function of autophagy in tumor prevention [81].

Autophagy, on the other hand, can contribute as a pro-survival mechanism after the onset of cancer. Cancer cells subjected to stress stimuli, such as nutrient deprivation, hypoxia, DNA damage, chemotherapy, and radiation, show a significant increase in autophagy. Here, the autophagic response helps cancer cells adapt to metabolic stress and acquire resistance to chemotherapeutic drugs, promoting tumor growth and survival [16]. Autophagy is, therefore, an intriguing therapeutic target in this example, and medicines that selectively block this metabolic process may improve chemosensitivity and tumor cell death. Hence, the only autophagy inhibitors authorized for clinical use by the US Food and Drug Administration are chloroquine and hydroxychloroquine at the moment [77].

Acute Myeloid Leukemia (AML)

Acute leukemias are a heterogeneous category of malignant hematological diseases defined by the uncontrolled growth of clonal neoplastic cells of the myeloid or lymphoid lineages. Acute leukemias are distinguished by their fast progression and inevitable BM failure, which results in severe anemia, leukopenia, and thrombocytopenia. When compared to nonleukemic cells or AML cells induced towards differentiation, primary AML blasts have reduced autophagy gene levels [40, 105]. Additionally, during neutrophil differentiation in acute promyelocytic leukemia (APL) cells, enhanced expression of autophagy receptor SQSMT1/p62 has been observed, which prevents ubiquitinated protein accumulation. SQSMT1/p62 functions as a prosurvival cellular pathway during the terminal differentiation of APL cells and is also required for cell proliferation and mitochondrial integrity. Defects in SQSMT1/p62 hinder myeloid leukemia development and mitophagy in this kind of malignant tumor [66].

Autophagy's function in leukemia progression differs depending on the kind of oncogene involved in the disease progression. A major kinase in AML is the RET proto-oncogene, which is a tyrosine kinase [88]. RET pathway activation reduces autophagy and the stability of leukemia-causing proteins such as mutant FMS-like tyrosine kinase 3 (FLT3). Hence, RET inhibition reduces FLT3 levels via autophagy. Protease inhibitors, on the other hand, promote FLT3 internal tandem duplication (FLT3/ITD) degeneration via autophagy. Inhibition of the FLT3-ITD mutation in AML cells, on the other hand, promotes a high level of basal autophagy by affecting autophagy-dependent proliferation. The amount of ATF4 transcription determines the degree of FLT3-ITD-dependent autophagy [57].

In leukemic mice, inhibiting autophagy by deleting ATG7 increases ROS production, which leads to a decrease in leukemia-initiating cells (LICs), as well as an increase in mitochondrial activity, cell death, and improved survival. As a result of enhanced apoptosis, the number of blasts in the peripheral blood falls. In mixed-lineage leukemia–eleven nineteen lysine-rich leukemia (MLL-ENL) animal model, suppression of autophagy by deletion of ATG7 or ATG5 resulted in more aggressive leukemia. Furthermore, reduced autophagy in MLL-ENL cells results in abnormal mitochondrial activity, including proliferation and transformation [94].

Because AML is such a diverse illness, autophagy may both promote and inhibit tumor growth depending on the subtype.

Myelodysplastic Syndrome (MDS)

MDS is a multifaceted disease caused by defective and inefficient hematopoiesis, aging, clonal hematopoiesis expansion, and a higher risk of secondary AML transformation. HSCs develop a sequence of recurrent genetic alterations prior to disease onset, giving them a proliferative edge over healthy HSCs. MDS pathogenesis is associated with autophagy and mitochondrial dysfunction. In high-risk MDS, mutations in autophagy genes are more common. The deletion of Atg7 in HSCs causes an MDS-like phenotype [68]. In addition, HSPCs from MDS patients frequently contain mitochondrial DNA abnormalities and abnormal mitochondrial respiration, which contributes to their susceptibility to normoxia. MDS is characterized by anemia, which is the most prevalent symptom [102]. As an ultrastructural indicator of improved mitophagy, erythroblasts from individuals with low-risk MDS have more autophagosomes and lysosomes containing mitochondria. In high-risk MDS, however, there is a buildup of irregularly structured, larger, and malfunctioning mitochondria, as well as mitochondrial iron deposits. In individuals with high-risk MDS, nuclear RBCs have lower LC3B levels and more mitochondrial inadequacy; moreover, lower LC3B levels in these patients are associated with lower hemoglobin levels [39]. ATG3 expression is also lower in MDS patients than in healthy individuals, and overexpression of ATG3 in the SKM-1 MDS cell line promotes caspase-dependent autophagy and cell death [113]. In erythroid precursor cells from high-risk MDS patients, NIX expression is reduced, which is linked to increased mitochondrial mass, loss of mitochondrial membrane potential, increased ROS, downregulation of ULK1 and AMPK, and activation of mTOR signaling.

Chronic Myeloid Leukemia (CML)

CML is characterized by the displacement of t(22;9) (q34;q11) and expression of the Breakpoint cluster region protein–Abelson murine leukemia viral oncogene homolog 1 (BCR-ABL) fusion protein, which has excessive tyrosine kinase activity, accounting for 15% of all forms of leukemia in adults. BCR-ABL1 promotes cell transformation by inducing MAPK15-dependent autophagy [79]. Several Atg genes, including the ATG4 family members, ATG5, and BECN1, are upregulated in CML $CD34^+$ HSC and HSPCs. BIM1 inhibition increases the expression of Cyclin-G2 (CCNG2) and reduces the tumor suppressor response in autophagy, which speeds CML progression to acute stages [7]. Lys05, a second-generation autophagy inhibitor, causes CML stem cells to go into dormancy and slows their proliferation. When used in combination with a tyrosine kinase inhibitor, Lys05 or PIK-III (PI3P class III inhibitor) decreases the number of primary CML LSCs [6].

Lymphoid Malignancies

Lymphomas are neoplastic diseases of the B-cell, T-cell, or NK-cell lineage that originate in the bone marrow from lymphoid progenitor cells. Acute lymphoid leukemia (ALL) mainly occurs in B-cell and thymocyte precursors.

The ETV6-RUNX1 (or TEL-AML1) fusion protein is seen in 25% of pediatric patients with B-cell precursor acute lymphoblastic leukemia and is the initial event in leukemogenesis. Although autophagy can help cells survive under stressful situations, it can also lead to cell death due to cellular consumption. Vps34, Beclin-1, and Vps15 make up the core autophagy-regulating complex. ETV6-RUNX1 was shown to activate and upregulate Vps34, an essential autophagy regulator, in ETV6-RUNX1-positive leukemic cells. In these cells, deletion of Vps34 significantly decreased proliferation and survival [80]. Autophagy is required for survival and leukemic transformation in hematopoietic cells expressing the BCR-ABL kinase. BCR-ABL cells had modest basal levels of autophagy in a study by Altman et al. but they were overly dependent on it for survival [3]. These cells rapidly underwent apoptosis when autophagy was impaired by Atg3 deletion or treatment with pharmacologic autophagy inhibitors. In addition, Atg3 deletion reduced BCR-ABL-mediated leukemogenesis in vivo experiments [3]. Autophagy has also been demonstrated to play a pro-survival effect in CLL cells. Suppression of autophagy in CLL patients' peripheral blood mononuclear cells (PBMCs) utilizing RNA interference targeting critical autophagy genes or chloroquine or by using 3-methyladenine reduced CLL cell survival. On the contrary, Gade et al. found that inhibiting death-associated protein kinase 1 (DAPK1), an autophagy-associated gene, decreases autophagy and increases CLL proliferation [23].

Final Remarks

Autophagy is an evolutionarily conserved cellular process that, under normal physiological conditions, plays an essential role in the maintenance of homeostasis and stemness in embryonic as well as in adult stem cells. It also plays an important role in the initiation and progression of human diseases that have been shown to be associated with maintenance of disease-specific stem cell compartments [25, 115] It is widely recognized that autophagy is a critical regulator of metabolic reprogramming during normal hematopoiesis from its development to aging [13, 53]. It is not surprising, therefore, that dysregulated autophagy is a common feature of several cancers and developmental disorders related to HSCs [16].

The intrinsic and extrinsic mechanisms regulate quiescence, self-renewal and differentiation of HSCs [8, 99, 114]. The intrinsic mechanisms include intracellular signaling pathways, epigenetic mechanism, factors that cause DNA damage, etc., whereas, the BM niche consisting of MSCs, ECM and other secretory factors, etc., constitute the extrinsic mechanisms [99, 114]. Several reports have pointed out that along with these mechanisms, autophagy also plays a very important role in governing the fate of HSCs [51]. Expanding our understanding of how autophagy, whether directly or indirectly, exploits the intrinsic and extrinsic mechanisms to cause aging or dysregulation of HSCs is, therefore, imperative. Hence, HSC fate regulation via autophagy modulation is a major focus of research interest for several groups [8, 51, 52, 65, 98].

Till now, several rejuvenation strategies involving the use of cell cycle inhibitors, signaling pathway inhibitors, epigenetic modifiers, NO donors [38, 51, 62], etc., that focus on modulating either the intrinsic or the extrinsic mechanisms of HSCs have been explored. These strategies either rejuvenate aged and dysregulated HSCs or rejuvenate their tissue microenvironment [38, 51, 62]. A recent report has shown that mere rejuvenation of aged HSCs does not protect them from the deleterious effects of the aged niche [29]. Thus, a combinatorial strategy that targets both the intrinsic and extrinsic mechanisms—ultimately translating into a "combinatorial rejuvenating stem cell therapy" that simultaneously rejuvenates both the HSCs and the microenvironment in which they reside seems promising. Such an approach would aid in developing and optimizing strategies that perhaps also rewires the autophagic machinery, having tremendous implications on improving the health span of people from birth until old age.

Acknowledgements The authors would like to thank Symbiosis Centre for Research & Innovation (SCRI), Symbiosis International (Deemed University) for providing Senior Research fellowship to

PB, Junior Research Fellowship (JRF) to PT, infrastructural support, and financial support (Grant number: SIU/SCRI/MJRP/19-20/ 1516-C).

Compliance with Ethical Standards

Funding This study was funded by Symbiosis Centre for Research & Innovation (SCRI), Symbiosis International (Deemed University).

Conflict of Interest The authors declare no conflict of interest.

Ethical Approval This article does not contain any studies with human participants performed by any of the authors. This article does not contain any studies with animals performed by any of the authors.

References

1. Alexander A, Kim J, Walker CL (2010) ATM engages the TSC2/mTORC1 signaling node to regulate autophagy. Autophagy 6:672–673. https://doi.org/10.4161/auto.6.5.12509
2. Alfaro IE, Albornoz A, Molina A, Moreno J, Cordero K, Criollo A, Budini M (2019) Chaperone mediated autophagy in the crosstalk of neurodegenerative diseases and metabolic disorders. Front Endocrinol (Lausanne) 9:778. https://doi.org/10.3389/fendo.2018.00778
3. Altman BJ, Jacobs SR, Mason EF, Michalek RD, MacIntyre AN, Coloff JL, Ilkayeva O, Jia W, He Y-W, Rathmell JC (2011) Autophagy is essential to suppress cell stress and to allow BCR-Abl-mediated leukemogenesis. Oncogene 30:1855–1867. https://doi.org/10.1038/onc.2010.561
4. Arias E, Koga H, Diaz A, Mocholi E, Patel B, Cuervo AM (2015) Lysosomal mTORC2/PHLPP1/Akt regulate chaperone-mediated autophagy. Mol Cell 59:270–284. https://doi.org/10.1016/j.molcel.2015.05.030
5. Ashford TP, Porter KR (1962) Cytoplasmic components in hepatic cell lysosomes. J Cell Biol 12:198–202. https://doi.org/10.1083/jcb.12.1.198
6. Baquero P, Dawson A, Mukhopadhyay A, Kuntz EM, Mitchell R, Olivares O, Ianniciello A, Scott MT, Dunn K, Nicastri MC, Winkler JD, Michie AM, Ryan KM, Halsey C, Gottlieb E, Keaney EP, Murphy LO, Amaravadi RK, Holyoake TL, Helgason GV (2019) Targeting quiescent leukemic stem cells using second generation autophagy inhibitors. Leukemia 33:981–994. https://doi.org/10.1038/s41375-018-0252-4
7. Bonnet D, Dick JE (1997) Human acute myeloid leukemia is organized as a hierarchy that originates from a primitive hematopoietic cell. Nat Med 3:730–737. https://doi.org/10.1038/nm0797-730
8. Budgude P, Kale V, Vaidya A (2021) Pharmacological Inhibition of p38 MAPK Rejuvenates bone marrow derived-mesenchymal stromal cells and boosts their hematopoietic stem cell-supportive ability. Stem Cell Rev Reports. https://doi.org/10.1007/s12015-021-10240-9
9. Cabukusta B, Neefjes J (2018) Mechanisms of lysosomal positioning and movement. Traffic 19:761–769. https://doi.org/10.1111/tra.12587
10. Canu G, Ruhrberg C (2021) First blood: the endothelial origins of hematopoietic progenitors. Angiogenesis 24:199–211. https://doi.org/10.1007/s10456-021-09783-9
11. Cerletti M, Jang YC, Finley LWS, Haigis MC, Wagers AJ (2012) Short-term calorie restriction enhances skeletal muscle stem cell function. Cell Stem Cell 10:515–519. https://doi.org/10.1016/j.stem.2012.04.002
12. Chang NC (2020) Autophagy and stem cells: self-eating for self-renewal. Front Cell Dev Biol 8:138. https://doi.org/10.3389/fcell.2020.00138

13. Chen C, Liu Y, Liu Y, Zheng P (2009) mTOR regulation and therapeutic rejuvenation of aging hematopoietic stem cells. Sci Signal 2:ra75–ra75. https://doi.org/10.1126/scisignal.2000559
14. Chen T, Shen L, Yu J, Wan H, Guo A, Chen J, Long Y, Zhao J, Pei G (2011) Rapamycin and other longevity-promoting compounds enhance the generation of mouse induced pluripotent stem cells. Aging Cell 10:908–911. https://doi.org/10.1111/j.1474-9726.2011.00722.x
15. Chiang H-L, Plant CP, Dice JF (1989) A role for a 70-kilodalton heat shock protein in lysosomal degradation of intracellular proteins. Science 80(246):382–385. https://doi.org/10.1126/science.2799391
16. Das CK, Mandal M, Kögel D (2018) Pro-survival autophagy and cancer cell resistance to therapy. Cancer Metastasis Rev 37:749–766. https://doi.org/10.1007/s10555-018-9727-z
17. Deretic V (2021) Autophagy in inflammation, infection, and immunometabolism. Immunity 54:437–453. https://doi.org/10.1016/j.immuni.2021.01.018
18. Dice JF (1990) Peptide sequences that target cytosolic proteins for lysosomal proteolysis. Trends Biochem Sci 15:305–309. https://doi.org/10.1016/0968-0004(90)90019-8
19. Diwan A, Koesters AG, Odley AM, Pushkaran S, Baines CP, Spike BT, Daria D, Jegga AG, Geiger H, Aronow BJ (2007) Unrestrained erythroblast development in Nix−/− mice reveals a mechanism for apoptotic modulation of erythropoiesis. Proc Natl Acad Sci 104:6794–6799. https://doi.org/10.1073/pnas.0610666104
20. Eisenstein M (2014) Molecular biology: remove, reuse, recycle. Nature 514:S2–S4. https://doi.org/10.1038/514S2a
21. Eliasson P, Jönsson J (2010) The hematopoietic stem cell niche: low in oxygen but a nice place to be. J Cell Physiol 222:17–22. https://doi.org/10.1002/jcp.21908
22. Feng Y, He D, Yao Z, Klionsky DJ (2014) The machinery of macroautophagy. Cell Res 24:24–41. https://doi.org/10.1038/cr.2013.168
23. Gade P, Kimball AS, DiNardo AC, Gangwal P, Ross DD, Boswell HS, Keay SK, Kalvakolanu DV (2016) Death-associated protein kinase-1 expression and autophagy in chronic lymphocytic leukemia are dependent on activating transcription factor-6 and CCAAT/enhancer-binding protein-β. J Biol Chem 291:22030–22042. https://doi.org/10.1074/jbc.M116.725796
24. García-Prat L, Martínez-Vicente M, Muñoz-Cánoves P (2016a) Autophagy: a decisive process for stemness. Oncotarget 7:12286. https://doi.org/10.18632/oncotarget.7766
25. García-Prat L, Martínez-Vicente M, Perdiguero E, Ortet L, Rodríguez-Ubreva J, Rebollo E, Ruiz-Bonilla V, Gutarra S, Ballestar E, Serrano AL, Sandri M, Muñoz-Cánoves P (2016) Autophagy maintains stemness by preventing senescence. Nature 529:37–42. https://doi.org/10.1038/nature16187
26. Gomez-Puerto MC, Folkerts H, Wierenga ATJ, Schepers K, Schuringa JJ, Coffer PJ, Vellenga E (2016) Autophagy Proteins ATG5 and ATG7 are essential for the maintenance of human CD34+ hematopoietic stem-progenitor Cells. Stem Cells 34:1651–1663. https://doi.org/10.1002/stem.2347
27. Guan J-L, Simon AK, Prescott M, Menendez JA, Liu F, Wang F, Wang C, Wolvetang E, Vazquez-Martin A, Zhang J (2013) Autophagy in stem cells. Autophagy 9:830–849. https://doi.org/10.4161/auto.24132
28. Gubas A, Dikic I (2021) A guide to the regulation of selective autophagy receptors. FEBS J. https://doi.org/10.1111/febs.15824
29. Guidi N, Marka G, Sakk V, Zheng Y, Florian MC, Geiger H (2021) An aged bone marrow niche restrains rejuvenated hematopoietic stem cells. Stem Cells. https://doi.org/10.1002/stem.3372
30. Harada K, Kotani T, Kirisako H, Sakoh-Nakatogawa M, Oikawa Y, Kimura Y, Hirano H, Yamamoto H, Ohsumi Y, Nakatogawa H (2019) Two distinct mechanisms target the autophagy-related E3 complex to the pre-autophagosomal structure. Elife 8:e43088. https://doi.org/10.7554/eLife.43088
31. Hashimoto M, Umemoto T, Nakamura-Ishizu A, Matsumura T, Yokomizo T, Sezaki M, Takizawa H, Suda T (2021) Autophagy is dispensable for the maintenance of hematopoietic stem cells in neonates. Blood Adv 5:1594–1604. https://doi.org/10.1182/bloodadvances.2020002410

32. Herpin A, Lescat L, Bobe J, Jenny A, Seiliez I (2020) Lighting chaperone-mediated autophagy (CMA) evolution with an ancient LAMP: the existence of a functional CMA activity in fish. Autophagy 16:1918–1920. https://doi.org/10.1080/15548627.2020.1797344
33. Ho TT, Warr MR, Adelman ER, Lansinger OM, Flach J, Verovskaya EV, Figueroa ME, Passegué E (2017) Autophagy maintains the metabolism and function of young and old stem cells. Nature 543:205–210. https://doi.org/10.1038/nature21388
34. Honda S, Arakawa S, Nishida Y, Yamaguchi H, Ishii E, Shimizu S (2014) Ulk1-mediated Atg5-independent macroautophagy mediates elimination of mitochondria from embryonic reticulocytes. Nat Commun 5:1–13. https://doi.org/10.1038/ncomms5004
35. Ianniciello A, Rattigan KM, Helgason GV (2018) The ins and outs of autophagy and metabolism in hematopoietic and leukemic stem cells: food for thought. Front cell Dev Biol 6:120. https://doi.org/10.3389/fcell.2018.00120
36. Ito K, Suda T (2014) Metabolic requirements for the maintenance of self-renewing stem cells. Nat Rev Mol Cell Biol 15:243–256. https://doi.org/10.1038/nrm3772
37. Ito K, Turcotte R, Cui J, Zimmerman SE, Pinho S, Mizoguchi T, Arai F, Runnels JM, Alt C, Teruya-Feldstein J (2016) Self-renewal of a purified Tie2+ hematopoietic stem cell population relies on mitochondrial clearance. Science 80(354):1156–1160. https://doi.org/10.1126/science.aaf5530
38. Jalnapurkar S, Moirangthem RD, Singh S, Limaye L, Kale V (2019) Microvesicles secreted by nitric oxide-primed mesenchymal stromal cells boost the engraftment potential of hematopoietic stem cells. Stem Cells 37:128–138. https://doi.org/10.1002/stem.2912
39. Jiang H, Yang L, Guo L, Cui N, Zhang G, Liu C, Xing L, Shao Z, Wang H (2018) Impaired mitophagy of nucleated erythroid cells leads to anemia in patients with myelodysplastic syndromes. Oxid Med Cell Longev 2018.https://doi.org/10.1155/2018/6328051
40. Jin J, Britschgi A, Schläfli AM, Humbert M, Shan-Krauer D, Batliner J, Federzoni EA, Ernst M, Torbett BE, Yousefi S, Simon H-U, Tschan MP (2018) Low Autophagy (ATG) gene expression is associated with an immature AML blast cell phenotype and can be restored during AML differentiation therapy. Oxid Med Cell Longev 2018:1482795. https://doi.org/10.1155/2018/1482795
41. Jung HE, Shim YR, Oh JE, Oh DS, Lee HK (2019) The autophagy protein Atg5 plays a crucial role in the maintenance and reconstitution ability of hematopoietic stem cells. Immune Netw 19.https://doi.org/10.4110/in.2019.19.e12
42. Kale VP (2019) Application of "primed" Mesenchymal Stromal cells in hematopoietic stem cell transplantation: current status and Future prospects. Stem Cells Dev 28:1473–1479. https://doi.org/10.1089/scd.2019.0149
43. Kang Y-A, Sanalkumar R, O'geen H, Linnemann AK, Chang C-J, Bouhassira EE, Farnham PJ, Keles S, Bresnick EH (2012) Autophagy driven by a master regulator of hematopoiesis. Mol Cell Biol 32:226–239.https://doi.org/10.1128/MCB.06166-11
44. Kaushik S, Cuervo AM (2018) The coming of age of chaperone-mediated autophagy. Nat Rev Mol Cell Biol 19:365–381. https://doi.org/10.1038/s41580-018-0001-6
45. Kim B-W, Jin Y, Kim J, Kim JH, Jung J, Kang S, Kim IY, Kim J, Cheong H, Song HK (2018) The C-terminal region of ATG101 bridges ULK1 and PtdIns3K complex in autophagy initiation. Autophagy 14:2104–2116. https://doi.org/10.1080/15548627.2018.1504716
46. Kiššová IB, Salin B, Schaeffer J, Bhatia S, Manon S, Camougrand N (2007) Selective and non-selective autophagic degradation of mitochondria in yeast. Autophagy 3:329–336. https://doi.org/10.4161/auto.4034
47. Klionsky DJ (2008) Autophagy revisited: a conversation with Christian de Duve. Autophagy 4:740–743. https://doi.org/10.4161/auto.6398
48. Klionsky DJ, Codogno P (2013) The mechanism and physiological function of macroautophagy. J Innate Immun 5:427–433. https://doi.org/10.1159/000351979
49. Kobayashi T, Piao W, Takamura T, Kori H, Miyachi H, Kitano S, Iwamoto Y, Yamada M, Imayoshi I, Shioda S (2019) Enhanced lysosomal degradation maintains the quiescent state of neural stem cells. Nat Commun 10:1–14. https://doi.org/10.1038/s41467-019-13203-4

50. Koschade SE, Brandts CH (2020) Selective autophagy in normal and malignant hematopoiesis. J Mol Biol 432:261–282. https://doi.org/10.1016/j.jmb.2019.06.025
51. Kulkarni R, Bajaj M, Ghode S, Jalnapurkar S, Limaye L, Kale VP (2018) Intercellular transfer of microvesicles from young mesenchymal stromal cells rejuvenates aged murine hematopoietic stem cells. Stem Cells 36:420–433. https://doi.org/10.1002/stem.2756
52. Kulkarni RS, Bajaj MS, Kale VP (2019) Induction and detection of autophagy in aged hematopoietic stem cells by exposing them to microvesicles secreted by HSC-supportive mesenchymal stromal cells. Methods Mol Biol 1854:21–34. https://doi.org/10.1007/7651_2018_166
53. Kuma A, Hatano M, Matsui M, Yamamoto A, Nakaya H, Yoshimori T, Ohsumi Y, Tokuhisa T, Mizushima N (2004) The role of autophagy during the early neonatal starvation period. Nature 432:1032–1036. https://doi.org/10.1038/nature03029
54. Kumar S, Geiger H (2017) HSC Niche Biology and HSC Expansion Ex Vivo. Trends Mol Med 23:799–819. https://doi.org/10.1016/j.molmed.2017.07.003
55. Kundu M, Lindsten T, Yang C-Y, Wu J, Zhao F, Zhang J, Selak MA, Ney PA, Thompson CB (2008) Ulk1 plays a critical role in the autophagic clearance of mitochondria and ribosomes during reticulocyte maturation. Blood J Am Soc Hematol 112:1493–1502. https://doi.org/10.1182/blood-2008-02-137398
56. Lacombe J, Herblot S, Rojas-Sutterlin S, Haman A, Barakat S, Iscove NN, Sauvageau G, Hoang T (2010) Scl regulates the quiescence and the long-term competence of hematopoietic stem cells. Blood 115:792–803. https://doi.org/10.1182/blood-2009-01-201384
57. Larrue C, Saland E, Boutzen H, Vergez F, David M, Joffre C, Hospital M-A, Tamburini J, Delabesse E, Manenti S, Sarry JE, Récher C (2016) Proteasome inhibitors induce FLT3-ITD degradation through autophagy in AML cells. Blood 127:882–892. https://doi.org/10.1182/blood-2015-05-646497
58. Lee Y-K, Lee J-A (2016) Role of the mammalian ATG8/LC3 family in autophagy: differential and compensatory roles in the spatiotemporal regulation of autophagy. BMB Rep 49:424. https://doi.org/10.5483/bmbrep.2016.49.8.081
59. Leeman DS, Hebestreit K, Ruetz T, Webb AE, McKay A, Pollina EA, Dulken BW, Zhao X, Yeo RW, Ho TT, Mahmoudi S, Devarajan K, Passegué E, Rando TA, Frydman J, Brunet A (2018) Lysosome activation clears aggregates and enhances quiescent neural stem cell activation during aging. Science (80) 359:1277 LP—1283. https://doi.org/10.1126/science.aag3048
60. Levine B, Kroemer G (2019) Biological functions of autophagy genes: a disease perspective. Cell 176:11–42. https://doi.org/10.1016/j.cell.2018.09.048
61. Li W, Li J, Bao J (2012) Microautophagy: lesser-known self-eating. Cell Mol life Sci 69:1125–1136. https://doi.org/10.1007/s00018-011-0865-5
62. Li X, Zeng X, Xu Y, Wang B, Zhao Y, Lai X, Qian P, Huang H (2020) Mechanisms and rejuvenation strategies for aged hematopoietic stem cells. J Hematol Oncol 13:31. https://doi.org/10.1186/s13045-020-00864-8
63. Liu F, Lee JY, Wei H, Tanabe O, Engel JD, Morrison SJ, Guan J-L (2010) FIP200 is required for the cell-autonomous maintenance of fetal hematopoietic stem cells. Blood J Am Soc Hematol 116:4806–4814. https://doi.org/10.1182/blood-2010-06-288589
64. Liu WJ, Ye L, Huang WF, Guo LJ, Xu ZG, Wu HL, Yang C, Liu HF (2016) p62 links the autophagy pathway and the ubiqutin–proteasome system upon ubiquitinated protein degradation. Cell Mol Biol Lett 21:29. https://doi.org/10.1186/s11658-016-0031-z
65. Lyu Z-S, Cao X-N, Wen Q, Mo X-D, Zhao H-Y, Chen Y-H, Wang Y, Chang Y-J, Xu L-P, Zhang X-H, Kong Y, Huang X-J (2020) Autophagy in endothelial cells regulates their haematopoiesis-supporting ability. EBioMedicine 53:102677. https://doi.org/10.1016/j.ebiom.2020.102677
66. Meenhuis A, van Veelen PA, de Looper H, van Boxtel N, van den Berge IJ, Sun SM, Taskesen E, Stern P, de Ru AH, van Adrichem AJ, Demmers J, Jongen-Lavrencic M, Löwenberg B, Touw IP, Sharp PA, Erkeland SJ (2011) MiR-17/20/93/106 promote hematopoietic cell expansion by targeting sequestosome 1-regulated pathways in mice. Blood 118:916–925. https://doi.org/10.1182/blood-2011-02-336487

67. Mizushima N, Levine B (2010) Autophagy in mammalian development and differentiation. Nat Cell Biol 12:823–830. https://doi.org/10.1038/ncb0910-823
68. Mortensen M, Ferguson DJP, Edelmann M, Kessler B, Morten KJ, Komatsu M, Simon AK (2010) Loss of autophagy in erythroid cells leads to defective removal of mitochondria and severe anemia in vivo. Proc Natl Acad Sci 107:832–837. https://doi.org/10.1073/pnas.0913170107
69. Mortensen M, Simon AK (2010) Nonredundant role of Atg7 in mitochondrial clearance during erythroid development. Autophagy 6:423–425. https://doi.org/10.4161/auto.6.3.11528
70. Mortensen M, Soilleux EJ, Djordjevic G, Tripp R, Lutteropp M, Sadighi-Akha E, Stranks AJ, Glanville J, Knight S, Jacobsen W, S-E, (2011) The autophagy protein Atg7 is essential for hematopoietic stem cell maintenance. J Exp Med 208:455–467. https://doi.org/10.1084/jem.20101145
71. Müller O, Sattler T, Flötenmeyer M, Schwarz H, Plattner H, Mayer A (2000) Autophagic tubes: vacuolar invaginations involved in lateral membrane sorting and inverse vesicle budding. J Cell Biol 151:519–528. https://doi.org/10.1083/jcb.151.3.519
72. Nakamura-Ishizu A, Ito K, Suda T (2020) Hematopoietic stem cell metabolism during development and aging. Dev Cell 54:239–255. https://doi.org/10.1016/j.devcel.2020.06.029
73. Nassar D, Blanpain C (2016) Cancer stem cells: basic concepts and therapeutic implications. Annu Rev Pathol Mech Dis 11:47–76. https://doi.org/10.1146/annurev-pathol-012615-044438
74. Noguchi M, Hirata N, Tanaka T, Suizu F, Nakajima H, Chiorini JA (2020) Autophagy as a modulator of cell death machinery. Cell Death Dis 11:1–12. https://doi.org/10.1038/s41419-020-2724-5
75. Nomura N, Ito C, Ooshio T, Tadokoro Y, Kohno S, Ueno M, Kobayashi M, Kasahara A, Takase Y, Kurayoshi K, Si S, Takahashi C, Komatsu M, Yanagawa T, Hirao A (2021) Essential role of autophagy in protecting neonatal haematopoietic stem cells from oxidative stress in a p62-independent manner. Sci Rep 11:1666. https://doi.org/10.1038/s41598-021-81076-z
76. Parzych KR, Klionsky DJ (2014) An overview of autophagy: morphology, mechanism, and regulation. Antioxid Redox Signal 20:460–473. https://doi.org/10.1089/ars.2013.5371
77. Pérez-Hernández M, Arias A, Martínez-García D, Pérez-Tomás R, Quesada R, Soto-Cerrato V (2019) Targeting autophagy for cancer treatment and tumor Chemosensitization. Cancers (Basel) 11.https://doi.org/10.3390/cancers11101599
78. Phadwal K, Watson AS, Simon AK (2013) Tightrope act: autophagy in stem cell renewal, differentiation, proliferation, and aging. Cell Mol Life Sci 70:89–103. https://doi.org/10.1007/s00018-012-1032-3
79. Piragyte I, Clapes T, Polyzou A, Klein Geltink RI, Lefkopoulos S, Yin N, Cauchy P, Curtis JD, Klaeylé L, Langa X, Beckmann CCA, Wlodarski MW, Müller P, Van Essen D, Rambold A, Kapp FG, Mione M, Buescher JM, Pearce EL, Polyzos A, Trompouki E (2018) A metabolic interplay coordinated by HLX regulates myeloid differentiation and AML through partly overlapping pathways. Nat Commun 9:3090. https://doi.org/10.1038/s41467-018-05311-4
80. Polak R, Bierings MB, van der Leije CS, Sanders MA, Roovers O, Marchante JRM, Boer JM, Cornelissen JJ, Pieters R, den Boer ML, Buitenhuis M (2019) Autophagy inhibition as a potential future targeted therapy for ETV6-RUNX1-driven B-cell precursor acute lymphoblastic leukemia. Haematologica 104:738–748. https://doi.org/10.3324/haematol.2018.193631
81. Puissant A, Robert G, Fenouille N, Luciano F, Cassuto J-P, Raynaud S, Auberger P (2010) Resveratrol promotes autophagic cell death in chronic myelogenous leukemia cells via JNK-Mediated p62/SQSTM1 expression and AMPK activation. Cancer Res 70:1042 LP—1052. https://doi.org/10.1158/0008-5472.CAN-09-3537
82. Rabanal-Ruiz Y, Otten EG, Korolchuk VI (2017) mTORC1 as the main gateway to autophagy. Essays Biochem 61:565–584. https://doi.org/10.1042/EBC20170027
83. Revuelta M, Matheu A (2017) Autophagy in stem cell aging. Aging Cell 16:912–915. https://doi.org/10.1111/acel.12655
84. Reya T, Morrison SJ, Clarke MF, Weissman IL (2001) Stem cells, cancer, and cancer stem cells. Nature 414:105–111. https://doi.org/10.1038/35102167

85. Riffelmacher T, Clarke A, Richter FC, Stranks A, Pandey S, Danielli S, Hublitz P, Yu Z, Johnson E, Schwerd T (2017) Autophagy-dependent generation of free fatty acids is critical for normal neutrophil differentiation. Immunity 47:466–480. https://doi.org/10.1016/j.immuni.2017.08.005
86. Riffelmacher T, Simon A (2017) Mechanistic roles of autophagy in hematopoietic differentiation. FEBS J 284:1008–1020. https://doi.org/10.1111/febs.13962
87. Rogov VV, Suzuki H, Marinković M, Lang V, Kato R, Kawasaki M, Buljubašić M, Šprung M, Rogova N, Wakatsuki S, Hamacher-Brady A, Dötsch V, Dikic I, Brady NR, Novak I (2017) Phosphorylation of the mitochondrial autophagy receptor Nix enhances its interaction with LC3 proteins. Sci Rep 7:1131. https://doi.org/10.1038/s41598-017-01258-6
88. Rudat S, Pfaus A, Cheng YY, Holtmann J, Ellegast JM, Bühler C, Di MD, Martinez E, Göllner S, Wickenhauser C, Müller-Tidow C, Lutz C, Bullinger L, Milsom MD, Sykes SM, Fröhling S, Scholl C (2018) RET-mediated autophagy suppression as targetable co-dependence in acute myeloid leukemia. Leukemia 32:2189–2202. https://doi.org/10.1038/s41375-018-0102-4
89. Sahu R, Kaushik S, Clement CC, Cannizzo ES, Scharf B, Follenzi A, Potolicchio I, Nieves E, Cuervo AM, Santambrogio L (2011) Microautophagy of cytosolic proteins by late endosomes. Dev Cell 20:131–139. https://doi.org/10.1016/j.devcel.2010.12.003
90. Sandoval H, Thiagarajan P, Dasgupta SK, Schumacher A, Prchal JT, Chen M, Wang J (2008) Essential role for Nix in autophagic maturation of erythroid cells. Nature 454:232–235. https://doi.org/10.1038/nature07006
91. Schüler SC, Gebert N, Ori A (2020) Stem cell aging: the upcoming era of proteins and metabolites. Mech Ageing Dev 190:111288. https://doi.org/10.1016/j.mad.2020.111288
92. Seita J, Weissman IL (2010) Hematopoietic stem cell: self-renewal versus differentiation. Wiley Interdiscip Rev Syst Biol Med 2:640–653. https://doi.org/10.1002/wsbm.86
93. Simsek T, Kocabas F, Zheng J, DeBerardinis RJ, Mahmoud AI, Olson EN, Schneider JW, Zhang CC, Sadek HA (2010) The distinct metabolic profile of hematopoietic stem cells reflects their location in a hypoxic niche. Cell Stem Cell 7(3):380–390
94. Sumitomo Y, Koya J, Nakazaki K, Kataoka K, Tsuruta-Kishino T, Morita K, Sato T, Kurokawa M (2016) Cytoprotective autophagy maintains leukemia-initiating cells in murine myeloid leukemia. Blood 128:1614–1624. https://doi.org/10.1182/blood-2015-12-684696
95. Takahashi K, Tanabe K, Ohnuki M, Narita M, Ichisaka T, Tomoda K, Yamanaka S (2007) Induction of pluripotent stem cells from adult human fibroblasts by defined factors. Cell 131:861–872. https://doi.org/10.1016/j.cell.2007.11.019
96. Takano-Ohmuro H, Mukaida M, Kominami E, Morioka K (2000) Autophagy in embryonic erythroid cells: its role in maturation. Eur J Cell Biol 79:759–764. https://doi.org/10.1078/0171-9335-00096
97. Takubo K, Nagamatsu G, Kobayashi CI, Nakamura-Ishizu A, Kobayashi H, Ikeda E, Goda N, Rahimi Y, Johnson RS, Soga T, Hirao A, Suda T (2013) Regulation of glycolysis by Pdk functions as a metabolic checkpoint for cell cycle quiescence in hematopoietic stem cells. Cell Stem Cell 12(1):49–61
98. Teli P, Kale V, Vaidya A (2021) Extracellular vesicles isolated from mesenchymal stromal cells primed with neurotrophic factors and signaling modifiers as potential therapeutics for neurodegenerative diseases. Curr Res Transl Med 69:103286. https://doi.org/10.1016/j.retram.2021.103286
99. Trowbridge JJ (2019) Intrinsic and extrinsic factors driving hematopoietic stem cell aging and bone marrow failure. Blood 134:SCI-35-SCI-35. https://doi.org/10.1182/blood-2019-121044
100. Udroiu I, Sgura A (2021) Growing and aging of hematopoietic stem cells. World J Stem Cells 13:594. https://doi.org/10.4252/wjsc.v13.i6.594
101. Unwin RD, Smith DL, Blinco D, Wilson CL, Miller CJ, Evans CA, Baldwin SA, Barnes K, Pierce A, Spooncer E, Whetton AD (2006) Quantitative proteomics reveals posttranslational control as a regulatory factor in primary hematopoietic stem cells. Blood 107(12):4687–4694
102. Visconte V, Nawrocki ST, Przychodzen BP, Kelly KR, Yin M, Parker Y, Clemente MJ, Barnard J, Advani AS, Carraway HE, Sekeres MA, Lindner D, Rogers HJ, Maciejewski JP, Carew JS

(2016) Targeting autophagy in myelodysplastic syndromes. Blood 128:4295. https://doi.org/10.1182/blood.V128.22.4295.4295
103. Wang S, Xia P, Ye B, Huang G, Liu J, Fan Z (2013) Transient activation of autophagy via Sox2-mediated suppression of mTOR Is an important early step in reprogramming to pluripotency. Stem Cell 13:617–625. https://doi.org/10.1016/j.stem.2013.10.005
104. Warr MR, Binnewies M, Flach J, Reynaud D, Garg T, Malhotra R, Debnath J, Passegué E (2013) FOXO3A directs a protective autophagy program in haematopoietic stem cells. Nature 494:323–327. https://doi.org/10.1038/nature11895
105. Watson AS, Riffelmacher T, Stranks A, Williams O, De Boer J, Cain K, MacFarlane M, McGouran J, Kessler B, Khandwala S, Chowdhury O, Puleston D, Phadwal K, Mortensen M, Ferguson D, Soilleux E, Woll P, Jacobsen SEW, Simon AK (2015) Autophagy limits proliferation and glycolytic metabolism in acute myeloid leukemia. Cell Death Discov 1:15008. https://doi.org/10.1038/cddiscovery.2015.8
106. White E (2012) Deconvoluting the context-dependent role for autophagy in cancer. Nat Rev Cancer 12:401–410. https://doi.org/10.1038/nrc3262
107. Xi Y, Dhaliwal JS, Ceizar M, Vaculik M, Kumar KL, Lagace DC (2016) Knockout of Atg5 delays the maturation and reduces the survival of adult-generated neurons in the hippocampus. Cell Death Dis 7:e2127–e2127. https://doi.org/10.1038/cddis.2015.406
108. Xie R, Nguyen S, McKeehan WL, Liu L (2010) Acetylated microtubules are required for fusion of autophagosomes with lysosomes. BMC Cell Biol 11:1–12. https://doi.org/10.1186/1471-2121-11-89
109. Yazdankhah M, Farioli-Vecchioli S, Tonchev AB, Stoykova A, Cecconi F (2014) The autophagy regulators Ambra1 and Beclin 1 are required for adult neurogenesis in the brain subventricular zone. Cell Death Dis 5:e1403–e1403. https://doi.org/10.1038/cddis.2014.358
110. Young LN, Goerdeler F, Hurley JH (2019) Structural pathway for allosteric activation of the autophagic PI 3-kinase complex I. Proc Natl Acad Sci 116:21508–21513. https://doi.org/10.1073/pnas.1911612116
111. Yu L, Chen Y, Tooze SA (2018) Autophagy pathway: cellular and molecular mechanisms. Autophagy 14:207–215. https://doi.org/10.1080/15548627.2017.1378838
112. Zhang Y, Morgan MJ, Chen K, Choksi S, Liu Z (2012) Induction of autophagy is essential for monocyte-macrophage differentiation. Blood J Am Soc Hematol 119:2895–2905. https://doi.org/10.1182/blood-2011-08-372383
113. Zhuang L, Ma Y, Wang Q, Zhang J, Zhu C, Zhang L, Xu X (2016) Atg3 overexpression enhances bortezomib-induced cell death in SKM-1 cell. PLoS ONE 11:e0158761. https://doi.org/10.1371/journal.pone.0158761
114. Zon LI (2008) Intrinsic and extrinsic control of haematopoietic stem-cell self-renewal. Nature 453:306–313. https://doi.org/10.1038/nature07038
115. Zuo R, Wang Y, Li J, Wu J, Wang W, Li B, Sun C, Wang Z, Shi C, Zhou Y (2019) Rapamycin induced autophagy inhibits inflammation-mediated endplate degeneration by enhancing Nrf2/Keap1 signaling of cartilage endplate stem cells. Stem Cells 37:828–840. https://doi.org/10.1002/stem.2999

Chapter 6
Autophagy in Muscle Stem Cells

Shulei Li, Romina L. Filippelli, Alice Jisoo Nam, and Natasha C. Chang

Abstract Muscle stem cells, also known as satellite cells, are responsible for the regenerative capacity of adult muscle tissue in response to stress and injury. Upon regenerative stimuli, satellite cells are activated and undergo myogenic commitment. Myogenic progenitors, which are termed myoblasts, undergo rapid proliferation, propagation, and differentiation into myocytes, which then fuse with each other to form new myotubes or to a pre-existing myotube. This process of myogenic differentiation is metabolically demanding and involves cellular remodeling of organelles and cellular architecture. Autophagy, a catabolic mechanism involving the sequestration of cellular contents into double membrane autophagosome vesicles, is strongly implicated at various stages during myogenesis; from the satellite stem cell to the mature muscle tissue. Moreover, aberrant autophagy (both the overstimulation and inhibition of autophagy) in both satellite cells and mature muscle cells can be detrimental for muscle health and physiology. This chapter outlines the importance of autophagy in maintaining skeletal muscle tissue homeostasis and satellite cell regenerative capacity.

Keywords Autophagy · Muscle stem cell · Myoblast · Myogenesis · Satellite cell · Skeletal muscle

Abbreviations

AICAR	5-Aminoimidazole-4-carboxamide ribonucleotide
AMPK	AMP-activated protein kinase

Shulei Li, Romina L. Filippelli, Alice Jisoo Nam: These authors contributed equally to this work.

S. Li · R. L. Filippelli · A. J. Nam · N. C. Chang (✉)
Faculty of Medicine and Health Sciences, Department of Biochemistry, McGill University, Montréal, Québec, Canada
e-mail: natasha.chang@mcgill.ca

B. V. Shravage and K. Turksen (eds.), *Autophagy in Stem Cell Maintenance and Differentiation*, Stem Cell Biology and Regenerative Medicine 73,
https://doi.org/10.1007/978-3-031-17362-2_6

BAG3	BAG cochaperone 3
CASA	Chaperone-assisted selective autophagy
DM1	Myotonic dystrophy type 1
ECM	Extracellular matrix
ER	Endoplasmic reticulum
FACS	Fluorescence-activated cell sorting
HSPA8/HSC70	Heat shock 70 kDa protein 8
LD	Lipid droplet
LSM	Lipid storage myopathy
MFN1/2	Mitofusin 1/2
miRNA	MicroRNA
mTOR	Mammalian target of rapamycin
mTORC1	MTOR complex 1
NAF-1	Nutrient-deprivation autophagy factor-1
NMJ	Neuromuscular junction
p62/SQSTM1	Autophagy receptor p62/sequestome 1
PGC-1α	Proliferator-activated receptor-γ coactivator-1α
ROS	Reactive oxygen species
SIRT1	Sirtuin-1
TALEN	Transcription activator-like effector nuclease
T tubules	Transverse tubules

Introduction

Skeletal muscle is a highly organized contractile tissue making up roughly 40% of human whole body lean mass [106]. The human musculoskeletal system is responsible for allowing movement, the maintenance of posture, body position and body temperature [43]. While both skeletal and cardiac muscle are forms of striated muscle, cardiac muscle functions as a self-stimulating and non-fatiguing group of muscle cells [102]. In contrast, the motor activity of skeletal muscle is voluntary and the tissue itself exhibits fatigue and has high energy requirements [102].

Skeletal Muscle Architecture

Skeletal muscles are supported by the cytoskeleton network composed of bundles of fascicles, which consist of bundles of muscle fibers, termed myofibers [102]. Single myofibers are multinucleated and are of variable lengths and shapes containing several myofibrils arranged in parallel and units of sarcomere, the basic contractile unit of muscles, arranged in series [129]. Single myofibers are encased by the sarcolemma, which acts as the muscle plasma membrane and is directly involved in synaptic transmission, action potential propagation, and excitation–contraction coupling in response to stimulation [43]. The sarcolemma is connected to the extracellular matrix (ECM) that surrounds skeletal muscle fibers and is encapsulated by the basal lamina, an outer membrane layer that defines the anatomical length and boundary of a myofiber [43]. The sarcoplasmic reticulum, which is the main area of calcium storage in skeletal muscles, forms a network with transverse (T) tubules that surrounds the myofibrils [128]. The connection between T tubules and the sarcoplasmic reticulum is critical for the release of calcium leading to muscle contraction in response to the action potential generated by motor neurons [128]. The ECM surrounding the muscle fibers is composed of various types of collagens, laminins, fibronectin, and proteoglycans, providing mechanical support to the myofibers during contraction [128].

Satellite Cells: Muscle-Resident Stem Cells

Adult skeletal muscles are stable under normal conditions and have remarkable capacity for regeneration after injury due to the presence of satellite cells, which are muscle resident somatic stem cells accounting for 3–6% of total myonuclei [165]. Satellite cells were first discovered in 1961 by Alexander Mauro upon examination of the peripheral region of myofibers dissected from the tibialis anterior muscles of the frog by electron microscopy [95]. Notably, while fused myonuclei are scattered across the myofibers, satellite cells reside along host myofibers directly above the sarcolemma and under the basal lamina [95, 102]. Satellite cells can be identified by their unique anatomical location with electron microscopy, which also reveals their morphological characteristics: a large nuclear-to-cytoplasmic ratio, few organelles, small nucleus, and condensed interphase chromatin [95, 136]. This morphology supports the notion that most satellite cells in healthy, unstressed muscles are mitotically quiescent and transcriptionally inactive [165]. Satellite cells can also be identified by immunofluorescence using antibodies specific to satellite cell markers such as the paired box transcription factor, PAX7 [136, 138]. In situ hybridization analyses in skeletal muscle tissues demonstrated that *Pax7* mRNA is expressed exclusively in satellite cells [138]. Moreover, *Pax7* is expressed in proliferating myoblasts derived from the satellite cell lineage and its expression is downregulated during myogenic differentiation [138].

Cell surface protein markers such as ITGA7, ITGB1, caveolin1, CD34, M-cadherin, CXCR4, N-CAM, syndecan -3 and -4 and VCAM-1 can be used for identifying satellite cells [35]. Fluorescence-activated cell sorting (FACS) is a widely used method to isolate satellite cells from freshly harvested muscle [90]. Using *Pax7-zsGreen* transgenic reporter mice, FACS with the cell surface markers ITGA7, ITGB1, CXCR4, and CD34 were shown to allow for successful prospective isolation of *Pax7*-expressing satellite cells [90]. A combination of CD34 and ITGA7 enables the isolation of a more restricted, namely quiescent, subset of satellite cells [90].

In contrast, VCAM-1 is expressed in satellite cells from young and old mice during quiescence and upon injury, and may be used for isolating activated satellite cells [87].

The Role of Satellite Cells

Satellite cells are responsible for postnatal growth of skeletal muscles [49]. During this period, satellite cells differentiate and contribute to muscle growth at varying rates; they exist in heterogenous pools comprised of 80% fast-dividing and 20% slow-dividing populations [135]. The slow-dividing population spends more time in G_0-phase between divisions and serves as a source for resident stem cells [135]. Flow cytometry analyses assessing various myogenic markers through stages of postnatal development demonstrated that postnatal growth is accompanied by waves of satellite cell commitment and differentiation [49]. Specifically, by assessing markers of a cycling signature $PAX7^+/Ki67^+$, non-cycling $PAX7^+/Ki67^-$, and the quiescence marker CD34, it was found that the levels of cycling satellite cells is highest in the early postnatal stage and becomes less prevalent as the muscles develop until adulthood when cycling satellite cells are absent and satellite cells express CD34 [49].

While satellite cell contribution to muscle is most apparent during early muscle development, satellite cells continue to contribute to muscle throughout life [71]. Genetic lineage tracing experiments have demonstrated that satellite cells in adult muscles contribute to muscle homeostasis in sedentary conditions [71]. *Pax7*-expressing satellite cells were labeled upon tamoxifen induction in $Pax7^{CreERT2}$:$Rosa^{mTmG}$ mice, allowing visualization of membrane-bound GFP following Cre-mediated recombination [71]. All myofibers examined over a 12- or 20-month period were GFP-positive indicating that satellite cells have contributed to these myofibers despite the lack of muscle injury or stress [71]. These results demonstrate that satellite cells actively participate in maintaining normal steady-state muscle homeostasis.

Myogenesis: Quiescence, Activation, Proliferation, Differentiation, and Self-renewal

In resting muscles, satellite cells are kept in a G_0-quiescent and mononucleated state within their niche [165]. Maintenance of the quiescent state is highly regulated and dependent on the expression of specific quiescence genes and post-transcriptional regulation of differentiation genes (Fukada et al. 2007). Indeed, microRNA (miRNA) pathways play a critical role in maintaining the quiescent state [30]. Microarray analysis of quiescent satellite cells revealed 22 highly expressed quiescence-specific miRNAs [30]. Specifically, miRNA-489 retains satellite cell quiescence by supressing activation and its overexpression inhibited muscle regeneration [30]. Additionally, microarray analysis of quiescent satellite cells revealed 507 genes, including cell cycle down-regulators and myogenic inhibitory factors that were highly expressed in the quiescent state [45]. Analysis of histone modifications via chromatin immunoprecipitation followed by sequencing have demonstrated that quiescent satellite cells are primed for rapid activation rather than staying strictly dormant [86]. An overwhelming majority of genes in quiescent satellite cells are marked by tri-methylated histone H3 lysine K4 (H3K4), a marker of active transcription, while only a few are marked with tri-methylated histone H3 lysine K27 (H3K27), a marker of gene repression, at the transcription start site [12, 86, 161].

Further to maintaining muscle homeostasis, satellite cells are responsible for the regenerative capacity of adult muscle tissue in response to stress and injury [165]. Using *Pax7*^{CreERT2} mice to genetically label satellite cells and characterize their response to muscle injury, Murphy and colleagues found that all regenerated muscle arose from *Pax7*-expressing satellite cells [105]. Importantly, genetic ablation of satellite cells via Cre-mediated expression of diphtheria toxin A results in a complete and irreversible loss of muscle regeneration [105]. Similar conclusions were drawn using an alternative mouse model with a different *Pax7*^{CreERT2} allele [82]. In a separate and complementary approach to deplete satellite cells that were genetically engineered to express the human diptheria toxin receptor following local intramuscular injection of diptheria toxin, it was confirmed that elimination of satellite cells resulted in loss of muscle tissue and a failure to regenerate damaged muscle [131]. Moreover, expression of *Pax7* in satellite cells is required for satellite cell function and regenerative myogenesis [157]. These studies altogether corroborate that satellite cells and *Pax7* expression are essential for muscle regeneration.

Following stress or injury to the muscle, satellite cells become activated and are recruited to the cell cycle [124]. Activated satellite cells undergo commitment to become myogenic progenitors, known as myoblasts, that are capable of undergoing rapid proliferation, propagation, and differentiation [115]. Myoblasts may fuse with pre-existing muscle fibers or fuse with each other to form new myotubes [115]. The activation and commitment of satellite cells to myogenesis are dependent and regulated by a hierarchy of transcription factors including PAX7 and the myogenic regulatory factors (MRFs), which include MYF5, MYOD, MRF4, and myogenin [57, 138]. Notably, *Myf5* is transcribed in quiescent satellite cells but *Myf5* transcripts are sequestered within mRNP granules along with their antagonist miRNA-31 [33]. Upon activation, mRNP granules dissociate and release *Myf5* mRNA, thus allowing its translation and the rapid accumulation of MYF5 protein to promote myogenesis [33]. *Myod* mRNA in quiescent satellite cells is regulated by the mRNA decay factor tristetraprolin which binds to the 3' UTR of the transcript, promoting its decay [55]. Upon activation, *Myod* transcripts are stabilized by the inactivation of tristetraprolin by the p38α/β MAP kinase [55]. Thus, unlike satellite stem cells, myoblast progenitors are characterized by their abundant expression of MYF5 and MYOD transcription factors [165]. MYF5 drives proliferation while MYOD drives early differentiation [130]. During myoblast differentiation, MYF5 and MYOD are downregulated, followed by enhanced expression of myogenin and MEF2, while MRF4 expression marks terminally differentiated and fused myotubes [165].

Recent studies indicate that the satellite cell population within the adult muscle exists as a heterogeneous population where some satellite cells are in a more committed state compared to others in a more stem cell-like state [150]. Upon examination of isolated satellite cells from *Myf5-nLacZ* mice it was found that 13% of quiescent satellite cells are β-Gal$^-$, and do not express *Myf5* in comparison to the majority of β-Gal$^+$/*Myf5*$^+$ satellite cells [77]. Upon activation, these *Myf5*$^-$ satellite cells undergo either symmetric expansion, wherein two identical *Myf5*$^-$ daughter cells are generated, or undergo asymmetric division, yielding one *Myf5*$^-$ satellite cell and one committed *Myf5*$^+$ satellite cell [77]. Satellite cell heterogeneity can also be characterized by differential PAX7 expression where satellite cells with high levels of PAX7 are less primed for differentiation and require more time for activation in comparison to satellite cells that express low levels of PAX7 [126]. Similarly, satellite cells vary in their expression of CD34, as quiescent satellite cells with low levels of CD34 exist in a more committed state compared to those with high levels of CD34 that exist in a more stem cell-like state [48].

Satellite Cells Contribute to Muscle Health

Muscle satellite cells are critical for both muscle homeostasis in the resting state as well as muscle regenerative capacity upon injury [165]. Disruptions in satellite cell functions are associated with impairments in the ability of the muscle to launch an effective regenerative response as observed in muscle wasting during aging and disease [14]. Age-related muscle deterioration, known as sarcopenia, is characterized by a decrease in muscle mass and strength and contributes significantly to a decrease in the quality of life and morbidity in the elderly [69]. During aging, muscle stem cells are numerically and functionally compromised while their niche also becomes less supportive [11]. Immunofluorescence staining against PAX7 in freshly isolated myofibers harvested from young and old mice indicated an age-associated decrease in satellite cell number [140]. Moreover, aged satellite cells break quiescence under homeostatic conditions, further depleting the muscle resident satellite cell pool [23]. In addition to aging, satellite cell function is altered in muscle degenerative diseases such as Duchenne muscular dystrophy (DMD), which is a progressive and fatal neuromuscular disease resulting from the loss of dystrophin [40, 166]. In contrast to aging, satellite cell numbers are elevated in dystrophic muscles, however, defective regulation of stem cell commitment and other cellular abnormalities contribute to the reduced regenerative capacity of dystrophic satellite cells [26, 75].

Maintenance of the satellite stem cell population is essential to sustain muscle homeostasis and tissue plasticity in response to movement, exercise, injury, stress, aging, and disease. One well known cellular pathway that contributes to the health and fitness of tissues and cells is autophagy. Autophagy is a catabolic program that is responsible for the degradation and recycling of cellular components in a lysosome-dependent manner [97]. Macroautophagy, one of the main types of autophagy, involves the sequestration of cytoplasmic constituents into double membraned structures known as autophagosomes [97]. Autophagosomes subsequently fuse with lysosomes, where their contents are broken down by lysosomal enzymes and the resulting macromolecules are released back to the cytoplasm for utilization [97]. In this chapter, we highlight the contribution of autophagy in muscle health and satellite stem cell function.

Autophagy in Skeletal Muscle

The daily voluntary movements of skeletal muscle place a high demand for energy production on the mitochondria of muscle cells [134]. Muscle contractions cause mechanical and metabolic alterations of proteins and organelles within muscle cells [134]. Additionally, this process results in an accumulation of reactive oxygen species (ROS), such as peroxidases, superoxides, and hydroxyl radicals [106, 118]. The presence of such ROS not only inhibits the phosphatidylinositol-3-kinase/Protein kinase B and the mammalian target of rapamycin (mTOR) signaling pathway, thereby suppressing protein synthesis, but may also result in the damage of cellular components [106, 118]. Thus, the skeletal muscle machinery is equipped with a waste removal mechanism to eliminate misfolded proteins and dysfunctional organelles [134]. Autophagy contributes to this quality control assurance in skeletal muscle through the sequestration of aberrant entities within autophagosomes that are subsequently delivered to the lysosome for degradation [134].

Basal Autophagy is Required to Maintain Muscle Mass

Autophagy occurs at basal levels in all eukaryotic cells [83]. With respect to skeletal muscle, autophagy is required to clear dysfunctional organelles and other forms of cellular waste, as their accumulation leads to the activation of catabolic pathways, resulting in muscle atrophy and consequential muscle weakness [94]. The role of autophagy in skeletal muscle has been examined through the deletion of autophagy genes in mouse models. As shown in mice harboring a muscle-specific deletion of *Atg7*, an essential autophagy gene, the loss of autophagy causes an accumulation of abnormally large mitochondria and a dilated sarcoplasmic reticulum [94]. Ultimately, this leads to an unfolded protein response, which suppresses protein synthesis alongside a simultaneous production of ROS from the dysfunctional mitochondria and, finally, cell death via apoptosis [94]. Phenotypically, disrupting autophagy in muscle manifests as a loss of muscle mass and muscle strength [94]. Similarly, in a model of attenuated autophagy via reduced expression of *Atg16l1*, a gene that is important for autophagosome biogenesis, muscle fibers of hypomorphic *Atg16l1* mice were smaller than their wild-type counterparts [109]. Moreover, the recovery and regeneration from a muscle injury is significantly slower in *Atg16l1* mice [109].

On the other hand, the loss of a negative autophagy regulator, which results in enhanced constitutive autophagy, also has detrimental effects on skeletal muscle health. Nutrient-deprivation autophagy factor-1 (NAF-1, also known as CISD2) is an endoplasmic reticulum (ER) BCL-2 interacting protein that promotes the ability of BCL-2 to antagonize Beclin 1-dependent autophagy [24, 25]. Of note, homozygous mutations of *NAF-1* cause Wolfram syndrome type 2, an autosomal recessive neurodegenerative disease [1]. The skeletal muscles of *Naf-1* null mice exhibit signs of degeneration, a dramatic reduction in force-generating capacity, dysregulated calcium flux, and elevated levels of autophagy [24, 29]. Furthermore, the mitochondria of *Naf-1* deficient muscle tissue and myoblasts are enlarged, suggesting an adaptative response to augmented autophagy [24, 51]. The brain and muscle tissues of *Naf-1* knockout mice exhibit mitochondrial breakdown and dysfunction as well as autophagic cell death [29]. The mitochondrial dysfunction in *Naf-1* null mice exacerbates with age and is accompanied by increased autophagy, which is characteristic of premature aging, thus implicating NAF-1 as a longevity factor [29].

Altogether these studies demonstrate that both the inhibition and augmentation of basal autophagy in skeletal muscle contribute to myofiber damage and may underlie certain muscle disorders [94]. Thus, maintaining a critical level of autophagy is essential for muscle homeostasis and health.

Role of Selective Autophagy in Muscle Homeostasis

In addition to general macroautophagy, selective autophagy has also been shown to play critical roles in maintaining muscle homeostasis. Chaperone-assisted selective autophagy (CASA) is a form of macroautophagy that mediates the specific degradation of ubiquitinated protein aggregates [70]. CASA involves encapsulation of protein aggregates within autophagosomes via chaperones and cochaperones that interact with the autophagy receptor p62/sequestosome 1 (SQSTM1) [70]. BAG cochaperone 3 (BAG3) and heat shock 70 kDa protein 8 (HSPA8/HSC70) chaperones induce CASA through direct binding of protein aggregates [70]. Mutations in BAG3 (Starvin in *Drosophila*), which colocalizes with the Z-disk marker α-actinin in adult fly muscle fibers, are associated with childhood muscle dystrophy [3]. BAG3, together with the dual-function co-chaperone/ubiquitin ligase CHIP, p62, HSC70 and the small heat shock protein HSPB8, promote the degradation of damaged Z-disk protein components including filamin to preserve muscle integrity [3]. Impairment of CASA-mediated proteostasis leads to Z-disk disintegration and progressive muscle weakness [3]. These findings demonstrate that CASA, which contributes to cellular proteostasis, is required for Z-disk maintenance in muscle [3].

Lipophagy is the selective autophagic degradation of lipid droplets (LDs), a eukaryotic intracellular lipid organelle responsible for the storage of triacyclglycerols, cholesterol esters, and retinyl esters [76]. The lipids stored within LDs are utilized for the synthesis of macromolecules, lipid membrane components such as phospholipids, and, most relevantly, for energy production [76]. Upon energetic demand, lipids can be accessed via lipolysis which is the degradation of lipids within LDs by cytosolic lipases [76]. Lipophagy is another method for the cell to access LD content using one or more cargo adaptors, such as p62 [76]. Lipid accumulation is characteristic of a variety of pathologies related to metabolic disorders [76]. Notably, lipid storage myopathy (LSM) is a group of clinically heterogenous diseases that are characterized by an accumulation of LDs in skeletal muscle, often adjacent to the mitochondria [2]. Lipophagy is critical in reducing this accumulation in LSM patients [2]. Interestingly, a major consequence of the treatment for insulin resistance, such as bariatric surgery, is the disappearance of LDs from skeletal muscle [81, 152, 168]. While the mechanism behind the reduction in LDs is unknown, it has been proposed that p62-mediated lipophagy is responsible for the breakdown of LDs [79].

Given the high energetic demands of skeletal muscle tissue, maintenance of mitochondrial homeostasis is critical for ATP biogenesis and muscle function [151]. Mitophagy, the selective autophagic degradation of mitochondria, helps to maintain the mitochondrial pool in muscle cells while releasing a relatively low amount of ROS [151]. Upon signs of mitochondrial dysfunction, such as high ROS production, loss of membrane potential, or respiratory impairment, mitochondria are targeted for degradation within the lysosome [151]. This degradative process is balanced with mitochondrial biogenesis, allowing for efficient organelle turnover that maintains the metabolic needs of muscle tissue [60, 151]. Thus, forms of selective autophagy, including CASA, lipophagy and mitophagy, contribute to muscle maintenance and health.

Autophagy in Muscle Regeneration and Exercise

Exercise training throughout life has numerous benefits for the body. During exercise, cellular remodeling of the muscle tissue is activated to meet the exercise-induced elevation in energy demands [153]. This remodeling process involves the synthesis of new organelles and proteins to replace cellular components that may have become oxidized and damaged during exercise [153]. Specifically, this includes the activation of both of the major cellular proteolytic programs: the ubiquitin proteasome and autophagy pathways [153]. Autophagy is triggered by metabolic stresses including nutrient insufficiency, oxidative stress, and calcium imbalance [98, 123, 153]. These stresses are caused inadvertently as by-products of exercise, which may explain why exercise triggers autophagy [9, 153]. A singular instance of endurance training can change protein turnover markers in a training level-dependent manner [133]. On a cellular level, an elevated production of ROS, higher NAD + levels, and a general increase in the AMP-to-ATP ratio activate AMP-activated protein kinase (AMPK) in response to exercise [18, 54, 153]. AMPK induces autophagy in part by activating ULK1, a key player within the autophagy induction complex that is responsible for autophagosome formation [72, 153].

Interestingly, recent work has shown that the induction of autophagy in response to exercise and other forms of stress may proceed in two distinct phases [117, 153]. The first phase involves a rapid increase in autophagic flux within minutes or hours of exposure to the stressor and is mediated by post-translational protein modifications of stress-responsive factors already present within the cell [117]. Contrastingly, the second phase is delayed and relies on transcription factors, such as p53, NF-kB, and STAT3, that activate transcriptional programs to synthesize stress-responsive factors that ensure long-term adaptation to stress [117].

Importantly, not only does physical exercise induce autophagy, but autophagy is essential to support muscle plasticity in response to exercise as it ensures exercise-induced metabolic responses and skeletal muscle adaptation to exercise training, allowing for an improvement in physical performance [85, 133]. BCL-2, which regulates Beclin 1-dependent autophagy, also controls autophagy induced by exercise [56, 111]. Mutant *Bcl2* mice that are deficient in autophagy activation exhibit impaired endurance and glucose metabolism during exercise [56]. These results suggest that the beneficial metabolic effects of exercise may be due in part to exercise-induced autophagy. Of note, nutritional availability is important as it has direct effects on autophagy, and thus exercise-induced autophagy is more evident when exercise is performed in a fasted state [133].

During acute exercise, the transcriptional coactivator peroxisome proliferator-activated receptor-γ coactivator-1α (PGC-1α) orchestrates mitochondrial biogenesis to aid skeletal muscle energetically [154]. PGC-1α, which regulates muscle oxidative capacity, also promotes exercise-induced autophagy [53]. Additionally, exercise increases the phosphorylation of the mitochondrial fission protein DRP1, indicating an increase in mitochondrial fission, which triggers mitophagy [63, 146]. Thus, mitochondrial biogenesis and muscle homeostasis are balanced by mitochondria and protein turnover during exercise. Disruptions in this equilibrium, which is evident in muscle disuse and aging, have negative impacts on cellular mitochondrial health [60, 73, 93, 151, 169].

Autophagy also plays an important role in the regeneration of skeletal muscle tissue that follows strenuous exercise or muscle injury, during which damage to proteins and organelles occur [17]. Autophagic flux increases during muscle regeneration [17]. This indicates that myotubes acquire a higher metabolic capacity as they differentiate and suggests that autophagy is required to mediate remodeling of the mitochondrial network during regeneration [17, 34, 163]. After injury, two phases of mitochondrial remodeling occur: the first phase consists of mitochondrial degradation, where mitochondria appear fragmented and there is a clear absence of mitochondria around contractile units of the muscle, while the second phase involves reorganization of the mitochondrial network, where the network becomes reinstated despite the total mitochondrial content still being lower than the pre-injury state [17]. As shown in *Ulk1* knockout mice, *Ulk1* is not only required for mitophagy but also for mitochondrial remodeling during the maturation phase of regeneration as *Ulk1*-deficient mice exhibited delayed mitochondrial remodeling after induced injury [17, 78]. Despite the importance of autophagy in muscle regeneration, it is also possible that autophagy alone does not adequately clear the abundance of damaged proteins and organelles which result from muscle damage [16]. This creates an autophagy "bottleneck" wherein autophagosomes accumulate in injured skeletal muscle cells [16]. This effect is not exclusive to skeletal muscle, as it has also been reported in cardiomyocytes [89]. The accumulation of autophagosomes within the tissue have been proposed to have a negative effect on satellite cells and their regenerative potential and could contribute to muscle pathologies [16].

Autophagy in Muscle Health

Autophagy is under the tight regulation of many signaling pathways in skeletal muscle, where it is essential for both energy production and consumption, as well as for the clearance of waste products and the turnover of macromolecules [162]. Autophagy is critical for maintaining skeletal muscle integrity under physiological and stress conditions, as a basal level of autophagy is necessary for muscle homeostasis [162]. However, both deficient and excessive autophagy disturb this homeostasis which may contribute to cell damage, muscle weakness, and muscle atrophy [162]. Moreover, mutations in genes that mediate autophagy underlie several muscle pathologies [162]. Thus, autophagy is necessary for muscle health.

The loss of skeletal muscle throughout the course of aging, known as sarcopenia, is well-documented and inevitable [19, 110]. Initially, sarcopenia was thought to result from a general decline in the synthesis of proteins alongside an enhancement of protein degradation [19]. Sarcopenia manifests phenotypically as muscle fiber atrophy and degeneration, muscle weakness, dysfunctional mitochondria, and increased oxidative stress [19, 21]. For this reason, the elderly population experiences a lowered quality of life as they are pre-disposed to an increased risk of morbidity, disability, and mortality [19, 156]. Interestingly, this degeneration is not attributable to the loss of motor neurons in the brain or spinal cord [19, 22, 100]. Rather, during aging, the interaction between neuromuscular junctions (NMJs) and myofibers is altered, ultimately leading to a loss of muscle innervation [19, 22, 155]. Studies in aged mice show that aged NMJs exhibit axonal swelling, sprouting, synaptic detachment, withdrawal of axons from postsynaptic sites, and fragmentation of the postsynaptic specialization [155]. Furthermore, autophagy also declines during aging [19]. As observed in *Atg7*$^{-/-}$ autophagy-deficient mice, aged mice show higher levels of muscle atrophy, centrally-nucleated fibers, and inflammation when autophagy is inhibited [19]. Remarkably, the NMJs of *Atg7* knockout mice are more fragmented and unstable than age-matched control mice [19]. *Atg7* knockout mice also exhibit elevated levels of mitochondrial dysfunction and, consequentially, oxidative stress [19]. Therefore, autophagy, which declines with age, is required for proper muscle and nerve function, and maintenance of the integrity of NMJs [19].

Mitofusin 2 (MFN2), alongside mitofusin 1 (MFN1), are proteins located at the outer mitochondrial membrane that are important mediators of mitochondrial dynamics (i.e., fusion and fission), mitochondrial network architecture, and mitochondrial metabolism [4, 139]. MFN2 also regulates autophagy, mitophagy, the unfolded protein response, oxidative metabolism, and general cell proliferation [28, 52, 103, 107, 139]. During aging, MFN2 protein expression decreases [139]. Further, *Mfn2* deficiency in young mice impairs autophagy and reduces mitochondrial quality, leading to an aggravated state of premature sarcopenia and metabolic deficiency [139]. Similarly, muscle-specific loss of AMPK, which normally activates both autophagy and mitophagy, resulted in exacerbated age-related myopathy and mitochondrial dysfunction [15]. These studies suggest that inducing autophagy via the activation of AMPK may act to prevent mitochondrial disease, hypoglycemia, and myopathy during aging [15].

Overall, there is a clear contribution between impaired autophagy and the decline of skeletal muscle mass and function during aging. For this reason, it is not surprising that re-establishing autophagy in muscles is a potential treatment for sarcopenia both by pharmacological and exercise-induced modulation [68, 110, 167]. A partial inhibition of mTOR complex 1 (mTORC1), using rapamycin-related drugs known as rapalogs, counteracts sarcopenia in rats, as evidenced by an increase in muscle mass and fiber cross-sectional area [68]. Moreover, treatment with rapalogs led to enhanced levels of autophagy and the downregulation of senescence markers $p16^{INK4a}$ and $p21^{CIP1}$ [68]. Further, exercise-induced autophagy, such as through treadmill and resistance exercise, suppressed muscle mass loss, and enhanced mitochondrial function and AMPK phosphorylation to better modulate autophagic flux [167].

Another example that illustrates the importance of a regulated autophagy program in muscle health is Pompe's disease. Pompe's disease is a highly heterogeneous and devastating disorder caused by deficiencies in the *GAA* gene which encodes for alpha-glucosidase, a lysosomal enzyme that breaks down glycogen to glucose [122]. Pompe's disease manifests in both infants and adults, the former being the most severe form with symptoms including cardiomegaly, hypotonia, and death by cardiorespiratory failure within the first year of life [74, 122]. Pompe's disease also affects the autophagic pathway [122]. Muscle fibers isolated from the knockout mouse model of Pompe's disease exhibit large amounts of autophagic accumulation, especially in type II fibers [122]. Enzyme replacement therapy is now available to Pompe's disease patients [122]. It utilizes a recombinant human GAA and has allowed infantile patients to survive significantly longer than untreated patients and has improved cardiac function [122]. However, only a small percentage of clinical trial patients saw improvements in mortality and skeletal muscle function [122]. Moreover, Pompe's disease type II fibers are more therapeutically resistant than their wild-type counterparts [120, 122]. Interestingly, and conversely to sarcopenia, muscle-specific genetic suppression of autophagy in Pompe's mice mitigates the enhanced presence of autophagosomes that is characteristic of Pompe's muscle fibers [121].

To conclude, both the excess of autophagy, as is the case in Pompe's disease, and impaired autophagy, which causes a gradual decrease in skeletal muscle mass during aging, result in muscle pathology. Thus, maintaining a balanced and homeostatic control of autophagy is crucial for skeletal muscle health.

Autophagy in Muscle Stem Cells

Muscle satellite cells are key players in maintaining muscle tissue homeostasis and facilitating regeneration [124]. These tissue resident stem cells are kept in a quiescent state until stimulated by stress or damage to activate, enter the cell cycle, and either expand, differentiate, or self-renew [31, 32, 99, 165]. Satellite cells maintain their quiescence through cytoprotective and cellular quality control mechanisms that inhibit irreversible withdrawal from the cell cycle; these mechanisms are lost in advanced age, during which satellite cells switch to a senescence-like state [145]. As a result, the number and function of satellite cells decline with age and the regenerative capacity of the skeletal muscle is profoundly compromised [13, 62, 119].

Homeostatic Maintenance of Muscle Stem Cells

Satellite cell quiescence is preserved via active regulation of organelle and protein homeostasis, implicating basal autophagy as a cellular quality control mechanism in satellite stem cells [47]. By enabling the recycling of macromolecules to provide energy-rich metabolites as well as eliminate damaged proteins, organelles, and toxic compounds, autophagy allows cells and tissues to continually adapt to stress [67, 92]. Thus, the contribution of autophagy in satellite cells parallels the importance of autophagy in muscle tissue, which is essential for maintenance of muscle mass and integrity [134]. As autophagic activity declines with age or due to genetic impairment, toxic cellular waste accumulates and satellite cell functions are disturbed, accompanied by perturbations in mitochondrial function and ATP production, as well as entry into senescence [47, 147]. As with muscle tissue, a critical balance in the levels of basal autophagy is essential for satellite stem cell integrity, as excessive autophagy can also lead to stem cell defects [38].

Transcriptomic analysis of quiescent satellite cells compared to activated satellite cells uncovered that autophagy is the most prevalent pathway during quiescence and autophagic genes are downregulated in association with aging [47]. Aged satellite cells display common traits of deficient autophagy, including formation of p62 aggregates, accumulation of autophagic vesicles, ubiquitin-positive inclusions, and reduced accumulation of LC3-II upon treatment with bafilomycin A1 (an autophagy-flux inhibitor which prevents lysosomal degradation), which altogether indicate reduced capacity for autophagosome formation [47]. The block in autophagic flux increases progressively with age in mice from young (3 months) to old (20–24 months) to geriatric (over 28 months) [47]. Geriatric satellite cells exhibited increased co-localization of p62 and ubiquitin aggregates in non-degraded autophagosomes, thus indicating a block in autophagosome clearance [47].

Genetic impairment of autophagy in young mice by specific deletion of *Atg7* in quiescent satellite cells severely reduced satellite cell numbers, while the remaining satellite cells exhibited signs of premature aging such as induction of senescence genes $p16^{INK4a}$, $p21^{CIP1}$, and $p15^{INK4b}$, as well as evidence of DNA damage [47]. Young *Atg7*-deficient satellite cells and aged satellite cells shared similar phenotypes, including accumulation of mitochondria, lysosomes, and p62 and ubiquitin-positive aggregates, and a lower proportion of healthy mitochondria [47]. These results indicate that basal autophagy is required for maintaining satellite cell integrity and fitness and to preserve the pool of quiescent satellite cells [47]. Following muscle injury, satellite cells from these mice displayed reduced activation and proliferation capacity, evidence of cell-intrinsic regenerative failure, as well as accelerated entry into senescence [47]. Thus, the decline in autophagy in satellite cells may underlie the physical loss and functional exhaustion of muscle satellite cells associated with aging. These findings also implicate defective autophagy as a cause of senescence, rather than a consequence arising from senescence.

In addition, enhanced levels of ROS and mitochondrial dysfunction are a common phenotype observed in both geriatric satellite cells as well as *Atg7*-deficient satellite cells [47]. Notably, ROS inhibition in geriatric satellite cells with Trolox, a vitamin E analogue, prevented the induction of senescence markers and induced autophagic flux, resulting in a reduction of ubiquitin and p62 aggregates as well as mitochondria-ROS colocalization [47]. Moreover, treatment with Trolox restored satellite cell expansion, and rescued satellite cell defects in proliferation and regenerative capacity [47].

The reinstatement of basal autophagy using either genetic or pharmacological approaches in geriatric mice reversed senescence in satellite cells and rescued regenerative function [47]. Reactivation of autophagy with the mTOR inhibitor rapamycin or ectopic expression of *Atg7* in geriatric satellite cells prior to transplantation into pre-injured muscles of young recipient mice restored their capacity for expansion and engraftment and prevented senescence [47]. Of note, similar defects in protein and organelle clearance were observed in aged human satellite cells, alongside increased ROS levels and markers of senescence [47]. Restoration of autophagy and organelle homeostasis in aged human satellite cells with rapamycin was able to rescue cells from entering senescence, prevent abnormal mitochondrial content and ROS levels, as well as protein aggregation [47]. Thus, inducing autophagy may serve as a therapeutic avenue to improve aged satellite cell function and to prevent age-related regenerative decline in muscle.

Autophagy in Satellite Cells is Under Circadian Regulation

The day–night oscillation of genes that maintain tissue homeostasis have been observed in numerous tissues, suggesting that adult stem cells are subject to circadian control [64]. Circadian rhythms segregate cellular functions throughout the 24-h day to minimize potential exposure to harmful situations and maximize cellular performance and energetic efficiency [64]. Gene ontology analysis of whole transcriptome gene expression data from satellite cells revealed that adult quiescent satellite cells expressed many transcripts required for homeostasis in an oscillatory manner, such as myotube differentiation and cell proliferation [143]. This "rhythmic transcriptome" encompasses genes within the transforming growth factor-beta/bone morphogenetic protein and fibroblast growth factor signaling pathways, which regulate maintenance of satellite cell quiescence and readiness for activation [143]. Additionally, transcripts involved in DNA double-strand break repair, including *Rad23a*, *Ercc4*, and *Xpa*, were rhythmically expressed, consistent with previous findings indicating that quiescent satellite cells compared to differentiated muscle cells, are more predisposed to repairing this type of damage [37, 143]. Intriguingly during aging, the oscillatory transcriptome of satellite cells is dramatically reprogrammed [143]. Aged satellite cells exhibited a distinct program of oscillatory genes, including those involved in mitochondrial DNA repair, cytokine production, and inflammation [143].

Interestingly, Solanas et al. found that the expression of key autophagy-related genes, including *Becn1*, *Flcn*, *Atg13*, and *Svip*, was under rhythmic control in adult satellite cells [143]. The expression of these genes peaked late at night or early in the morning, resulting in higher levels of autophagic activity during the day [143]. In comparison, autophagy genes in aged satellite cells were not under circadian control and consequently, autophagy levels were significantly reduced throughout the day [143]. These findings suggest that aged satellite cells lose their capacity to rhythmically recycle damaged cellular components that are produced in the cell. Thus, the age-associated loss of rhythmic regulation of autophagy leads to an overall decline in autophagy, ultimately impairing the cell's intracellular quality control mechanism to sustain organelle and protein homeostasis, maintain quiescence, and preserve stemness [47, 143].

Quiescent Satellite Cells Exist in Two Distinct Metabolic States

Molecules and pathways responsible for regulating cellular energy status and metabolism, such as the nutrient sensing mTOR pathway, have been shown to influence different aspects of stem cell function, including pluripotency, differentiation, proliferation, and self-renewal [27, 41, 104, 132]. The downstream targets of these metabolic pathways include those relevant to the autophagic process and can act to induce autophagy during conditions of increased energetic demand and stress.

Satellite stem cells within the quiescent state can exist in two functional phases; G_0 and G_{Alert} [127]. Cells in G_{Alert} are considered to be in an intermediate "alert" phase while still in quiescence [127]. Metabolically, these cells have higher mitochondrial activity, larger cellular volumes, enhanced differentiation kinetics, and a higher propensity to cycle than those in G_0 [127]. The mTOR pathway, which inhibits autophagy, has been shown to be required for the transition of satellite cells from G_0

to G_{Alert} [127]. Thus, autophagy in this context may contribute to a "deeper" quiescent satellite cell, while the transition to G_{alert} requires active mTOR and concomitant inhibition of autophagy.

Contribution of Autophagy in Satellite Cell Activation

As stem cells activate and exit quiescence, different bioenergetic requirements exist during the differentiation process [41]. Quiescent stem cells initiate the activation process from a position characterized by low mitochondrial content and activity, metabolism, and translation rates. Thus, during activation, stem cells need to fulfill a high demand for energy and nutrients in order to support cellular growth [88]. An increase in ATP production is associated with progression through G_1 of the cell cycle as it is necessary to fuel DNA replication and cellular growth processes [41, 88]. During muscle regeneration, Fiacco and colleagues reported that autophagy is induced upon satellite cell activation following injury and that autophagy levels return to baseline by the end of the regeneration process [39]. Pharmacological induction of autophagy led to enhanced satellite cell activation and proliferation, while inhibiting autophagy resulted in impaired satellite cell regenerative capacity [39].

Another study by Tang and Rando found that autophagy in satellite cells was induced early during the activation process from quiescence [147]. Autophagy levels remained elevated during satellite cell proliferation and were subsequently reduced during self-renewal [147]. In contrast to the study conducted by Garcia-Prat and colleagues, Tang and Rando did not detect basal autophagic flux in quiescent satellite cells [47, 147]. Autophagy was detected in more than half of activated satellite cells 1.5 days following muscle injury, which increased to over 80% of activated satellite cells by 2.5 days post-injury [147]. These results illustrate the increase in autophagic flux required during satellite cell activation upon muscle injury in vivo [147]. Examination of single myofiber-associated satellite cells ex vivo revealed that autophagy was induced early in activated satellite cells prior to the initiation of DNA synthesis [147].

The inhibition of autophagy through chemical inhibition with either chloroquine, 3-methyladenine, or siRNA-mediated knockdown of essential autophagy genes *Atg5* and *Atg7*, resulted in delayed satellite cell activation [147]. Moreover, the ATP content was greatly reduced following autophagy inhibition, suggesting that autophagy contributes bioenergetic resources to facilitate satellite cell activation [147]. The delay in satellite cell activation was partially rescued upon supplementation with exogenous sodium pyruvate [147]. Sirtuin-1 (SIRT1), an NAD-dependent deacetylase that responds to changes in cellular metabolism, was required for autophagy induction during satellite cell activation [147]. In satellite cells, SIRT1 interacts with and deacetylates ATG7 [147]. Chemical inhibition of SIRT1 reduced autophagic flux and satellite cell-specific deletion of *Sirt1* resulted in delayed satellite cell activation [147]. Therefore, autophagic flux is upregulated during satellite cell activation to meet the necessary metabolic requirements for satellite cells to proceed through the myogenic program.

Autophagy Prevents Apoptosis in Satellite Cells

Activated cells whose energetic needs are not met by autophagy become susceptible to apoptosis, a form of programmed cell death [65]. During cell fate decisions, young satellite cells induce autophagy over apoptosis, whereas aged satellite cells are more likely to induce apoptosis [65]. Indeed, autophagy and apoptosis have opposing correlations related with satellite cell aging. Unlike autophagy that decreases with age, apoptosis is increased across the lifespan [137]. When compared with young and middle-aged satellite cells, old and geriatric satellite cells displayed enhanced levels of cleaved poly(ADP-ribose) polymerase (PARP), a marker of apoptosis, and were progressively positive for TUNEL and annexin V labelling [160]. Moreover, young and aged satellite cells exhibited differential susceptibility to apoptosis in the absence of autophagy [160]. Upon inhibition of autophagy mediated by *Atg5* knockdown, apoptosis was not induced in young satellite cells, but was increased two-fold in geriatric satellite cells resulting in enhanced cell death [160]. Apoptosis induced by the suppression of autophagy occurred via the canonical apoptotic pathway, as a pan-caspase inhibitor was able to prevent cell death and *Bcl2* overexpression reduced the onset of apoptosis [160]. Moreover, in young satellite cells, *Atg5* depletion delayed cell proliferation, while geriatric satellite cells failed to proliferate [160].

AMPK regulates autophagy and apoptosis in part through its ability to phosphorylate the cyclin inhibitor $p27^{Kip1}$ and has been shown to play an important role in satellite cell-mediated muscle regeneration [44, 84, 149]. $p27^{Kip1}$ can prevent apoptosis by inhibiting the activation of Cdk2 and the activity of the pro-apoptotic factor BAX [50, 58]. AMPK-dependent phosphorylation of $p27^{Kip1}$ at Thr198 promotes its stability and cytoplasmic translocation, leading to increased autophagy and decreased apoptosis [84, 160]. Both AMPK and $p27^{Kip1}$ phosphorylation were reduced in old mice and to a greater extent in geriatric mice [160]. Restoration of AMPK activity using the AMP analog, 5-aminoimidazole-4-carboxamide ribonucleotide (AICAR) in geriatric satellite cells prevented cell death [160]. Compared to control young satellite cells, which exhibited high transplantation efficiency, satellite cells in which AMPK activity was genetically suppressed prior to transplantation, showed a significant decline in engraftment [160]. In contrast, geriatric satellite cells, which displayed inherently poor transplantation efficiency compared to young cells, exhibited improved engraftment upon constitutive activation of AMPK with AICAR [160]. AICAR treatment of geriatric satellite cells led to reduced levels of senescence and expression of the senescence genes $p16^{INK4a}$ and $p21^{CIP1}$ [160]. Thus, the AMPK/$p27^{Kip1}$ signaling axis controls the autophagy/apoptosis balance in satellite stem cells, and activating this pathway may improve aged satellite cell function and muscle regeneration.

Satellite Cell Function and Regenerative Capacity

Calorie restriction is well known to have a positive effect on the lifespan and prevents age-related deterioration [61]. Calorie restriction is also a potent inducer of autophagy and is a non-genetic and non-chemical method used to stimulate autophagy in animal models [6]. Intriguingly, short-term calorie restriction in both young and old mice enhanced satellite cell numbers and improved muscle stem cell function [20]. Satellite cells from calorie restricted mice exhibited enhanced mitochondrial content, suggesting a switch to fatty acid oxidation and oxidative phosphorylation for energy production [20]. Moreover, satellite cells from calorie restricted mice exhibited increased expression of SIRT1 and FOXO3, both of which mediate autophagy [20]. Importantly, mice maintained on a calorie restricted diet displayed enhanced muscle repair in response to injury and improved satellite cell transplantation and engraftment efficiency [20].

Metformin is a drug that mimics calorie restriction and is used for the treatment of type 2 diabetes [159]. Metformin has been shown to activate autophagy through its ability to activate AMPK and inhibit mTOR signaling [141]. Treatment of immortalized C2C12 mouse myoblasts with metformin prevented terminal differentiation and permanent exit from the cell cycle [113]. In satellite cells, treatment with metformin resulted in a delay in satellite cell activation in association with delayed downregulation of PAX7 and differentiation [112]. Thus, metformin retains satellite cells in a more stem-like and pre-differentiation state. Upon muscle injury, metformin treatment delayed activation from quiescence [112]. This effect was attributed to a reduction in phosphorylation of ribosomal S6 kinase, a downstream target and readout of mTOR signaling, indicating an inhibition in mTOR activity and protein synthesis [112]. Altogether, these studies indicate that factors mediating metabolism and autophagy play an important role in stem cell function and regenerative capacity.

The Role of mTOR Signaling in Myogenesis

Upon the presence of an external stimuli, quiescent satellite cells undergo chronological stages of myogenesis; they re-enter the cell cycle and give rise to proliferative myoblasts, which subsequently differentiate into myocytes, and fuse to form myofibers [10]. Adult myogenesis, which relies on PAX7-positive satellite cells, recapitulates many mechanisms present during embryonic muscle development [35, 125]. Both embryonic and adult myogenesis are highly dependent on mTORC1, an established regulator of cellular growth [125]. Inactivation of mTORC1 via genetic deletion of *Raptor* (a component of mTORC1) in mouse embryonic muscle progenitors impaired muscle development and resulted in perinatal lethality [125]. *Raptor*-deficient embryos exhibited a 50% reduction in *Myf5* expression, indicating that the inactivation of mTORC1 directly influences the early stages of myogenesis [125]. In adult myogenesis, mTORC1 is activated in satellite cells following muscle injury and remains high during the proliferative phase of myogenesis [125]. In contrast, myocytes and myotubes do not exhibit phosphorylation of ribosomal S6 kinase, indicating low mTORC1 activity during late myogenesis [125]. Of note, mTORC2 appears to be dispensable for myogenesis [125]. Thus, early stages of embryonic and adult myogenesis are highly dependent on mTORC1 and inhibiting mTORC1 is detrimental to myogenesis [125].

Myotonic dystrophy type 1 (DM1) is a neuromuscular disease caused by mutated transcripts of the myotonic dystrophy protein kinase harboring expanded CTG repeats that results in the formation of nuclear RNA foci and disturb RNA-binding proteins [46, 80, 144]. DM1 manifests phenotypically as muscle atrophy and decreased skeletal muscle regeneration [46, 80, 144]. Moreover, DM1 satellite cells exhibit reduced proliferative capacity and enhanced levels of autophagy, which are thought to contribute to the muscle regeneration defect [144]. Intriguingly, satellite cells differentiated from induced pluripotent stem cells obtained from DM1 patients that were subsequently edited by transcription activator-like effector nucleases (TALENs) to target the CTG repeats showed reduced levels of autophagy, increased levels of phosphorylated mTOR, and enhanced cell proliferation rates [144]. Accordingly, this rescue in cell proliferation in DM1 TALEN-edited satellite cells was abrogated upon treatment with the mTOR inhibitor rapamycin [144]. Thus, the proliferation defect in DM1 satellite cells was rescued via the activation of mTOR and inhibition of autophagy [8, 144].

Coordination of Myogenic Differentiation and p53-Dependent Autophagy

In addition to the temporal coordination of myogenic proteins during differentiation, autophagy during myogenesis is also regulated in a differentiation stage-dependent manner [10]. While autophagy levels are reduced during myoblast proliferation, autophagy is required during the later stages of myogenic differentiation to protect myoblasts from apoptosis during differentiation [96]. The onset of autophagy during differentiation is mediated by an initial increase in apoptosis, which induces the activation of autophagy to subsequently prevent excessive apoptosis [66].

Myocytes, which are fully differentiated myoblast-derived cells that have not yet undergone fusion, exhibit decreased autophagic flux [42]. In contrast, autophagy is required for myocyte fusion. This is evidenced by an increase in the transcript levels of autophagy-related genes and positive immunofluorescent staining for autophagosome and lysosome proteins during fusion [42]. Moreover, when autophagy is genetically inactivated by siRNA silencing of Beclin 1 in vitro, the nuclear fusion index of myotubes, a quantifiable readout for myocyte fusion and thus differentiation efficiency, is reduced by 1.7-fold [42]. Accordingly, myotubes, but not myocytes, exhibit accumulated levels of autophagic LC3-II in the presence of lysosomal inhibitors [42]. Moreover, chemical inhibition of autophagy during differentiation resulted in reduced expression of myosin heavy chain and myogenin, which are markers of terminal differentiation, and delayed formation of myotubes [96].

p53 is a multifaceted protein which can shuffle between both the nucleus and cytoplasm to achieve different outcomes [101, 116]. In addition to its role as a tumor suppressor gene, it also acts as an activator or inhibitor of autophagy, dependent on its subcellular localization [91, 101, 148]. When p53 is localized to the nucleus, it activates autophagy induced by exogenous stress, leading to either a pro-death or pro-survival outcome [91, 148]. In contrast, cytoplasmic p53 inhibits autophagy induced by ER stress or nutrient deprivation [91, 148]. Myoblasts derived from *p53* null mice exhibited a reduction in basal autophagy and impaired ability to terminally differentiate into myotubes [42]. Moreover, mitochondrial biogenesis, which occurs during myogenic differentiation, is impaired in the absence of p53 [42]. Thus, p53-mediated autophagy is required for metabolic remodeling and myoblast fusion during differentiation [42]. Interestingly, p53 also imposes a quality control mechanism during differentiation and is activated upon genotoxic stress [164]. p53 binds directly to the myogenin promoter to repress myogenin expression and delay differentiation [164]. This ultimately protects terminally differentiated muscle cells from post-mitotic nuclear abnormalities [164]. Thus, p53-mediated autophagy and differentiation contribute to proper myotube fusion.

Mitophagy is Required for Mitochondrial Biogenesis and Myogenesis

Mitochondrial dynamics and mitophagy also play critical roles in regulating satellite cell quiescence and activation [7, 59]. Satellite cell activation is mediated through mitochondrial fragmentation, a process that is negatively regulated by the mitochondrial fusion protein optic atrophy 1 (OPA1) [7]. Satellite cell specific deletion of OPA1 drives quiescent satellite cells into a G_{alert} state [7]. On the other hand, satellite cell deletion of the mitochondrial fission protein dynamin-related protein 1 (DRP1) prevents satellite cell expansion and inhibits muscle regeneration [59]. Thus, mitochondrial dynamics play an integral role in controlling satellite stem cell fate. During myogenic differentiation, as myoblasts differentiate into myotubes, the mitochondrial network is also altered to adapt to different metabolic needs. Mitochondria are remodeled during differentiation, including alterations in their abundance, morphology, and functional properties [158]. Additionally, there is a switch from glycolysis, which serves as the main energy source for myoblasts, to oxidative phosphorylation at the terminal differentiation stage [158]. Thus, remodeling of the mitochondrial network, which includes their degradation and biogenesis, is under tight regulation to ensure the proper advancement of differentiation [158]. During the early stages of myoblast differentiation, an increase in mitochondrial fragmentation and removal of mitochondria via p62/SQSTM1-mediated mitophagy were observed [142]. Mitochondria biogenesis is subsequently upregulated via a PGC-1α-dependent pathway, resulting in a myotube freshly populated with new mitochondria that are better primed for oxidative phosphorylation [142]. Additionally, CRISPR-Cas9-mediated deletion of *Bnip3*, a member of the BCL-2 protein family that is upregulated during mitophagy and which modulates mitochondrial membrane permeability, impairs myoblast differentiation [5].

The E3 ubiquitin ligase Parkin is recruited to mitochondria upon the loss of mitochondrial membrane potential to initiate mitophagy in several cell types [108]. Interestingly, inducing Parkin-mediated mitophagy by uncoupling mitochondria during in vitro myogenesis resulted in excessive mitophagy and myotube atrophy [114]. Loss of Parkin function via siRNA-mediated knockdown resulted in impaired mitochondrial turnover as well as myotube atrophy [114]. Similarly, Parkin knockout mice exhibit enhanced fibrosis and decreased myofiber cross-sectional area following cardiotoxin injury, suggesting an impairment in regeneration [36]. In the absence of Parkin, satellite cells exhibit reduced differentiation capacity and increased proliferation [36]. Altogether, these studies indicate that mitophagy is an important contributor to the myogenic program to support the differentiation process and meet the metabolic needs of the tissue.

Conclusion

Autophagy is clearly important for muscle health and contributes to the maintenance of the integrity and plasticity of the muscle tissue as well as the regenerative capacity and fitness of muscle stem cells. Constitutive basal levels of autophagy are important for the homeostatic maintenance of cells and tissues, while induced autophagy mediates cellular response to stress and enhanced metabolic requirements. Impaired autophagy has detrimental effects on muscle health and is the underlying cause of age-related muscle decline and various myopathies including muscular dystrophies. Importantly, muscle stem cells, which contribute to both muscle homeostasis and muscle regenerative capacity throughout life, are also dependent on a dynamic autophagy program (Fig. 6.1). An interplay of metabolic sensing pathways that include regulators such as mTOR and AMPK, thus control autophagy to ensure that the level of autophagy is finely tuned to the metabolic needs of the satellite cell as it transitions through the myogenic differentiation process (Fig. 6.2). Ultimately, the ability of satellite cells to contribute to muscle repair and sustain future rounds of regeneration is an important determinant of muscle health.

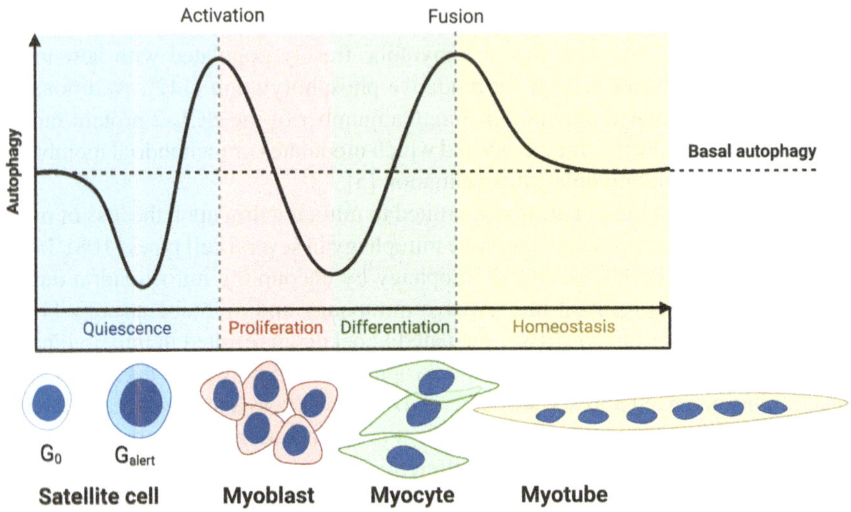

Fig. 6.1 Dynamic contribution of autophagy during myogenic differentiation. Basal autophagy is required to maintain satellite stem cell homeostasis, preserve stemness, and prevent the accumulation of factors that can lead to senescence. Quiescent satellite cells in G_{alert} are metabolically distinct from G_0 satellite cells, and are dependent on mTORC1, which inhibits autophagy. Following a regeneration stimulus, autophagy is upregulated in satellite cells to provide sufficient energy to facilitate the transition from quiescence to activation. Myoblasts exhibit low autophagic flux, as cells prioritize cell proliferation and growth to expand the progenitor population. Myocyte fusion requires autophagy for remodeling of the mitochondria network. Finally, mature muscle cells maintain a basal level of autophagy to ensure homeostatic maintenance of muscle mass and health

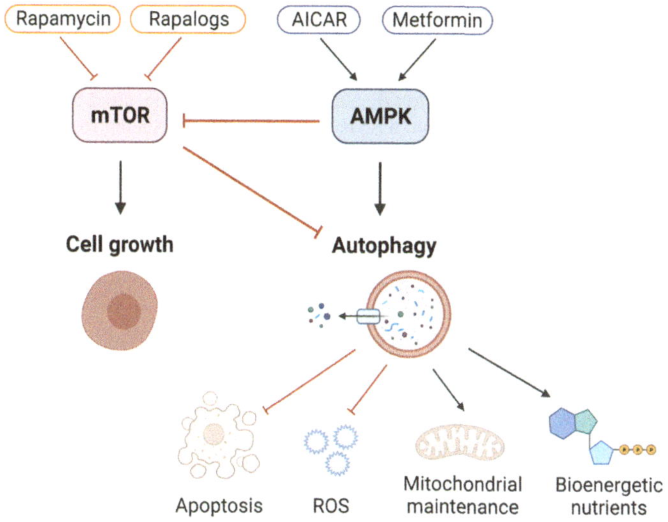

Fig. 6.2 Nutrient sensing pathways regulate autophagy. Autophagy is regulated by mTOR and AMPK protein kinase complexes. mTOR inhibits autophagy and promotes cell growth, while AMPK inhibits mTOR and induces autophagy. Compounds that modulate autophagy include rapamycin and rapologs that inhibit mTOR, as well as AICAR and metformin that activate AMPK. In muscle cells, autophagy inhibits apoptosis and prevents the accumulation of reactive oxygen species (ROS). Autophagy also maintains the mitochondrial network and provides metabolic bioenergetic nutrients

Acknowledgements Research in the Chang laboratory is supported by research grants from the Stem Cell Network, Defeat Duchenne Canada, Fonds de recherche du Québec—Santé (FRQS), and the Canadian Institutes of Health Research. SL is supported by a studentship from the Rosalind & Morris Goodman Cancer Institute. RLF is supported by a Master's training award from FRQS. NCC is supported by a Chercheurs-boursiers—Junior 1 award from FRQS. Figures were created with BioRender.com.

Compliance with Ethical Standards

Disclosure of Interests All authors declare they have no conflict of interest.

References

1. Amr S et al (2007) A homozygous mutation in a novel zinc-finger protein, ERIS, is responsible for Wolfram syndrome 2. Am J Hum Genet 81(4):673–683. https://doi.org/10.1086/520961
2. Angelini C, Nascimbeni AC, Cenacchi G, Tasca E (2016) Lipolysis and lipophagy in lipid storage myopathies. Biochim Biophys Acta 1862(7):1367–1373. https://doi.org/10.1016/j.bbadis.2016.04.008
3. Arndt V et al (2010) Chaperone-assisted selective autophagy is essential for muscle maintenance. Curr Biol 20(2):143–148. https://doi.org/10.1016/j.cub.2009.11.022
4. Bach D et al (2003) Mitofusin-2 determines mitochondrial network architecture and mitochondrial metabolism. A novel regulatory mechanism altered in obesity. J Biol Chem 278(19):17190–17197. https://doi.org/10.1074/jbc.M212754200

5. Baechler BL, Bloemberg D, Quadrilatero J (2019) Mitophagy regulates mitochondrial network signaling, oxidative stress, and apoptosis during myoblast differentiation. Autophagy 15(9):1606–1619. https://doi.org/10.1080/15548627.2019.1591672
6. Bagherniya M, Butler AE, Barreto GE, Sahebkar A (2018) The effect of fasting or calorie restriction on autophagy induction: a review of the literature. Ageing Res Rev 47:183–197. https://doi.org/10.1016/j.arr.2018.08.004
7. Baker N et al (2022) The mitochondrial protein OPA1 regulates the quiescent state of adult muscle stem cells. Cell Stem Cell 29(9):1315–1332.e9. https://doi.org/10.1016/j.stem.2022.07.010
8. Bargiela A, Cerro-Herreros E, Fernandez-Costa JM, Vilchez JJ, Llamusi B, Artero R (2015) Increased autophagy and apoptosis contribute to muscle atrophy in a myotonic dystrophy type 1 Drosophila model. Dis Model Mech 8(7):679–690. https://doi.org/10.1242/dmm.018127
9. Barry DW, Hansen KC, van Pelt RE, Witten M, Wolfe P, Kohrt WM (2011) Acute calcium ingestion attenuates exercise-induced disruption of calcium homeostasis. Med Sci Sports Exerc 43(4):617–623. https://doi.org/10.1249/MSS.0b013e3181f79fa8
10. Bentzinger CF, Wang YX, Rudnicki MA (2012) Building muscle: molecular regulation of myogenesis. Cold Spring Harb Perspect Biol 4(2). https://doi.org/10.1101/cshperspect.a008342
11. Blau HM, Cosgrove BD, Ho ATV (2015) The central role of muscle stem cells in regenerative failure with aging. Nat Med 21(8):854–862. https://doi.org/10.1038/nm.3918
12. Bogliotti YS, Ross PJ (2012) Mechanisms of histone H3 lysine 27 trimethylation remodeling during early mammalian development. Epigenetics 7(9):976–981. https://doi.org/10.4161/epi.21615
13. Brack AS, Bildsoe H, Hughes SM (2005) Evidence that satellite cell decrement contributes to preferential decline in nuclear number from large fibres during murine age-related muscle atrophy. J Cell Sci 118(Pt 20):4813–4821. https://doi.org/10.1242/jcs.02602
14. Brooks SV, Faulkner JA (1994) Skeletal muscle weakness in old age: underlying mechanisms. Med Sci Sports Exerc 26(4):432–439
15. Bujak AL et al (2015) AMPK activation of muscle autophagy prevents fasting-induced hypoglycemia and myopathy during aging. Cell Metab 21(6):883–890. https://doi.org/10.1016/j.cmet.2015.05.016
16. Call JA, Nichenko AS (2020) Autophagy: an essential but limited cellular process for timely skeletal muscle recovery from injury. Autophagy 16(7):1344–1347. https://doi.org/10.1080/15548627.2020.1753000
17. Call JA, Wilson RJ, Laker RC, Zhang M, Kundu M, Yan Z (2017) Ulk1-mediated autophagy plays an essential role in mitochondrial remodeling and functional regeneration of skeletal muscle. Am J Physiol Cell Physiol 312(6):C724–C732. https://doi.org/10.1152/ajpcell.00348.2016
18. Canto C et al (2009) AMPK regulates energy expenditure by modulating NAD+ metabolism and SIRT1 activity. Nature 458(7241):1056–1060. https://doi.org/10.1038/nature07813
19. Carnio S et al (2014) Autophagy impairment in muscle induces neuromuscular junction degeneration and precocious aging. Cell Rep 8(5):1509–1521. https://doi.org/10.1016/j.celrep.2014.07.061
20. Cerletti M, Jang YC, Finley LW, Haigis MC, Wagers AJ (2012) Short-term calorie restriction enhances skeletal muscle stem cell function. Cell Stem Cell 10(5):515–519
21. Cesari M, Landi F, Vellas B, Bernabei R, Marzetti E (2014) Sarcopenia and physical frailty: two sides of the same coin. Front Aging Neurosci 6:192. https://doi.org/10.3389/fnagi.2014.00192
22. Chai RJ, Vukovic J, Dunlop S, Grounds MD, Shavlakadze T (2011) Striking denervation of neuromuscular junctions without lumbar motoneuron loss in geriatric mouse muscle. PLoS ONE 6(12):e28090. https://doi.org/10.1371/journal.pone.0028090
23. Chakkalakal JV, Jones KM, Basson MA, Brack AS (2012) The aged niche disrupts muscle stem cell quiescence. Nature 490(7420):355–360. https://doi.org/10.1038/nature11438

24. Chang NC et al (2012) Bcl-2-associated autophagy regulator Naf-1 required for maintenance of skeletal muscle. Hum Mol Genet 21(10):2277–2287. https://doi.org/10.1093/hmg/dds048
25. Chang NC, Nguyen M, Germain M, Shore GC (2010) Antagonism of Beclin 1-dependent autophagy by BCL-2 at the endoplasmic reticulum requires NAF-1. EMBO J 29(3):606–618. https://doi.org/10.1038/emboj.2009.369
26. Chang NC et al (2018) The dystrophin glycoprotein complex regulates the epigenetic activation of muscle stem cell commitment. Cell Stem Cell 22(5):755-768.e756. https://doi.org/10.1016/j.stem.2018.03.022
27. Chen C, Liu Y, Liu R, Ikenoue T, Guan K-L, Liu Y, Zheng P (2008) TSC–mTOR maintains quiescence and function of hematopoietic stem cells by repressing mitochondrial biogenesis and reactive oxygen species. J Exp Med 205(10):2397–2408
28. Chen Y, Dorn GW 2nd (2013) PINK1-phosphorylated mitofusin 2 is a Parkin receptor for culling damaged mitochondria. Science 340(6131):471–475. https://doi.org/10.1126/science.1231031
29. Chen YF et al (2009) Cisd2 deficiency drives premature aging and causes mitochondria-mediated defects in mice. Genes Dev 23(10):1183–1194. https://doi.org/10.1101/gad.1779509
30. Cheung TH et al (2012) Maintenance of muscle stem-cell quiescence by microRNA-489. Nature 482(7386):524–528. https://doi.org/10.1038/nature10834
31. Cheung TH, Rando TA (2013) Molecular regulation of stem cell quiescence. Nat Rev Mol Cell Biol 14(6):329–340
32. Comai G, Tajbakhsh S (2014) Molecular and cellular regulation of skeletal myogenesis. Curr Top Dev Biol 110:1–73
33. Crist Colin G, Montarras D, Buckingham M (2012) Muscle satellite cells are primed for myogenesis but maintain quiescence with sequestration of Myf5 mRNA targeted by microRNA-31 in mRNP granules. Cell Stem Cell 11(1):118–126. https://doi.org/10.1016/j.stem.2012.03.011
34. Drake JC, Wilson RJ, Yan Z (2016) Molecular mechanisms for mitochondrial adaptation to exercise training in skeletal muscle. FASEB J 30(1):13–22. https://doi.org/10.1096/fj.15-276337
35. Dumont NA, Bentzinger CF, Sincennes MC, Rudnicki MA (2015) Satellite cells and skeletal muscle regeneration. Compr Physiol 5(3):1027–1059. https://doi.org/10.1002/cphy.c140068
36. Esteca MV et al (2020) Loss of Parkin results in altered muscle stem cell differentiation during regeneration. Int J Mol Sci 21(21). https://doi.org/10.3390/ijms21218007
37. Ferdousi LV et al (2014) More efficient repair of DNA double-strand breaks in skeletal muscle stem cells compared to their committed progeny. Stem Cell Res 13(3):492–507
38. Fernandes SA et al (2020) Altered in vitro muscle differentiation in X-linked myopathy with excessive autophagy. Dis Models Mech 13(2). https://doi.org/10.1242/dmm.041244
39. Fiacco E et al (2016) Autophagy regulates satellite cell ability to regenerate normal and dystrophic muscles. Cell Death Differ 23(11):1839–1849. https://doi.org/10.1038/cdd.2016.70
40. Filippelli RL, Chang NC (2021) Empowering muscle stem cells for the treatment of duchenne muscular dystrophy. Cells Tissues Organs 1–14. https://doi.org/10.1159/000514305
41. Folmes CD, Dzeja PP, Nelson TJ, Terzic A (2012) Metabolic plasticity in stem cell homeostasis and differentiation. Cell Stem Cell 11(5):596–606
42. Fortini P et al (2016) The fine tuning of metabolism, autophagy and differentiation during in vitro myogenesis. Cell Death Dis 7:e2168. https://doi.org/10.1038/cddis.2016.50
43. Frontera WR, Ochala J (2015) Skeletal muscle: a brief review of structure and function. Calcif Tissue Int 96(3):183–195. https://doi.org/10.1007/s00223-014-9915-y
44. Fu X, Zhu M-J, Dodson MV, Du M (2015) AMP-activated protein kinase stimulates Warburg-like glycolysis and activation of satellite cells during muscle regeneration. J Biol Chem 290(44):26445–26456
45. Fukada S-I et al (2007) Molecular signature of quiescent satellite cells in adult skeletal muscle. Stem Cells 25(10):2448–2459. https://doi.org/10.1634/stemcells.2007-0019

46. Gagnon C, Gallais B, Laberge L (2018) Myotonic dystrophy type 1: reasons to be OPTIMISTIC. Lancet Neurol 17(8):652–653. https://doi.org/10.1016/S1474-4422(18)30240-0
47. Garcia-Prat L et al (2016) Autophagy maintains stemness by preventing senescence. Nature 529(7584):37–42. https://doi.org/10.1038/nature16187
48. García-Prat L et al (2020) FoxO maintains a genuine muscle stem-cell quiescent state until geriatric age. Nat Cell Biol 22(11):1307–1318. https://doi.org/10.1038/s41556-020-00593-7
49. Gattazzo F, Laurent B, Relaix F, Rouard H, Didier N (2020) Distinct phases of postnatal skeletal muscle growth govern the progressive establishment of muscle stem cell quiescence. Stem Cell Rep 15(3):597–611. https://doi.org/10.1016/j.stemcr.2020.07.011
50. Gil-Gómez G, Berns A, Brady HJ (1998) A link between cell cycle and cell death: Bax and Bcl-2 modulate Cdk2 activation during thymocyte apoptosis. EMBO J 17(24):7209–7218
51. Gomes LC, Di Benedetto G, Scorrano L (2011) During autophagy mitochondria elongate, are spared from degradation and sustain cell viability. Nat Cell Biol 13(5):589–598
52. Hailey DW, Rambold AS, Satpute-Krishnan P, Mitra K, Sougrat R, Kim PK, Lippincott-Schwartz J (2010) Mitochondria supply membranes for autophagosome biogenesis during starvation. Cell 141(4):656–667. https://doi.org/10.1016/j.cell.2010.04.009
53. Halling JF, Ringholm S, Nielsen MM, Overby P, Pilegaard H (2016) PGC-1alpha promotes exercise-induced autophagy in mouse skeletal muscle. Physiol Rep 4(3). https://doi.org/10.14814/phy2.12698
54. Hardie DG (2011) Energy sensing by the AMP-activated protein kinase and its effects on muscle metabolism. Proc Nutr Soc 70(1):92–99. https://doi.org/10.1017/S0029665110003915
55. Hausburg MA et al (2015) Post-transcriptional regulation of satellite cell quiescence by TTP-mediated mRNA decay. Elife 4:e03390
56. He C et al (2012) Exercise-induced BCL2-regulated autophagy is required for muscle glucose homeostasis. Nature 481(7382):511–515. https://doi.org/10.1038/nature10758
57. Hernández-Hernández JM, García-González EG, Brun CE, Rudnicki MA (2017) The myogenic regulatory factors, determinants of muscle development, cell identity and regeneration. Semin Cell Dev Biol 72:10–18. https://doi.org/10.1016/j.semcdb.2017.11.010
58. Hiromura K, Pippin JW, Fero ML, Roberts JM, Shankland SJ (1999) Modulation of apoptosis by the cyclin-dependent kinase inhibitor p27 Kip1. J Clin Investig 103(5):597–604
59. Hong X et al (2022) Mitochondrial dynamics maintain muscle stem cell regenerative competence throughout adult life by regulating metabolism and mitophagy. Cell Stem Cell 29(9):1298–1314.e10. https://doi.org/10.1016/j.stem.2022.07.009
60. Hood DA, Memme JM, Oliveira AN, Triolo M (2019) Maintenance of skeletal muscle mitochondria in health, exercise, and aging. Annu Rev Physiol 81:19–41. https://doi.org/10.1146/annurev-physiol-020518-114310
61. Hursting SD, Lavigne JA, Berrigan D, Perkins SN, Barrett JC (2003) Calorie restriction, aging, and cancer prevention: mechanisms of action and applicability to humans. Annu Rev Med 54(1):131–152
62. Hwang AB, Brack AS (2018) Muscle stem cells and aging. Curr Top Dev Biol 126:299–322. https://doi.org/10.1016/bs.ctdb.2017.08.008
63. Jamart C, Francaux M, Millet GY, Deldicque L, Frere D, Feasson L (2012) Modulation of autophagy and ubiquitin-proteasome pathways during ultra-endurance running. J Appl Physiol (1985) 112(9):1529–1537. https://doi.org/10.1152/japplphysiol.00952.2011
64. Janich P, Meng Q-J, Benitah SA (2014) Circadian control of tissue homeostasis and adult stem cells. Curr Opin Cell Biol 31:8–15
65. Jejurikar SS, Henkelman EA, Cederna PS, Marcelo CL, Urbanchek MG, Kuzon WM (2006) Aging increases the susceptibility of skeletal muscle derived satellite cells to apoptosis. Exp Gerontol 41(9):828–836. https://doi.org/10.1016/j.exger.2006.06.053
66. Jiang A, Guo H, Wu W, Liu H (2021) The crosstalk between autophagy and apoptosis is necessary for myogenic differentiation. J Agric Food Chem 69(13):3942–3951. https://doi.org/10.1021/acs.jafc.1c00140

67. Jiang P, Mizushima N (2014) Autophagy and human diseases. Cell Res 24(1):69–79. https://doi.org/10.1038/cr.2013.161
68. Joseph GA et al (2019) Partial inhibition of mTORC1 in aged rats counteracts the decline in muscle mass and reverses molecular signaling associated with sarcopenia. Mol Cell Biol 39(19). https://doi.org/10.1128/MCB.00141-19
69. Karakelides H, Nair KS (2005) Sarcopenia of aging and its metabolic impact. In: Current topics in developmental biology, vol 68. Academic Press, pp 123–148. https://doi.org/10.1016/S0070-2153(05)68005-2
70. Kaushik S, Cuervo AM (2012) Chaperones in autophagy. Pharmacol Res 66(6):484–493. https://doi.org/10.1016/j.phrs.2012.10.002
71. Keefe AC et al (2015) Muscle stem cells contribute to myofibres in sedentary adult mice. Nat Commun 6:7087–7087. https://doi.org/10.1038/ncomms8087
72. Kim J, Kundu M, Viollet B, Guan K-L (2011) AMPK and mTOR regulate autophagy through direct phosphorylation of Ulk1. Nat Cell Biol 13(2):132–141. https://doi.org/10.1038/ncb2152
73. Kim Y, Triolo M, Hood DA (2017) Impact of aging and exercise on mitochondrial quality control in skeletal muscle. Oxid Med Cell Longev 2017:3165396. https://doi.org/10.1155/2017/3165396
74. Kishnani PS et al (2006) Pompe disease diagnosis and management guideline. Genet Med 8(5):267–288. https://doi.org/10.1097/01.gim.0000218152.87434.f3
75. Kottlors M, Kirschner J (2010) Elevated satellite cell number in Duchenne muscular dystrophy. Cell Tissue Res 340(3):541–548. https://doi.org/10.1007/s00441-010-0976-6
76. Kounakis K, Chaniotakis M, Markaki M, Tavernarakis N (2019) Emerging roles of lipophagy in health and disease. Front Cell Dev Biol 7:185. https://doi.org/10.3389/fcell.2019.00185
77. Kuang S, Kuroda K, Le Grand F, Rudnicki MA (2007) Asymmetric self-renewal and commitment of satellite stem cells in muscle. Cell 129(5):999–1010. https://doi.org/10.1016/j.cell.2007.03.044
78. Kundu M et al (2008) Ulk1 plays a critical role in the autophagic clearance of mitochondria and ribosomes during reticulocyte maturation. Blood 112(4):1493–1502. https://doi.org/10.1182/blood-2008-02-137398
79. Lam T et al (2016) Reversal of intramyocellular lipid accumulation by lipophagy and a p62-mediated pathway. Cell Death Discov 2:16061. https://doi.org/10.1038/cddiscovery.2016.61
80. Lee JE, Cooper TA (2009) Pathogenic mechanisms of myotonic dystrophy. Biochem Soc Trans 37(Pt 6):1281–1286. https://doi.org/10.1042/BST0371281
81. Leichman JG et al (2008) Dramatic reversal of derangements in muscle metabolism and left ventricular function after bariatric surgery. Am J Med 121(11):966–973. https://doi.org/10.1016/j.amjmed.2008.06.033
82. Lepper C, Partridge TA, Fan C-M (2011) An absolute requirement for Pax7-positive satellite cells in acute injury-induced skeletal muscle regeneration. Development 138(17):3639–3646. https://doi.org/10.1242/dev.067595
83. Levine B, Kroemer G (2008) Autophagy in the pathogenesis of disease. Cell 132(1):27–42. https://doi.org/10.1016/j.cell.2007.12.018
84. Liang J et al (2007) The energy sensing LKB1–AMPK pathway regulates p27 kip1 phosphorylation mediating the decision to enter autophagy or apoptosis. Nat Cell Biol 9(2):218–224. https://doi.org/10.1038/ncb1537
85. Lira VA et al (2013) Autophagy is required for exercise training-induced skeletal muscle adaptation and improvement of physical performance. FASEB J 27(10):4184–4193. https://doi.org/10.1096/fj.13-228486
86. Liu L et al (2013) Chromatin modifications as determinants of muscle stem cell quiescence and chronological aging. Cell Rep 4(1):189–204. https://doi.org/10.1016/j.celrep.2013.05.043
87. Liu L, Cheung TH, Charville GW, Rando TA (2015) Isolation of skeletal muscle stem cells by fluorescence-activated cell sorting. Nat Protoc 10(10):1612–1624. https://doi.org/10.1038/nprot.2015.110

88. Lunt SY, Vander Heiden MG (2011) Aerobic glycolysis: meeting the metabolic requirements of cell proliferation. Annu Rev Cell Dev Biol 27:441–464. https://doi.org/10.1146/annurev-cellbio-092910-154237
89. Ma X, Liu H, Foyil SR, Godar RJ, Weinheimer CJ, Hill JA, Diwan A (2012) Impaired autophagosome clearance contributes to cardiomyocyte death in ischemia/reperfusion injury. Circulation 125(25):3170–3181. https://doi.org/10.1161/CIRCULATIONAHA.111.041814
90. Maesner CC, Almada AE, Wagers AJ (2016) Established cell surface markers efficiently isolate highly overlapping populations of skeletal muscle satellite cells by fluorescence-activated cell sorting. Skeletal Muscle 6(1):35. https://doi.org/10.1186/s13395-016-0106-6
91. Maiuri MC, Galluzzi L, Morselli E, Kepp O, Malik SA, Kroemer G (2010) Autophagy regulation by p53. Curr Opin Cell Biol 22(2):181–185. https://doi.org/10.1016/j.ceb.2009.12.001
92. Mariño G, Madeo F, Kroemer G (2011) Autophagy for tissue homeostasis and neuroprotection. Curr Opin Cell Biol 23(2):198–206. https://doi.org/10.1016/j.ceb.2010.10.001
93. Marzetti E, Calvani R, Cesari M, Buford TW, Lorenzi M, Behnke BJ, Leeuwenburgh C (2013) Mitochondrial dysfunction and sarcopenia of aging: from signaling pathways to clinical trials. Int J Biochem Cell Biol 45(10):2288–2301. https://doi.org/10.1016/j.biocel.2013.06.024
94. Masiero E et al (2009) Autophagy is required to maintain muscle mass. Cell Metab 10(6):507–515. https://doi.org/10.1016/j.cmet.2009.10.008
95. Mauro A (1961) Satellite cell of skeletal muscle fibers. J Biophys Biochem Cytol 9(2):493–495. https://doi.org/10.1083/jcb.9.2.493
96. McMillan EM, Quadrilatero J (2014) Autophagy is required and protects against apoptosis during myoblast differentiation. Biochem J 462(2):267–277. https://doi.org/10.1042/BJ20140312
97. Mizushima N (2007) Autophagy: process and function. Genes Dev 21(22):2861–2873. https://doi.org/10.1101/gad.1599207
98. Mofarrahi M et al (2013) Autophagic flux and oxidative capacity of skeletal muscles during acute starvation. Autophagy 9(10):1604–1620. https://doi.org/10.4161/auto.25955
99. Montarras D, L'honoré A, Buckingham M (2013) Lying low but ready for action: the quiescent muscle satellite cell. FEBS J 280(17):4036–4050. https://doi.org/10.1111/febs.12372
100. Morrison JH, Hof PR (1997) Life and death of neurons in the aging brain. Science 278(5337):412–419. https://doi.org/10.1126/science.278.5337.412
101. Mrakovcic M, Frohlich LF (2018) p53-mediated molecular control of autophagy in tumor cells. Biomolecules 8(2). https://doi.org/10.3390/biom8020014
102. Mukund K, Subramaniam S (2020) Skeletal muscle: A review of molecular structure and function, in health and disease. Wiley Interdiscip Rev Syst Biol Med 12(1):e1462–e1462. https://doi.org/10.1002/wsbm.1462
103. Munoz JP et al (2013) Mfn2 modulates the UPR and mitochondrial function via repression of PERK. EMBO J 32(17):2348–2361. https://doi.org/10.1038/emboj.2013.168
104. Murakami M et al (2004) mTOR is essential for growth and proliferation in early mouse embryos and embryonic stem cells. Mol Cell Biol 24(15):6710–6718
105. Murphy MM, Lawson JA, Mathew SJ, Hutcheson DA, Kardon G (2011) Satellite cells, connective tissue fibroblasts and their interactions are crucial for muscle regeneration. Development 138(17):3625–3637. https://doi.org/10.1242/dev.064162
106. Neel BA, Lin Y, Pessin JE (2013) Skeletal muscle autophagy: a new metabolic regulator. Trends Endocrinol Metab 24(12):635–643. https://doi.org/10.1016/j.tem.2013.09.004
107. Ngoh GA, Papanicolaou KN, Walsh K (2012) Loss of mitofusin 2 promotes endoplasmic reticulum stress. J Biol Chem 287(24):20321–20332. https://doi.org/10.1074/jbc.M112.359174
108. Panicker N, Dawson VL, Dawson TM (2017) Activation mechanisms of the E3 ubiquitin ligase parkin. Biochem J 474(18):3075–3086. https://doi.org/10.1042/BCJ20170476
109. Paolini A et al (2018) Attenuation of autophagy impacts on muscle fibre development, starvation induced stress and fibre regeneration following acute injury. Sci Rep 8(1):9062. https://doi.org/10.1038/s41598-018-27429-7

110. Park SS, Seo YK, Kwon KS (2019) Sarcopenia targeting with autophagy mechanism by exercise. BMB Rep 52(1):64–69. https://doi.org/10.5483/BMBRep.2019.52.1.292
111. Pattingre S et al (2005) Bcl-2 antiapoptotic proteins inhibit Beclin 1-dependent autophagy. Cell 122(6):927–939. https://doi.org/10.1016/j.cell.2005.07.002
112. Pavlidou T et al (2019) Metformin delays satellite cell activation and maintains quiescence. Stem Cells Int 2019:5980465. https://doi.org/10.1155/2019/5980465
113. Pavlidou T, Rosina M, Fuoco C, Gerini G, Gargioli C, Castagnoli L, Cesareni G (2017) Regulation of myoblast differentiation by metabolic perturbations induced by metformin. PLoS ONE 12(8):e0182475. https://doi.org/10.1371/journal.pone.0182475
114. Peker N, Donipadi V, Sharma M, McFarlane C, Kambadur R (2018) Loss of Parkin impairs mitochondrial function and leads to muscle atrophy. Am J Physiol Cell Physiol 315(2):C164–C185. https://doi.org/10.1152/ajpcell.00064.2017
115. Petrany MJ, Millay DP (2019) Cell fusion: merging membranes and making muscle. Trends Cell Biol 29(12):964–973. https://doi.org/10.1016/j.tcb.2019.09.002
116. Phelan A, Elliott G, O'Hare P (1998) Intercellular delivery of functional p53 by the herpesvirus protein VP22. Nat Biotechnol 16(5):440–443. https://doi.org/10.1038/nbt0598-440
117. Pietrocola F, Izzo V, Niso-Santano M, Vacchelli E, Galluzzi L, Maiuri MC, Kroemer G (2013) Regulation of autophagy by stress-responsive transcription factors. Semin Cancer Biol 23(5):310–322. https://doi.org/10.1016/j.semcancer.2013.05.008
118. Powers SK, Ji LL, Kavazis AN, Jackson MJ (2011) Reactive oxygen species: impact on skeletal muscle. Compr Physiol 1(2):941–969. https://doi.org/10.1002/cphy.c100054
119. Price FD et al (2014) Inhibition of JAK-STAT signaling stimulates adult satellite cell function. Nat Med 20(10):1174–1181. https://doi.org/10.1038/nm.3655
120. Raben N et al (2005) Replacing acid alpha-glucosidase in Pompe disease: recombinant and transgenic enzymes are equipotent, but neither completely clears glycogen from type II muscle fibers. Mol Ther 11(1):48–56. https://doi.org/10.1016/j.ymthe.2004.09.017
121. Raben N et al (2008) Suppression of autophagy in skeletal muscle uncovers the accumulation of ubiquitinated proteins and their potential role in muscle damage in Pompe disease. Hum Mol Genet 17(24):3897–3908. https://doi.org/10.1093/hmg/ddn292
122. Raben N, Roberts A, Plotz PH (2007) Role of autophagy in the pathogenesis of Pompe disease. Acta Myol 26(1):45–48
123. Rahman M, Mofarrahi M, Kristof AS, Nkengfac B, Harel S, Hussain SN (2014) Reactive oxygen species regulation of autophagy in skeletal muscles. Antioxid Redox Signal 20(3):443–459. https://doi.org/10.1089/ars.2013.5410
124. Relaix F, Zammit PS (2012) Satellite cells are essential for skeletal muscle regeneration: the cell on the edge returns centre stage. Development 139(16):2845–2856. https://doi.org/10.1242/dev.069088
125. Rion N, Castets P, Lin S, Enderle L, Reinhard JR, Eickhorst C, Ruegg MA (2019) mTOR controls embryonic and adult myogenesis via mTORC1. Development 146(7). https://doi.org/10.1242/dev.172460
126. Rocheteau P, Gayraud-Morel B, Siegl-Cachedenier I, Blasco MA, Tajbakhsh S (2012) A subpopulation of adult skeletal muscle stem cells retains all template DNA strands after cell division. Cell 148(1–2):112–125. https://doi.org/10.1016/j.cell.2011.11.049
127. Rodgers JT et al (2014) mTORC1 controls the adaptive transition of quiescent stem cells from G0 to GAlert. Nature 510(7505):393–396. https://doi.org/10.1038/nature13255
128. Rossi D, Barone V, Giacomello E, Cusimano V, Sorrentino V (2008) The sarcoplasmic reticulum: an organized patchwork of specialized domains. Traffic 9(7):1044–1049. https://doi.org/10.1111/j.1600-0854.2008.00717.x
129. Roy RR, Edgerton VR (2009) Skeletal muscle architecture. In: Binder MD, Hirokawa N, Windhorst U (eds) Encyclopedia of neuroscience. Springer Berlin Heidelberg, pp 3702–3707. https://doi.org/10.1007/978-3-540-29678-2_5425
130. Rudnicki MA, Le Grand F, McKinnell I, Kuang S (2008) The molecular regulation of muscle stem cell function. Cold Spring Harb Symp Quant Biol 73:323–331. https://doi.org/10.1101/sqb.2008.73.064

131. Sambasivan R et al (2011) Pax7-expressing satellite cells are indispensable for adult skeletal muscle regeneration. Development 138(17):3647–3656. https://doi.org/10.1242/dev.067587
132. Sampath P, Pritchard DK, Pabon L, Reinecke H, Schwartz SM, Morris DR, Murry CE (2008) A hierarchical network controls protein translation during murine embryonic stem cell self-renewal and differentiation. Cell Stem Cell 2(5):448–460. https://doi.org/10.1016/j.stem.2008.03.013
133. Sanchez AM, Bernardi H, Py G, Candau RB (2014) Autophagy is essential to support skeletal muscle plasticity in response to endurance exercise. Am J Physiol Regul Integr Comp Physiol 307(8):R956-969. https://doi.org/10.1152/ajpregu.00187.2014
134. Sandri M (2010) Autophagy in skeletal muscle. FEBS Lett 584(7):1411–1416. https://doi.org/10.1016/j.febslet.2010.01.056
135. Schultz E (1996) Satellite cell proliferative compartments in growing skeletal muscles. Dev Biol 175(1):84–94. https://doi.org/10.1006/dbio.1996.0097
136. Schultz E, Gibson MC, Champion T (1978) Satellite cells are mitotically quiescent in mature mouse muscle: an EM and radioautographic study. J Exp Zool 206(3):451–456. https://doi.org/10.1002/jez.1402060314
137. Schultz E, Lipton BH (1982) Skeletal muscle satellite cells: changes in proliferation potential as a function of age. Mech Ageing Dev 20(4):377–383. https://doi.org/10.1016/0047-6374(82)90105-1
138. Seale P, Sabourin LA, Girgis-Gabardo A, Mansouri A, Gruss P, Rudnicki MA (2000) Pax7 Is Required for the Specification of Myogenic Satellite Cells. Cell 102(6):777–786. https://doi.org/10.1016/S0092-8674(00)00066-0
139. Sebastian D et al (2016) Mfn2 deficiency links age-related sarcopenia and impaired autophagy to activation of an adaptive mitophagy pathway. EMBO J 35 (15):1677–1693. https://doi.org/10.15252/embj.201593084
140. Shefer G, Van de Mark DP, Richardson JB, Yablonka-Reuveni Z (2006) Satellite-cell pool size does matter: defining the myogenic potency of aging skeletal muscle. Dev Biol 294(1):50–66. https://doi.org/10.1016/j.ydbio.2006.02.022
141. Shi W et al (2012) Therapeutic metformin/AMPK activation blocked lymphoma cell growth via inhibition of mTOR pathway and induction of autophagy. Cell Death Dis 3(3):e275–e275. https://doi.org/10.1038/cddis.2012.13
142. Sin J et al (2016) Mitophagy is required for mitochondrial biogenesis and myogenic differentiation of C2C12 myoblasts. Autophagy 12(2):369–380. https://doi.org/10.1080/15548627.2015.1115172
143. Solanas G et al. (2017) Aged stem cells reprogram their daily rhythmic functions to adapt to stress. Cell 170(4):678–692.e620. https://doi.org/10.1016/j.cell.2017.07.035
144. Song KY et al (2020) MBNL1 reverses the proliferation defect of skeletal muscle satellite cells in myotonic dystrophy type 1 by inhibiting autophagy via the mTOR pathway. Cell Death Dis 11(7):545. https://doi.org/10.1038/s41419-020-02756-8
145. Sousa-Victor P et al (2014) Geriatric muscle stem cells switch reversible quiescence into senescence. Nature 506(7488):316–321. https://doi.org/10.1038/nature13013
146. Tanaka A, Cleland MM, Xu S, Narendra DP, Suen D-F, Karbowski M, Youle RJ (2010) Proteasome and p97 mediate mitophagy and degradation of mitofusins induced by Parkin. J Cell Biol 191(7):1367–1380
147. Tang AH, Rando TA (2014) Induction of autophagy supports the bioenergetic demands of quiescent muscle stem cell activation. EMBO J 33(23):2782–2797. https://doi.org/10.15252/embj.201488278
148. Tasdemir E et al (2008) A dual role of p53 in the control of autophagy. Autophagy 4(6):810–814. https://doi.org/10.4161/auto.6486
149. Theret M et al (2017) AMPK α1-LDH pathway regulates muscle stem cell self-renewal by controlling metabolic homeostasis. EMBO J 36(13):1946–1962. https://doi.org/10.15252/embj.201695273
150. Tierney MT, Sacco A (2016) Satellite cell heterogeneity in skeletal muscle homeostasis. Trends Cell Biol 26(6):434–444. https://doi.org/10.1016/j.tcb.2016.02.004

151. Triolo M, Hood DA (2021) Manifestations of age on autophagy, mitophagy and lysosomes in skeletal muscle. Cells 10(5). https://doi.org/10.3390/cells10051054
152. Unger RH, Orci L (2001) Diseases of liporegulation: new perspective on obesity and related disorders. FASEB J 15(2):312–321. https://doi.org/10.1096/fj.00-0590
153. Vainshtein A, Hood DA (2016) The regulation of autophagy during exercise in skeletal muscle. J Appl Physiol (1985) 120(6):664–673. https://doi.org/10.1152/japplphysiol.00550.2015
154. Vainshtein A, Tryon LD, Pauly M, Hood DA (2015) Role of PGC-1alpha during acute exercise-induced autophagy and mitophagy in skeletal muscle. Am J Physiol Cell Physiol 308(9):C710-719. https://doi.org/10.1152/ajpcell.00380.2014
155. Valdez G, Tapia JC, Kang H, Clemenson GD Jr, Gage FH, Lichtman JW, Sanes JR (2010) Attenuation of age-related changes in mouse neuromuscular synapses by caloric restriction and exercise. Proc Natl Acad Sci U S A 107(33):14863–14868. https://doi.org/10.1073/pnas.1002220107
156. Visser M, Schaap LA (2011) Consequences of sarcopenia. Clin Geriatr Med 27(3):387–399. https://doi.org/10.1016/j.cger.2011.03.006
157. von Maltzahn J, Jones AE, Parks RJ, Rudnicki MA (2013) Pax7 is critical for the normal function of satellite cells in adult skeletal muscle. Proc Natl Acad Sci 110(41):16474. https://doi.org/10.1073/pnas.1307680110
158. Wagatsuma A, Sakuma K (2013) Mitochondria as a potential regulator of myogenesis. Sci World J 2013:593267. https://doi.org/10.1155/2013/593267
159. Wang YW, He SJ, Feng X, Cheng J, Luo YT, Tian L, Huang Q (2017) Metformin: a review of its potential indications. Drug Des Devel Ther 11:2421–2429. https://doi.org/10.2147/DDDT.S141675
160. White JP, Billin AN, Campbell ME, Russell AJ, Huffman KM, Kraus WE (2018) The AMPK/p27Kip1 axis regulates autophagy/apoptosis decisions in aged skeletal muscle stem cells. Stem Cell Rep 11(2):425–439. https://doi.org/10.1016/j.stemcr.2018.06.014
161. Wysocka J et al (2006) A PHD finger of NURF couples histone H3 lysine 4 trimethylation with chromatin remodelling. Nature 442(7098):86–90. https://doi.org/10.1038/nature04815
162. Xia Q, Huang X, Huang J, Zheng Y, March ME, Li J, Wei Y (2021) The role of autophagy in skeletal muscle diseases. Front Physiol 12:638983. https://doi.org/10.3389/fphys.2021.638983
163. Yan Z, Lira VA, Greene NP (2012) Exercise training-induced regulation of mitochondrial quality. Exerc Sport Sci Rev 40(3):159–164. https://doi.org/10.1097/JES.0b013e3182575599
164. Yang ZJ et al (2015) p53 suppresses muscle differentiation at the myogenin step in response to genotoxic stress. Cell Death Differ 22(4):560–573. https://doi.org/10.1038/cdd.2014.189
165. Yin H, Price F, Rudnicki MA (2013) Satellite cells and the muscle stem cell niche. Physiol Rev 93(1):23–67. https://doi.org/10.1152/physrev.00043.2011
166. Yiu EM, Kornberg AJ (2015) Duchenne muscular dystrophy. J Paediatr Child Health 51(8):759–764. https://doi.org/10.1111/jpc.12868
167. Zeng Z, Liang J, Wu L, Zhang H, Lv J, Chen N (2020) Exercise-induced autophagy suppresses sarcopenia through Akt/mTOR and Akt/FoxO3a signal pathways and AMPK-mediated mitochondrial quality control. Front Physiol 11:583478. https://doi.org/10.3389/fphys.2020.583478
168. Zhou YT et al (2000) Lipotoxic heart disease in obese rats: implications for human obesity. Proc Natl Acad Sci USA 97(4):1784–1789. https://doi.org/10.1073/pnas.97.4.1784
169. Ziaaldini MM, Marzetti E, Picca A, Murlasits Z (2017) Biochemical pathways of sarcopenia and their modulation by physical exercise: a narrative review. Front Med (Lausanne) 4:167. https://doi.org/10.3389/fmed.2017.00167

Chapter 7
Autophagy in the Intestinal Stem Cells

Ebru Goncu

Abstract Intestinal epithelial cells that are exposed to damage caused by the contents of the lumen die, and therefore a continuous and rapid turnover takes place in the epithelial cells of the digestive tract to maintain optimal function of the intestinal epithelium throughout life. Studies in different animal groups have shown that this process is mediated by a population of intestinal stem cells (ISC) that can self-sustain over extended periods and potentially differentiate into enterocytes, goblet cells, enteroendocrine cells, and paneth cells. More than 300 million new epithelial cells must be produced per day in the small intestine to compensate for the high cell death rate in the villi, so cell losses in the tissue must be balanced by the proliferation of stem cells. While the decrease in the intestinal stem cell population causes tissue dysfunction, excessive stem cell proliferation may cause tumor formation. Autophagy is an evolutionarily conserved pathway that maintains cellular homeostasis by degrading long-lived proteins and damaged organelles in the cytosol. At basal levels, autophagy removes damaged components that threaten cellular homeostasis. It is also a particularly important mechanism that performs functions ranging from providing metabolic nutrients to cells under starvation stress to defending against microbial attacks. Autophagic mechanisms are highly active in ISCs and are critical for the survival and function of these cells. Deficiencies in autophagy in intestinal stem cells are associated with various pathological conditions such as ulcerative colitis, Crohn's disease, and gastrointestinal cancers. Understanding the roles of autophagy in the survival, stemness, proliferation, and differentiation processes of intestinal stem cells is important for elucidating the cellular mechanisms of gastrointestinal diseases and for the development of new treatment options.

Keywords Autophagy · Autophagy-related genes · Differentiation · Intestine · Pathologies · Proliferation · Stem cell

E. Goncu (✉)
Faculty of Science, Department of Biology, Ege University, Bornova, Izmir, Turkey
e-mail: ebru.goncu@ege.edu.tr

B. V. Shravage and K. Turksen (eds.), *Autophagy in Stem Cell Maintenance and Differentiation*, Stem Cell Biology and Regenerative Medicine 73,
https://doi.org/10.1007/978-3-031-17362-2_7

Abbreviations

AKT	Protein kinase B
AMPK	5-AMP-activated protein kinase
Atg	Autophagy-related
Atg16L	Atg16-like protein
Atg5BD	Atg5-binding motif
CCD	Helix-coil domain
CD	Crohn's disease
CHK1	Checkpoint kinase 1
EB	Enteroblast
EC	Enterocyte
EE	Enteroendocrine
EGF	Epidermal growth factor
EGFR	Epidermal growth factor receptor
ERK	Extracellular signal-regulated kinase
FIP200	Focal adhesion kinase family interacting protein of 200-kDa
FLNA	Filamin A, alpha
IBD	Inflammatory bowel disease
IM	Isolation membrane
KEAP1	Kelch-like ECH-associated protein 1
LC3	Microtubule-associated protein light chain 3
Lgr5	Leu-rich, repeat-containing G protein-coupled receptor 5
MDP	Muramil dipeptide
mTOR	Mammalian target of rapamycin
mTORC1	Mammalian target of rapamycin (mTOR) complex 1
NOD2	Nucleotide- oligomerization domain protein 2
Nrf2	Nuclear factor (erythroid-derived-2)-like 2
OXPHOS	Mitochondrial oxidative phosphorylation
PE	Phosphatidylethanolamine
PI3K	Class III phosphatidylinositol 3-kinase
Pre-EE	Pre-enteroendocrine
RAB19	Ras-associated binding 19
Rheb	Ras homolog enriched in brain
ROS	Reactive oxygen species
S6K	S6 protein kinase
SNX	Sorting Nexin
TA	Transit-amplifying
TSC2	Tuberous sclerosis complex 2
ULK1	Unc-51 like autophagy activating kinase 1
WD40	WD40 repeat
WNT	Wingless/Int-1 family of proteins

Introduction

The intestine is crucial for performing many vital functions throughout life, such as food digestion, nutrient absorption, glucose homeostasis, and energy maintenance. The intestinal epithelium consists of two distinct parts, the proliferative crypts of Lieberkühn and the long projections called villi. Villi are continuous structures that feature a mature epithelial cell and contain a single layer of differentiated cells that can no longer divide. Differentiated epithelial cells fill the villi and are classified according to their function. Enterocytes absorb nutrients, goblet cells secrete a protective mucus barrier, and enteroendocrine cells produce gastrointestinal hormones [54, 98]. Paneth cells, on the other hand, are located in the lower part of the proliferation chamber and perform two important functions both by joining the stem cell niche and by secreting antibacterial peptides [19, 87, 98]. The proliferative compartment contains undifferentiated and highly proliferating cells that populate the Lieberkühn crypts. The epithelium in this region is responsible for maintaining the enormous cell turnover and providing a protective niche for stem cells. Stem cells exist in a niche of these epithelial and mesenchymal cells and extracellular substrates that provide an optimal microenvironment for their maintenance, proliferation, and differentiation [7, 121]. Paneth cells are in close association with stem cells and are an important source of several niche factors, including epidermal growth factor (EGF), WNT, Wingless/Int-1 family of protein 3A (WNT3A), and Notch ligand [90]. The amount of stem cells in the crypt varies from species to species, but ranges make up 0.4–60% of the crypt [61] and are capable of self-renewal and differentiation into mature cell types to preserve the integrity of the intestinal epithelium [54].

The intestinal epithelium is constantly renewed depending on food content, exposure to toxins, and pathogens [40, 77]. Tissue homeostasis in continuously regenerating tissues is regulated by the tightly controlled proliferation and differentiation of multipotent somatic stem cells, which have the potential to differentiate into different cell types of a given tissue. The critical balance between cell production and cell loss must be strictly maintained [71]. In mammals, the monolayer intestinal epithelium is renewed every 2–5 days [86]. This regeneration process involves rapid and sustained proliferation of epithelial stem cells at the base of the crypt, followed by migration of these cells along the crypt-villi axis [98]. Both at the normal homeostatic level and under stressful conditions, ISC divisions are regulated by multiple signaling pathways [5, 38, 54, 65]. The regulatory mechanism at the stem cell level allows stem cells to divide sparingly, maintain their long-term potential, rapidly restore tissue homeostasis, and repair injured tissues.

During regeneration, small populations of ISCs at the base of the crypt regularly divide to produce highly proliferating progenitors known as transit-amplifying (TA) cells. The newly formed TA cells divide 2–3 times and begin to differentiate into absorptive or secretory cells as they migrate upward from the base of the villus. When differentiated cells exit the crypt, cell proliferation stops, and these epithelial cells then continue to migrate upwards through the villi [6]. The Paneth cell cycle is the only exception to this rapid self-renewal. These cells are renewed every 3–6 weeks by differentiation from specialized secretory cell progenitors that follow a downward migration route from the crypt floor [9, 36]. The first identified marker for ISCs was the leu-rich, repeat-containing G protein-coupled receptor 5 (Lgr5). This receptor has been identified as a WNT target gene that is selectively expressed at the base of adult intestinal crypts. Stem cell-specific expression of Lgr5 has been confirmed using mouse models [60]. To maintain intestinal homeostasis throughout life, the functional and genomic integrity of Lgr5+ ISCs must be maintained under different stress conditions such as infection, physical or chemical stress. Thus, there are unique mechanisms to ensure the protection and survival of these cells [96]. The fate of adult stem cells depends on several factors such as growth factors, cellular niche, metabolic pathways, calcium homeostasis, and autophagy [15]. Growth factors are essential signaling molecules that support the proliferation and differentiation of adult stem cells. These factors regulate cell–cell contacts and cell–matrix adhesions to create a microenvironment that regulates the survival and fate of stem cells [10].

Autophagy is a catabolic process involved in the destruction of cells' own intracellular proteins, lipids, and organelles [66]. Autophagy-mediated recycling of cellular components is a critical process in cellular homeostasis and tissue remodeling during development [14]. Three basic types of autophagy have been identified in mammalian cells: microautophagy, chaperone-mediated autophagy, and macroautophagy [78]. The common form of autophagy is macroautophagy and an evolutionarily highly conserved catabolic mechanism involved in various physiological processes regulated by cellular signaling pathways, including normal development, growth, and immunity.

More than 30 autophagy-related (Atg) genes have been identified in yeast and these genes have also been shown to be evolutionarily conserved in many species. The proteins encoded by these genes form a series of complexes that participate in distinct stages of the autophagic process, including induction of autophagy, autophagosome formation, and autophagosome-lysosome fusion [74, 102]. Autophagy induction is regulated by the mammalian target of rapamycin (mTOR) complex 1 (mTORC1) [102]. Under starvation conditions, after inhibition of mTORC1, autophagy is activated by the formation of the ULK complex, which includes Atg13, Unc-51 like autophagy activating kinase 1 (ULK1), an interacting protein of the focal adhesion kinase family (FIP200) and Atg101. In addition, 5-AMP-activated protein kinase (AMPK) is involved in autophagy activation by suppressing mTOR [37]. Class III phosphatidylinositol 3-kinase (PI3K) complex I migrates to the endoplasmic reticulum (ER) after the formation of the ULK complex. ULK and PI3K complexes promote nucleation of the isolation membrane (IM) to initiate autophagosome formation. Atg9 is recruited to the IM by the ULK complex to perform its function to transport membrane components used for IM expansion. Next, two ubiquitin-like conjugation systems, the Atg12 conjugation system (Atg12-Atg5-Atg16) and the microtubule-associated protein light chain 3 (LC3) system, allow the IM to expand and mature to form autophagosome [26]. Later, the mature autophagosome fuses with lysosomes or endosomes to form the autolysosome. The contents of the autophagosome are digested by lysosomal enzymes and the resulting components are returned to the cytoplasm for reuse [102].

Autophagy has important roles in maintaining homeostasis and in disease processes, including metabolic diseases, cancer, and autoimmune diseases [46, 114, 115]. There are numerous studies on the role of autophagy in embryogenesis and development, as deletion of various Atg genes results in typical phenotypes or death [1]. Furthermore, autophagy interacts with important developmental pathways regulated by signals such as Wnt, Sonic hedgehog, transforming growth factor-b, and fibroblast growth factor [41, 47, 59, 119]. These findings suggest that autophagy may regulate cell fate decisions such as differentiation and proliferation.

A growing number of studies in recent years have shown that autophagy plays an important role in maintaining homeostasis in both embryonic stem cells and adult stem cells under physiological conditions [14, 29]. The contribution of autophagy to the maintenance, proliferation, and differentiation of stem cells has been studied in a variety of adult stem cell types, including ISCs. Autophagic mechanisms are highly active in ISC and are critical for the survival and function of these cells, as well as in the initiation and progression of pathologies that occur in the gut [3, 96]. In conclusion, understanding the role of autophagy in maintaining intestinal homeostasis in stem cells will make an important contribution to the understanding of the pathogenesis of intestinal diseases and the development of treatment strategies.

The Role of Autophagy in Intestinal Stem Cell Maintenance and Proliferation

Maintaining the survival and functionality of stem cells is a requirement for the remaining life of the organism. Therefore, functional quality control of stem cells is critical. Because the cellular transformation in the intestinal tissue is rapid, tightly controlling the transition between the dormant and active states of the ISC is especially important in terms of maintaining the dormant state of the stem cells and allowing them to re-enter the cell cycle when necessary [81, 99]. Increasing evidence suggests that autophagy has a vital role in maintaining stem cell function through simultaneous regulation of cell remodeling and metabolism, while also serving as an essential quality control mechanism [24]. In addition, autophagy is important in regulating the mitochondrial quality of stem cells, protecting them from oxidative damage, regulating energy homeostasis, maintaining differentiation potential, and controlling proliferation [12, 51].

Extensive studies on various stem cells show that their differentiated progeny have different metabolic requirements [97]. Similarly, ISCs must efficiently adapt their metabolism to meet the energy demands of division and differentiation. The presence of a significantly broad mitochondrial network and high mitochondrial membrane potential in ISCs suggests that they meet their energy requirements via oxidative phosphorylation and electron transport chain. Metabolic analysis of adult intestinal Lgr5+ ISCs and Paneth cells revealed the importance of glycolysis in Paneth cells and mitochondrial oxidation in ISCs in energy metabolism [83]. The study showed that Paneth cells create a metabolic niche to produce lactate, which is converted to pyruvate in ISCs and fuels mitochondrial oxidative phosphorylation (OXPHOS) during homeostasis. In addition to Paneth cells, lactate can also be produced by the microbiota to support ISC proliferation [23]. During mitochondrial OXPHOS, besides energy, reactive oxygen species (ROS) are also produced as a by-product. Small amounts of mitochondria-derived reactive ROS are required for ISC maintenance, proliferation, and differentiation, but excessive ROS production can cause oxidative stress, cellular dysfunction, and p53-mediated ISC death [57].

Studies have shown that autophagy reduces excess reactive oxygen species for the maintenance of ISC. Reducing ROS levels helps maintain the viability of ISCs, thereby providing a positive effect on epithelial regeneration because of sudden stress conditions such as infection, exposure to food toxins, nutritional factors, and xenobiotics [55]. In mouse intestinal epithelium, specific deletion of the autophagy-related gene Atg5 in Lgr5 ISCs was found to result in increased intracellular ROS and decreased ISC number [3]. Atg5 is involved in the early stages of autophagosome formation by playing a role in conjugation with Atg12 and the lipidation processes of Atg8 (LC3) (Fig. 7.1a). It has been determined to be an important gene, especially for development, cell differentiation, maintenance of homeostasis, and regulation of immunity [112]. Given that the primary source of intracellular ROS is mitochondria, the autophagy defect resulting from the absence of Atg5 has also resulted in various cellular dysfunctions associated with the accumulation of damaged mitochondria in the ISC [3, 122].

It has also been determined that the guts of these mice have reduced regeneration capacity after an injury, due to a lack of intrinsic autophagy in ISCs. In the intestines of Atg5-deficient mice, impaired gut regeneration following irradiation has been associated with the accumulation of dysfunctional mitochondria and increasing ROS levels due to insufficient autophagy in ISCs [3]. The results showed that functional autophagy in cells can limit the amount of ROS produced through the removal of damaged mitochondria and maintain cellular homeostasis (Fig. 7.1b).

In addition to all these effects, loss of autophagy function may disrupt cellular energy homeostasis, resulting in impaired cellular energy production and decreased cell proliferation.

The role of autophagy in maintaining intestinal stem cells has also been demonstrated by the function of another autophagy-related gene, Atg7. In different Atg7-focused studies, autophagy defects have been reported to occur in Atg7-deficient cells [17, 49, 105]. Atg7 is a protein involved in autophagosome formation and expansion. The phagophore membrane formed during autophagy is enriched with phosphatidylethanolamine (PE), a phospholipid reported to positively regulate autophagic activity [82]. PE acts as an anchor for the recruitment of cytosolic Atg8 to the formed phagophore membrane. Lipidation of Atg8 with PE is a multistep process driven by the E1-like enzymatic activity of homodimeric Atg7. Atg8 is first processed by Atg4, a cysteine protease, to expose a glycine residue at its C-terminus. C-terminal

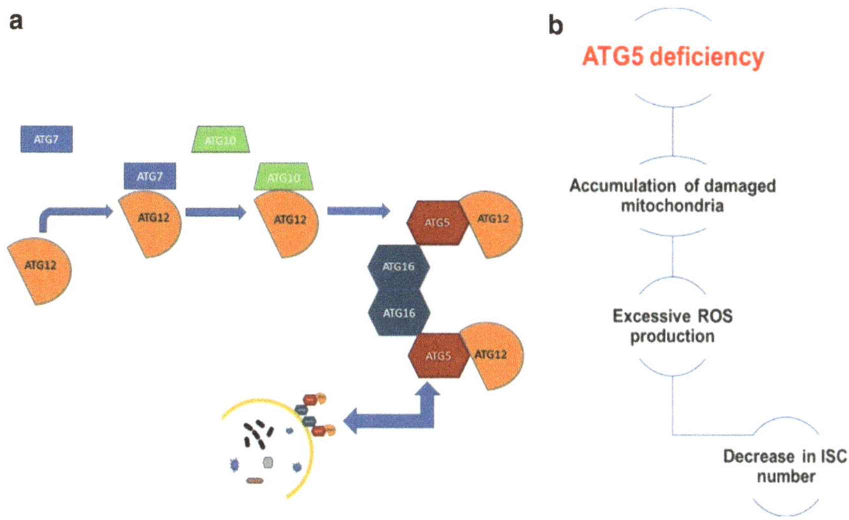

Fig. 7.1 ATG5 plays a role in the expansion of the autophagosome membrane by forming a complex with ATG 12 and ATG16 (**a**) and ATG 5 deficiency causes reduced ISC number (**b**)

processed Atg8 is then activated by Atg7 via adenylation and transferred to Atg3 to be conjugated with PE. After lipidation, Atg8 localizes to both inner and outer autophagosomal membranes and is subsequently degraded by autolysosome formation. Atg7 is also involved in a second autophagy conjugation reaction that promotes Atg8 lipidation. During this reaction, Atg12 is adenylated by Atg7, then transferred to Atg5 via the E2-like enzyme Atg10, producing Atg5-Atg12 conjugates (Fig. 7.2). The resulting complex localizes only to the outer autophagosomal membrane and leaves the membrane before autophagosome closure. Atg5–Atg12 promotes lipidation of the complex Atg8, which forms a complex with Atg16L [26]. Studies have shown that deletion of Atg7 prevents Atg8 lipidation [42, 53].

Changes in ISCs because of Atg7 deficiency have revealed both autophagy-independent and autophagy-related roles of the gene [96]. Increasing evidence for Atg proteins suggests that they may have functions that do not overlap with autophagy, including apoptosis, modulation of cell trafficking, protein secretion, transcription, translation, and membrane reorganization [25, 92]. Disruptions in DNA repair responses of Lgr5+ ISCs have been reported in Atg7-deficient mice, in contrast to differentiated cells and TA progenitors. The long-term presence of stem cells in tissues increases the risk of accumulation of mutations in these cells. Therefore, it has been suggested that the capacity of stem cells to repair DNA or to resist elimination when damaged may be greater than that of other cells [103]. Independent of its function in autophagy, Atg7 has been reported to regulate p53 activity to ensure cell survival by regulating the cell cycle during metabolic stress [52]. It has been shown that p53 activation, which occurs because of Atg7 deficiency changing the DNA repair mechanism, functions as a protection against the negative effects of

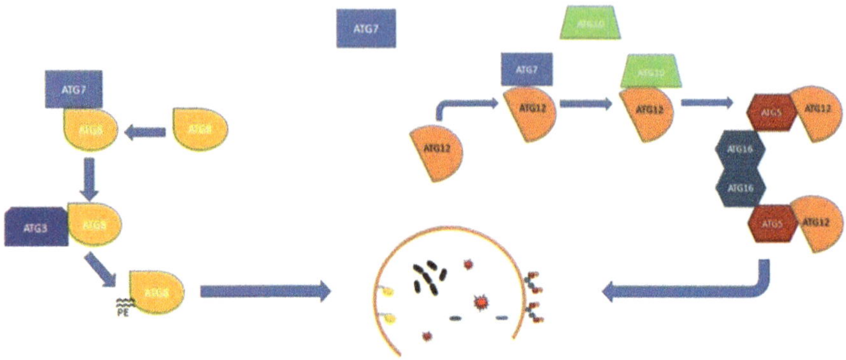

Fig. 7.2 ATG7 plays a role in both the Atg5–ATG12 conjugation system and the ATG8 lipidation process during the expansion and maturation process of the autophagosome

Atg7 inhibition. Subsequent DNA repair processes maintain intestinal homeostasis by protecting Lgr5+ ISC. Increased p53 activation induced by Atg7 deficiency may also contribute to tumor suppression, whereas loss of p53 promotes DNA damage and tumorigenesis in Atg7-deficient Lgr5+ ISC. These results revealed that Atg7 acts in conjunction with p53 to maintain Lgr5+ ISC integrity, independent of its function in autophagy [96].

It has been determined that the autophagy pathway also plays a key role in maintaining the genomic integrity of Lgr5+ ISC by regulating DNA repair pathways. In vitro studies suggest that autophagy can directly control DNA damage repair by increasing proteasomal degradation of proteins that mediate homologous recombination repair, such as checkpoint kinase 1 (CHK1), RAD51, and filamin A, alpha (FLNA). It has been shown that loss of Atg7 increases double-strand break formation and apoptosis in Lgr5 + ISC, but activation of autophagy induced after 24 h of starvation re-increases the DNA repair capacity of Lgr5 + ISC and reduces apoptosis in these cells. Autophagy provides DNA damage repair pathways to cells in this process [96].

Oxidative stress causes aging and death by preventing the proliferation of cells [22]. In contrast to differentiated cells and TA progenitors, Atg7-deficient mice have been reported to cause a defective antioxidant response in Lgr5 + ISCs, resulting in the accumulation of both mitochondrial superoxide and cytoplasmic ROS in stem cells. Excessive production of ROS and DNA damage in the absence of Atg 7 resulted in the accumulation of cytotoxic damage and death of cells by p53-mediated apoptosis. The fact that autophagy is an important mechanism for the maintenance of intestinal stem cells by reducing excess reactive oxygen species is also supported by the protective role of nuclear factor (erythroid-derived-2)-like 2 (NRF2) in Atg 7-deficient ISCs [3, 110]. It is thought that NRF2 provides adaptive protection against oxidative and proteotoxic stress in cells and plays a role in maintaining cellular homeostasis [93]. Under normal conditions, the proteasome pathway degrades NRF2, the master regulator of the antioxidant defense system, via an E3 ubiquitin ligase Kelch-like ECH-associated protein 1 (KEAP1) [45]. An increase in ROS causes the release of NRF2 from KEAP1, resulting in the upregulation of the expression of several antioxidant genes. It was determined that NRF2 is specifically induced as a result of autophagy deficiency, is involved in the regulation of the autophagic process in response to the oxidative stress level, and functions in a feedback loop together with AMPK [44]. The protective function of NRF2 has been reported to be essential for the survival of mice with loss of Atg7. Because both Atg7 and NRF2 deficiency caused mice to die rapidly due to damage in the small intestine [111]. All these results demonstrate the critical importance of autophagy for the protection of ISCs from oxidative damage and the roles of Atg7 in maintaining robust antioxidant defenses and DNA repair to minimize the effects of intrinsic and environmental stresses on ISC homeostasis [96]. Demonstration of the role of NRF2 in maintaining stem cell viability in ATG7 loss [111] reveals its compensatory effect against autophagy loss [110]. It has also been determined that NRF2 provides a resistance mechanism in cancer cells whose essential autophagy genes have been deleted [95]. In conclusion, NRF2 limits ISC damage as a result of autophagy inhibition [110].

It has also been shown that the death of Atg7-deficient intestinal stem cells is dependent on the microbiota. The microbiota functions as an exogenous source of ROS, and excessive ROS production has also been shown to be responsible for stem cell death [43]. Paneth cell defects associated with Atg7 loss have not been found to affect the niche function of these cells but contribute to apoptosis of Lgr5 + ISC through an antimicrobial defense defect, indicating close links between the microbiota, Atg7, and Lgr5 + ISC. Treatment of these mice with broad-spectrum antibiotics reduces DNA damage and increases the survival rates of stem cells. It has been suggested that the production and secretion of an antimicrobial peptide thought to be involved in the direct interaction between bacteria and stem cells is impaired in Atg7 deficiency, and therefore, the microbiota may also contribute to the accumulation of intrinsic stress in Atg7-deficient stem cells. Muramil dipeptide (MDP) is an immunoreactive peptide located in the peptidoglycan motif encoding the "building blocks" of the bacterial cell wall and providing cytoprotective effects on stem cells by directly interacting with the autophagy mechanism through its intracellular receptor, nucleotide- oligomerization domain protein 2 (NOD2). Survival rates after delivery of MDP to Atg7-deficient organoids suggest that post-treatment autophagy may mediate the cytoprotective response involved in stem cells' response to certain microbial signals and function as an important protective mechanism [96]. The changes that occurred in Atg7-deficient ISCs are summarized in Fig. 7.3. All these results demonstrate the critical importance of autophagy for the protection of ISCs from oxidative damage and the roles of Atg7 to minimize the effects of intrinsic and environmental stresses on ISC homeostasis [96].

The function of the MDP-NOD2 signaling pathway in connection with autophagy in intestinal stem cells has also been demonstrated in a study by Levy et al. [55]. NOD2 has been identified as a component of the innate, nuclear factor-kB (NF-κB)-dependent, proinflammatory immune response against bacterial pathogens in immune cells such as macrophages and dendritic cells. In specific immune cell types

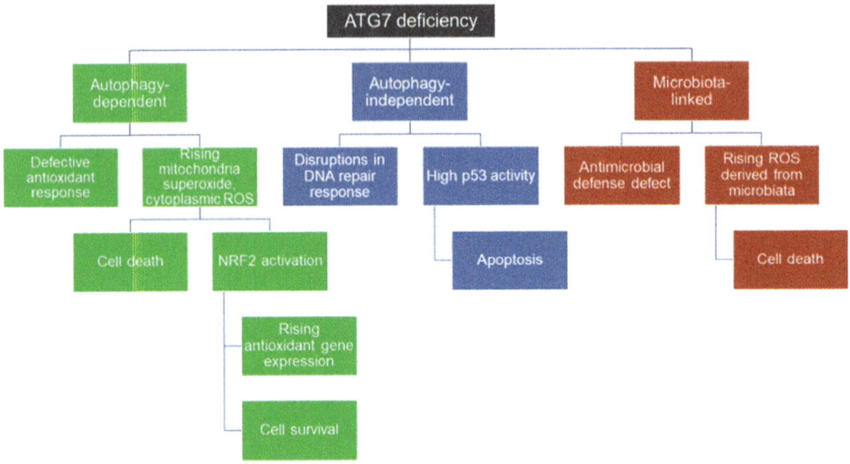

Fig. 7.3 Atg7 deficiency causes abnormalities in intestinal stem cells in diverse ways

such as dendritic cells, NOD2-mediated MDP can initiate autophagy [18]. Modulation of stem cell viability by MDP, including the NOD2 pathway, is crucial for the cytoprotective process associated with ROS production [55, 101]. Similarly, this signaling pathway plays a cytoprotective role in ISCs that express high levels of NOD2 transcripts [73]. MDP can activate many immunological signaling pathways through specific interaction with NOD2. This results in the activation of NF-κB a ubiquitous transcription factor that induces the expression of proinflammatory cytokines [76]. In the study by Levy et al. [55], they have determined that the MDP-NOD2 signaling pathway interacts with the autophagy-related protein Atg16L1 in ISCs.

Atg16 is an autophagy-related protein identified through its association with the autophagy protein Atg5. In the process of autophagy, Atg5 is covalently conjugated to Atg12, and the complex interacts with Atg16. Atg16 is required for the localization of the complex to the pre-autophagosome structure, where Atg8 catalyzes its lipidation (Fig. 7.2). The Atg16 protein identified in yeast weighs 17 kDa and consists of an Atg5-binding domain and a helix-helix domain. It has been determined that the genomes of higher eukaryotes encode larger proteins weighing 55–68 kDa and show weak homology with yeast Atg16 at their N-terminus. For this reason, they are called Atg16-like proteins (Atg16L) in mammals. Mammalian genomes encode two Atg16L proteins, designated Atg16L1 and Atg16L2. Atg16L1 is a core autophagy protein that targets the Atg5-Atg12 conjugate to the phagophore and interacts with Atg5-Atg12 to form a large multimeric complex that facilitates the conjugation of Atg8 proteins to PE [109]. The mammalian homolog of Atg16 was first cloned in mice in 2003, data from studies investigating its roles in cellular homeostasis, and the identification of a Crohn's disease susceptibility variant of its human homolog, Atg16L1, in 2007, have increased interest in this protein. Atg16L1 has been shown to have a central role in blocking NF-κB activity and stimulating mitophagy, which selectively removes mitochondria [63]. Findings demonstrated that activation of autophagy is involved in the MDP-NOD2 cytoprotection pathway and reduces mitochondrial ROS production in ISCs via Atg16L1 (Fig. 7.4a). The low survival rates of intestinal crypt organoids containing Atg16L1-deficient ISCs revealed the contribution of this process to stem cell homeostasis. The importance of MDP in ISC viability in relation to autophagy was demonstrated in another study. In this study, the use of MDP after irradiation in mice was found to have beneficial effects on autophagy in vivo and in vitro. MDP application caused an increase in the viability of stem cells and a decrease in the amount of active caspase 3 and mitochondria [62].

Slit proteins are highly conserved, secreted glycoproteins that mediate their function by binding to transmembrane receptors known as Robo-receptors and play a role in autophagy in intestinal stem cells [11, 108]. Slit/Robo signaling was first identified as an extracellular signature to direct axon guidance, promote axon branching, and control neuronal migration. Many reports have suggested that, in addition to axon guidance, the Slit/Robo pathway is involved in developmental processes and the regulation of various physiological processes. An abnormal Slit/Robo expression in cells can lead to cancer development, progression, and metastasis [94]. Intestinal stem cells have been found to express Slit2 and its single-pass type I membrane surface receptor Robo1 [120]. It has been determined that ISCs retain their self-renewal functions after sodium sulfate treatment in mice with Slit2 overexpression but reduce ISCs in partial Robo1/2 knockout mice. In the analysis of the molecular mechanism of this situation, it has been reported that in the presence of Slit2 overexpression, a normal autophagy process can occur with a small amount of p62 protein accumulation, while partial Robo1/2 degradation reduces autophagy and causes intense p62

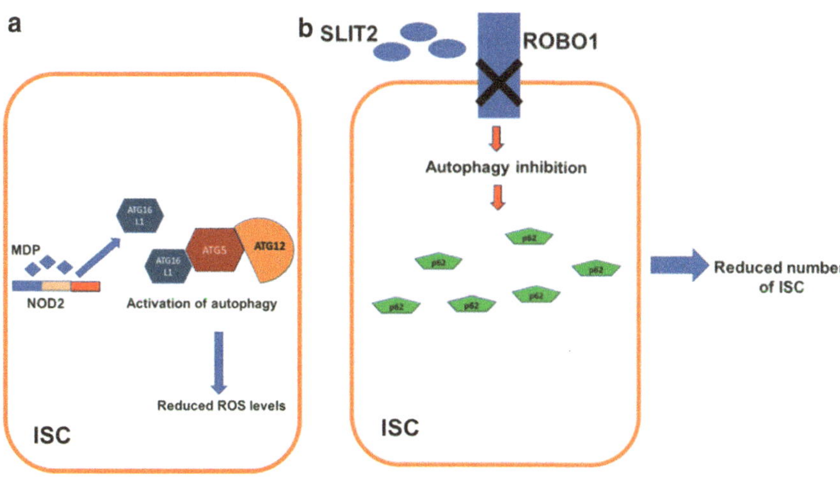

Fig. 7.4 Two different signaling pathways, the MDP-NOD2 (**a**) and Slit-Robo (**b**), are involved in the regulation of autophagy in ISCs. The MDP-NOD2 signaling pathway induces autophagy in ISCs through Atg16L1 (**a**). ROBO1 deficiency causes p62 accumulation due to inhibition of autophagy in ISC (**b**)

accumulation [109]. An autophagy substrate used as a marker of autophagy activity, p62 plays a crucial role in targeting ubiquitin-modified proteins to the proteasome or autophagic machinery. Activation of autophagy typically decreases protein expression of p62/SQSTM1, while pharmacological or genetic inhibition of autophagy increases intracellular p62/SQSTM1 levels [85]. All these results demonstrated that autophagy plays a significant role in maintaining Lgr5 + stem cell proliferation through Slit2/Robo1 signaling in ISCs (Fig. 7.4b).

Changes in nutrient availability and their effect on mTOR activity may play a role in ISC proliferation. During the aging process in mice, the number of crypt and TA cells decreases due to increased mTOR activation. It has been determined that inhibition of mTOR provides partial maintenance of crypt numbers and proliferating cells [32]. In experiments with caloric restriction in mice, an increase in ISCs and Paneth cell numbers was observed. It has been suggested that the reason for this is the increase in cyclic ADP ribose production, which stimulates ISC proliferation, in Paneth cells, due to mTORC1 inhibition [113]. Given that ISCs are affected by both mitochondrial functions and nutrient availability, autophagy is likely to affect their metabolic state.

Role of Autophagy in Intestinal Stem Cell Differentiation

In addition to acting as a quality control mechanism in differentiated cells, autophagy has also been shown to participate in differentiation as a cell remodeling mechanism that promotes morphological and structural changes in stem cells. In the differentiation process of stem cells, autophagy enables cell remodeling by removing unneeded cell components and structures and is therefore a necessary alteration factor in cell phenotype in the process of cell differentiation [64, 80].

Few studies demonstrated the role of autophagy in the differentiation process of intestinal stem cells. In one of these studies, it was determined that the ISCs of mice with Atg16L1 defects undergo an abnormal differentiation process due to excessive Notch signaling. It is known that the Notch signal has a very important role from embryogenesis to adulthood. Notch receptors and ligands are both transmembrane proteins. The maturation and activation of Notch require a series of proteolytic cleavage steps [50]. Notch signaling in the intestinal stem cell population is essential for the maintenance of the normal architecture of the intestinal epithelium [21]. Atg16L1-deficient mice showed impaired notch signaling compared to their normal counterparts, with a dramatic reduction in villus length associated with abnormal stem cell differentiation (Fig. 7.5). It has been demonstrated that high Notch levels in Atg16L1-deficient mice occur due to impaired autophagic degradation of Notch1, and that over-activated Notch signaling causes a developmental abnormality in intestinal villi [107].

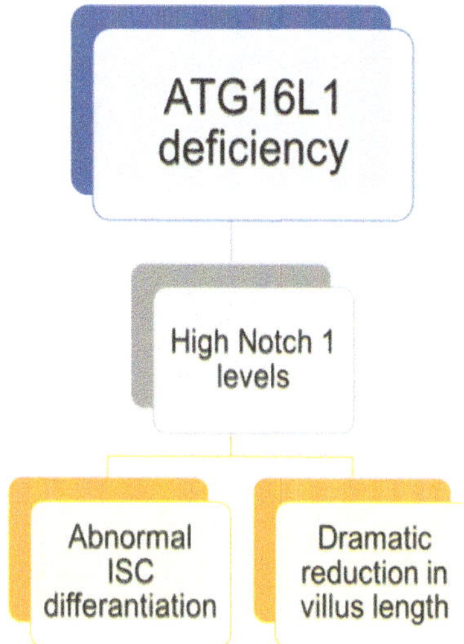

Fig. 7.5 ATG16L1 is involved in ISC differentiation

Insects as Model Organisms in Autophagy Studies in Intestinal Stem Cells

Insects, especially *Drosophila melanogaster*, are frequently used as model organisms in studies on the biology of intestinal stem cells. In *Drosophila*, intestinal homeostasis is maintained as in mammals by tightly controlled ISC proliferation and differentiation [38, 56]. Intestinal stem cells in *Drosophila* first produce progenitor cells called enteroblasts (EBs). Enteroblasts differentiate into enteroendocrine (EE) or enterocyte (EC) cell types without further division. *Drosophila* EE cells perform the functions of both EE and Paneth cells of humans [35].

In a study in the midgut of *Drosophila*, the autophagy-related protein Atg16 was determined to have a role in the differentiation of enteroendocrine cells from ISCs in association with the Slit/Robo pathway involved in developmental events [70]. Later studies in mice revealed the existence of a similar regulatory mechanism in mammalian intestinal stem cells [108]. Slit/Robo signaling has been identified as a key regulator of EE cell differentiation under homeostatic conditions in *Drosophila*. Fully differentiated EE cells secrete Slit, which acts on its receptor Robo2 expressed by both ISCs and differentiating precursor cells [8, 116]. Robo signaling in intestinal stem cells is also associated with Prospero, an evolutionarily conserved transcription factor and homeodomain-related protein with double subcellular localization that is involved in initiating the differentiation of progenitors in various tissues [16]. Activation of Robo signaling represses the homeodomain transcription factor, Prospero, leading to differentiation of EBs into ECs rather than EE cells (Fig. 7.6).

In the absence of Slit-Robo signaling, Ras-associated binding 19 (RAB19) protein can bind directly to Atg16 to promote gut secretory cell differentiation in *Drosophila*. The RAB family of small GTPases are master regulators of protein and lipid uptake

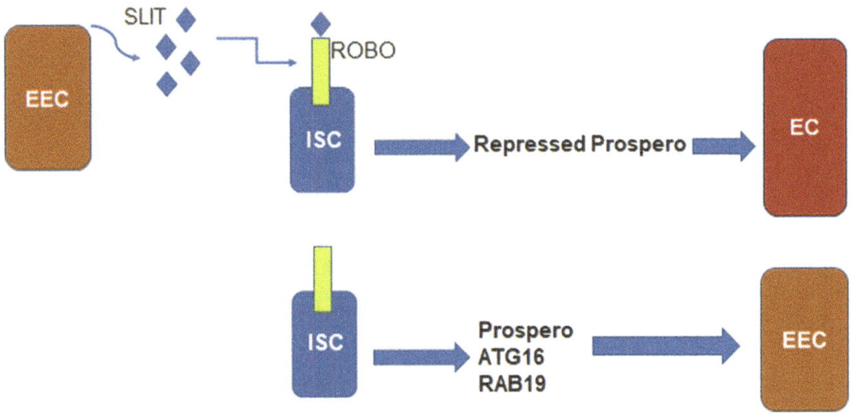

Fig. 7.6 Atg16 is involved in the differentiation process of ISCs into enteroendocrine cells in *Drosophila*

into cell locations at precise times. With this negative feedback loop, the number of mature EE cells in the gut is tightly controlled. The autophagy-related protein Atg16 usually contains 3 key domains. These are an Atg5-binding motif (Atg5BD), a helix-coil domain (CCD), and a WD repeat (WD40) domain. It has been determined that both Atg5BD and CCD are required for starvation-induced autophagy, but the WD40 domain mediates protein–protein interactions [75]. Pre-enteroendocrine (pre-EE) cells expressing Atg16 or mutant RAB19 lacking the WD40 domain have failed to differentiate into functional EE cells and accumulated in the gut of *Drosophil*a. It has been shown that loss of Atg16 or its binding partner RAB19 affects their differentiation by impairing slit production in pre-enteroendocrine cells, and the function of Atg16 in this process is independent of autophagy [70].

In recent years, *Drosophila* has also been used as a model organism in understanding the biology of the aging process and in studies related to diseases such as cancer, which are more common with age. Diseases such as tissue aging and cancer are associated with age-related changes in adult stem cells [33]. Metformin, a widely used drug for type 2 diabetes, has an anti-aging effect [67] and inhibits age and oxidative stress-related ISC abnormalities by suppressing the protein kinase B (AKT)/TOR pathway [39]. In a study investigating the effect of metformin on the aging-related phenotype of *Drosophila* ISC, it was shown that the drug inhibits the aging-related phenotype, but Atg6 is required for this effect [69]. The autophagy-related gene Atg6 (Autophagy-related gene 6 in mammals, Beclin1) is an essential component of endosome formation and has a crucial role in both autophagy and endocytosis. Loss of Atg6/Beclin1 is common in human breast, ovarian, and prostate cancer [13].

Atg6 and other autophagy-related genes are negatively regulated by the AKT/TOR pathway [100]. Allelic loss of Beclin1 is associated with activation of the DNA damage response in vitro. It has been determined that autophagy also prevents DNA damage in stem cells in the *Drosophila* gut [71]. Atg6 deletion in ISC induces ISC senescence phenotypes such as hyperproliferation, centrosome amplification, and

DNA damage accumulation, whereas EB-specific Atg6 depletion has no effect [69]. DNA damage has been shown to cause cell cycle arrest in these cells due to defective progression of the S-phase, which can ultimately lead to apoptotic elimination of cells [79]. Since the role of Atg6 in the maintenance of oxidative stress is known [106], cell cycle arrest and subsequent apoptosis as a result of autophagy inhibition resulting from Atg6 deficiency is likely the result of DNA damage resulting from excessive ROS production by unhealthy mitochondria [71]. Accumulation of CHK2, a kinase activated by DNA damage, in the nuclei of Atg6 knockout ISCs can also cause cell cycle arrest and apoptosis after the loss of autophagy. Loss of CHK2 has been reported to cause hyperproliferation of both normal and autophagy-deficient ISCs (Fig. 7.7). In addition, studies in mammals have reported that the anti-proliferative effect of metformin is partially or completely dependent on autophagy, and metformin stimulates Beclin1-triggered autophagy [58].

The scientists obtained important data to understand how autophagy regulates ISC behavior by manipulating autophagy-related genes in the *Drosophila* intestinal epithelium. In *Drosophila* ISC, for example, Atg1 activity has been determined to be required for the suppression of tumor-like hyperproliferation. Depletion of other autophagy-related genes, Atg5 and Atg14, was found to suppress ISC proliferation and differentiation [117]. These results suggest that autophagy-related genes are critical for ISC integrity. In addition, loss of UVRAG, which plays a role in autophagosome formation, has been reported to induce intestinal dysplasia independent of autophagy due to the defect in endocytic function [72].

Research has also been done on the role of autophagy in the division processes of stem cells in the *Drosophila* midgut. Sorting Nexins (SNX) are a conserved class of phospholipid-binding proteins and regulate a variety of intracellular membrane trafficking events. The *Drosophila* genome, SH3PX1, and its human ortholog, SNX18,

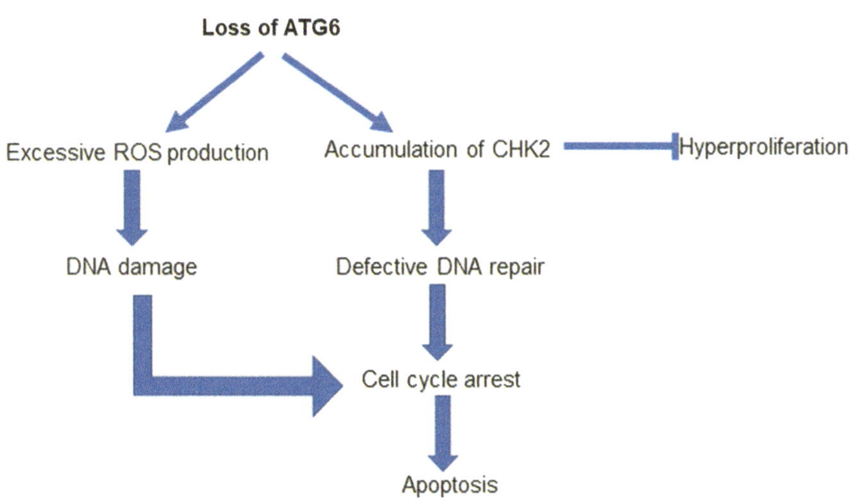

Fig. 7.7 Loss of Atg6 causes death of ISCs in different ways

have been shown to bind to PI(4,5)P2-containing endosomal membranes, form tubules and promote autophagosome formation [48]. Loss of SH3PX1 resulted in a dramatic increase in the proliferation of ISC. It has been determined that the epidermal growth factor receptor (EGFR)/ extracellular signal-regulated kinase (ERK) signaling pathway, which is an important cellular signaling pathway that regulates cell proliferation, differentiation, cell cycle, and migration, is directly activated by the loss of SH3PX1 function. Inhibition of this signaling pathway significantly blocked proliferation in ISCs. Under normal conditions, EGFR signaling in intestinal stem cells is kept low in ISCs through autophagy, contributing to the balance between ISC division rates. Upon disruption of autophagy as a result of SH3PX1 loss, enhanced cell-autonomous EGFR signaling increases ISC proliferation and results in hyperplasia [117] (Fig. 7.8a).

The role of TOR kinase, which is the main regulator of cell growth and an inhibitor of autophagy, has been demonstrated in Drosophila midgut stem cells [2, 91]. Ras homolog enriched in brain (Rheb) plays an important role in insulin/TOR/ signaling and regulates cell growth and cell cycle progression [4]. Rheb has been shown to interact with and activates TOR in vitro [89]. Overexpression of the TOR activator Rheb in flies has been reported to lead to loss of intestinal integrity and premature death (Fig. 7.8b). In addition, loss of the Rheb inhibitor tuberous sclerosis complex 2 (TSC2) impairs ISC maintenance and intestinal tissue homeostasis via TOR. Another study in Drosophila showed that loss of Atg9 and other autophagy-related proteins

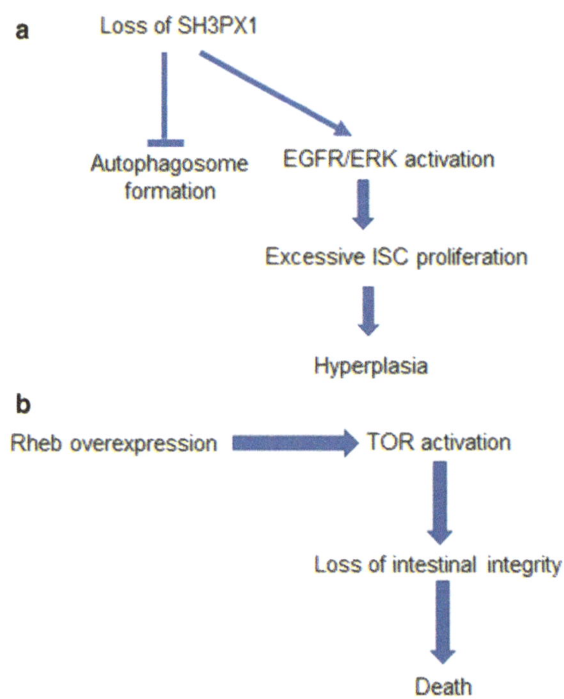

Fig. 7.8 SH3PX1-mediated autophagy controls stem cell proliferation (**a**) and excessive TOR activation reduces the number and function of ISC then causes abnormal intestinal structure (**b**) in *Drosophila*

resulted in increased Tor signaling and abnormally enlarged IECs, but no change in cell numbers. In contrast inhibition of TOR by rapamycin and consequent stimulation of autophagy reverses this phenotype [104]. Age-related reduction in both number and function in ISCs has been determined to be associated with TOR and autophagy. In *Drosophila*, it has been shown that autophagy induced in aged ISCs can partially alleviate impaired protein homeostasis [84].

In addition to *Drosophila melanogaster*, the silkworm *Bombyx mori* is another insect species that is frequently used as a model organism in physiology studies. During larval-pupal metamorphosis, the *Bombyx mori* midgut undergoes a remodeling process. In this process, while the larval midgut cells degenerate by programmed cell death, the pupal midgut is formed by the proliferation and differentiation of ISC [28]. It was determined that administration of autophagy inhibitor chloroquine to *Bombyx mori* larvae in the prepupal period did not prevent ISC proliferation, but differentiation was interrupted, and the healthy epithelial organization did not occur. Inhibition of autophagy also induced necrosis-like cell death in some areas of the epithelium. These results show that autophagy is required for normal differentiation of ISCs in *Bombyx mori* [30].

Pathologies Associated with Autophagy in Intestinal Stem Cells

As can be seen from studies on the role of autophagy in intestinal stem cells, it is a fundamental mechanism in stem cell maintenance, self-renewal, and differentiation processes. Numerous studies have shown that abnormalities in the proliferation and differentiation processes of the ISCs may lead to the development of certain pathologies in the gastrointestinal tract [27]. A study in *Drosophila* revealed that defects in the shi/dynamin, Rab5, Rab7, SH3PX1, Atg1, Atg5, Atg6, Atg7, Atg8a, Atg9, Atg12, Atg16, and Syx17 genes in the endocytosis-autophagy network result in excessive ISC proliferation. It has been determined that altering intracellular vesicle trafficking by suppressing macroautophagy/autophagy or endocytosis genes deregulates intestinal stem cell (ISC) proliferation and leads to severe intestinal hyperplasia affecting viability. Such disruptions in stem cell activities can then lead to colorectal cancer or intestinal atrophy [118].

In addition to its functions in digestive processes, the gastrointestinal (GI) tract of metazoans contributes to homeostasis by acting as a barrier that protects the body from exogenous elements, including pathogenic bacteria. Autophagy can regulate many aspects of gut physiology in the maintenance of intestinal epithelial architecture, metabolic regulation, the function of specific intestinal epithelial subsets, regulation of inflammatory pathways, and defense against infection. Damage to the intestinal barrier is characteristic of intestinal pathologies and especially chronic inflammation of the intestine known as inflammatory bowel disease (IBD) represented by Crohn's disease (CD) and ulcerative colitis (UC) [88]. The tight link between autophagy and intestine physiology was determined by the observation of the strong association between genes of the best autophagy pathway and susceptibility to CD [31]. The role of autophagy in the mucosal barrier has been demonstrated by the genetic relationship between Atg16L1 and Crohn's disease, an important form of inflammatory bowel disease [20]. It has been determined that IBD susceptibility-associated Atg16L1 (Atg16L1T300A) induces a caspase cleavage site then destabilizes the protein product and reduces autophagy [68]. As a result, decreased autophagy causes dysfunction of the intestinal barrier and the inflammatory process of IBD.

Concluding Remarks

Maintaining the cellular homeostasis and functionality of the digestive system is a necessity for the continuation of life, as demonstrated by studies in different animal groups, including humans and insects. The continuity of stem cells and the tight regulation of their proliferation and differentiation processes in the intestinal tissue, where the cellular cycle takes place continuously are essential for the maintenance of intestinal homeostasis and function. An increasing number of studies have proven that autophagy is an important cellular mechanism that plays a critical role in the development of organisms and adult life. As seen in the literature mentioned in this review, autophagy also plays a critical role in the homeostasis of intestinal stem cells. However, there are still many questions waiting to be answered. It is seen that there is relatively more known about the role of autophagy in the viability and proliferation processes of intestinal stem cells, but research on the possible roles of autophagy in the differentiation process is insufficient. Pathologies occurring in the intestine deeply affect human and animal health and can lead to death. The importance of stem cell functions in the formation and maintenance of healthy intestinal epithelium is well known. It is necessary to reveal the cellular mechanisms underlying intestinal diseases for the development of prevention, diagnosis, and treatment processes. Therefore, further studies can contribute to this aim by increasing the knowledge about the role of autophagy in the homeostasis of intestinal stem cells, self-renewal, proliferation, and differentiation processes.

Acknowledgements This work was supported by Ege University, Faculty of Science, Insect Physiology Research Laboratory. Figures were created with Microsoft Office PowerPoint.

Conflict of Interest The author declares that they have no conflict of interest.

Ethical Approval The author declares that this article does not contain any studies with human participants or animals.

References

1. Agnello M, Bosco L, Chiarelli, R, Martino, C, Roccheri, MC (2015) The role of autophagy and apoptosis during embryo development. In: Cell death-autophagy, apoptosis and necrosis. IntechOpen. https://doi.org/10.5772/61765
2. Amcheslavsky A, Ito N, Jiang J, Ip YT (2011) Tuberous sclerosis complex and Myc coordinate the growth and division of *Drosophila* intestinal stem cells. J Cell Biol 193(4):695–710. https://doi.org/10.1083/jcb.201103018
3. Asano J et al (2017) Intrinsic autophagy is required for the maintenance of intestinal stem cells and for irradiation-induced intestinal regeneration. Cell Rep 20(5):1050–1060. https://doi.org/10.1016/j.celrep.2017.07.019
4. Aspuria PJ, Tamanoi F (2004) The Rheb family of GTP-binding proteins. Cell Signal 16:1105–1112. https://doi.org/10.1016/j.cellsig.2004.03.019
5. Bach SP, Renehan AG, Potten CS (2000) Stem cells: the intestinal stem cell as a paradigm. Carcinogenesis 21(3):469–476. https://doi.org/10.1093/carcin/21.3.469
6. Barker N (2014) Adult intestinal stem cells: critical drivers of epithelial homeostasis and regeneration. Nat Rev Mol Cell Biol 15:19–33. https://doi.org/10.1038/nrm3721
7. Baulies A et al (2020) The transcription co-repressors MTG8 and MTG16 regulate exit of intestinal stem cells from their niche and differentiation into enterocyte vs secretory lineages. Gastroenterology 159(4):1328–1341. https://doi.org/10.1053/j.gastro.2020.06.012
8. Biteau B, Jasper H (2014) Slit/Robo signaling regulates cell fate decisions in the intestinal stem cell lineage of *Drosophila*. Cell Rep 7(6):1867–1875. https://doi.org/10.1016/j.celrep.2014.05.024
9. Bjerknes M, Cheng H (2005) Gastrointestinal stem cells. II. Intestinal stem cells. Am J Physiol Gastrointest Liver Physiol 289:381–387. https://doi.org/10.1152/ajpgi.00160.2005
10. Blahnova VH (2020) Combinations of growth factors for human mesenchymal stem cell proliferation and osteogenic differentiation. Bone Joint Res 9(7):412–420. https://doi.org/10.1302/2046-3758.97.BJR-2019-0183.R2
11. Blockus H, Chédotal A (2016) Slit-Robo signaling. Development 143:3037–3044. https://doi.org/10.1242/dev.132829
12. Cairns G, Thumiah-Mootoo M, Burelle Y, Khacho M (2020) Mitophagy: a new player in stem cell biology. Biology (Basel) 9(12):481. https://doi.org/10.3390/biology9120481
13. Cao Y, Klionsky DJ (2007) Physiological functions of Atg6/Beclin 1: a unique autophagy-related protein. Cell Res 17:839–849. https://doi.org/10.1038/cr.2007.78
14. Chang NC (2020) Autophagy and stem cells: self-eating for self-renewal. Front Cell Dev Biol 4(8):138. https://doi.org/10.3389/fcell.2020.00138
15. Chen S et al (2021) Role of autophagy in the maintenance of stemness in adult stem cells: a disease-relevant mechanism of action front cell. Dev Biol 9:715200. https://doi.org/10.3389/fcell.2021.715200
16. Choksi SP et al (2006) Prospero acts as a binary switch between self-renewal and differentiation in *Drosophila* neural stem cells. Dev Cell 11(6):775–789. https://doi.org/10.1016/j.devcel.2006.09.015
17. Collier JJ et al (2021) Emerging roles of ATG7in human healthand disease. EMBO Mol Med 13(12):e14824. https://doi.org/10.15252/emmm.202114824
18. Cooney R et al (2010) NOD2 stimulation induces autophagy in dendritic cells influencing bacterial handling and antigen presentation. Nat Med 16:90–97. https://doi.org/10.1038/nm.2069
19. Cray P, Sheahan BJ, Dekaney CM (2021) Secretory sorcery: paneth cell control of intestinal repair and homeostasis. Cell Mol Gastroenterol Hepatol 12:1239–1250. https://doi.org/10.1016/j.jcmgh.2021.06.006
20. Cummings JRF et al (2007) Confirmation of the role of ATG16L1 as a Crohn's disease susceptibility gene. Inflamm Bowel Dis 13(8):941–946. https://doi.org/10.1002/ibd.20162
21. Demitrack ES, Samuelson LC (2016) Notch regulation of gastrointestinal stem cells. J Physiol 594(17):4791–4803. https://doi.org/10.1113/JP271667

22. Denu RA, Hematti P (2016) Effects of oxidative stress on mesenchymal stem cell biology. Oxid Med Cell Longev 2016:2989076. https://doi.org/10.1155/2016/2989076
23. Dickson I (2017) A case of metabolic identity in the intestinal crypt. Nat Rev Gastroenterol Hepatol 14:259. https://doi.org/10.1038/nrgastro.2017.41
24. El-Gowily AH, Abosheasha MA (2021) Differential mechanisms of autophagy in cancer stem cells: emphasizing gastrointestinal cancers. Cell Biochem Funct 39:162–173. https://doi.org/10.1002/cbf.3552
25. Galluzzi L, Green DR (2019) Autophagy-independent functions of the autophagy machinery. Cell 177(7):1682–1699. https://doi.org/10.1016/j.cell.2019.05.026
26. Geng J, Klionsky DJ (2008) The Atg8 and Atg12 ubiquitin-like conjugation systems in macroautophagy. Protein modifications: beyond the usual suspects review series. EMBO Rep 9(9):859–864. https://doi.org/10.1038/embor.2008.163
27. Gersemann M, Stange EF, Wehkamp J (2011) From intestinal stem cells to inflammatory bowel diseases. World J Gastroenterol 17(27):3198–3203. https://doi.org/10.3748/wjg.v17.i27.3198
28. Goncu E, Uranlı R, Selek G, Parlak O (2016) Developmental expression of ecdysone-related genes associated with metamorphic changes during midgut remodeling of silkworm *Bombyx mori* (Lepidoptera:Bombycidae). J Insect Sci 16(1):86. https://doi.org/10.1093/jisesa/iew061
29. Guan JL et al (2013) Autophagy in stem cells. Autophagy 9(6):830–849. https://doi.org/10.4161/auto.24132
30. Gunay B, Goncu E (2021) Role of autophagy in midgut stem cells of silkworm *Bombyx mori*, during larval-pupal metamorphosis. Arch Insect Biochem Physiol 108:e21832. https://doi.org/10.1002/arch.21832
31. Haq S, Grondin J, Banskota S, Khan WI (2019) Autophagy: roles in intestinal mucosal homeostasis and inflammation. J Biomed Sci 26:19. https://doi.org/10.1186/s12929-019-0512-2
32. He D et al (2020) Gut stem cell aging is driven by mTORC1 via a p38 MAPK-p53 pathway. Nat Commun 11:37. https://doi.org/10.1038/s41467-019-13911-x
33. He Y, Jasper H (2014) Studying aging in *Drosophila*. Methods 68(1):129–133. https://doi.org/10.1016/j.ymeth.2014.04.008
34. Hochmuth CE, Biteau B, Jasper H (2011) Redox Regulation by Keap1 and Nrf2 Controls Intestinal Stem Cell Proliferation in Drosophila. Cell Stem Cell 8(2):188–199, S1934590910007010. https://doi.org/10.1016/j.stem.2010.12.006
35. Hung RJ et al (2020) A cell atlas of the adult *Drosophila* midgut. PNAS 117(3):1514–1523. https://doi.org/10.1073/pnas.1916820117
36. Ireland H, Houghton C, Howard L, Winton DJ (2005) Cellular inheritance of a cre-activated reporter gene to determine Paneth cell longevity in the murine small intestine. Dev Dyn 233(4):1332–1336. https://doi.org/10.1002/dvdy.20446
37. Jang M et al (2018) AMPK contributes to autophagosome maturation and lysosomal fusion. Sci Rep 8:12637. https://doi.org/10.1038/s41598-018-30977-7
38. Jiang H, Edgar BA (2011) Intestinal stem cells in the adult *Drosophila* midgut. Exp Cell Res 317(19):2780–2788. https://doi.org/10.1016/j.yexcr.2011.07.020
39. Jiang LL, Liu L (2020) Effect of metformin on stem cells: molecular mechanism and clinical prospect. World J Stem Cells 12(12):1455–1473. https://doi.org/10.4252/wjsc.v12.i12.1455
40. Jiang H et al (2009) Cytokine/Jak/Stat signaling mediates regeneration and homeostasis in the *Drosophila* midgut. Cell 137:1343–1355. https://doi.org/10.1016/j.cell.2009.05.014
41. Jimenez-Sanchez M et al (2012) The Hedgehog signalling pathway regulates autophagy. Nat Commun 13(3):1200. https://doi.org/10.1038/ncomms2212
42. Johansen T, Lamark T (2020) Selective autophagy: ATG8 family proteins, LIR motifs and cargo receptors. J Mol Biol 432:80–103. https://doi.org/10.1016/j.jmb.2019.07.016
43. Jones RM, Mercante JW, Neish AS (2012) Reactive oxygen production induced by the gut microbiota: pharmacotherapeutic implications. Curr Med Chem 19(10):1519–1529. https://doi.org/10.2174/092986712799828283

44. Kapuy O, Papp D, Vellai T, Banhegyi G, Korcsmaros T (2018) Systems-level feedbacks of NRF2 controlling autophagy upon oxidative stress response. Antioxidants 7(3):39. https://doi.org/10.3390/antiox7030039
45. Kensler TW, Wakabayashi N, Biswal S (2007) Cell survival responses to environmental stresses via the Keap1-Nrf2-ARE pathway. Ann Rev Pharmacol Toxicol 47:89–116. https://doi.org/10.1146/annurev.pharmtox.46.120604.141046
46. Kitada M, Koya D (2021) Autophagy in metabolic disease and ageing. Nat Rev Endocrinol 17(11):647–661. https://doi.org/10.1038/s41574-021-00551-9
47. Kiyono K et al (2009) Autophagy is activated by TGF-beta and potentiates TGF-beta-mediated growth inhibition in human hepatocellular carcinoma cells. Cancer Res 69:8844–8852. https://doi.org/10.1158/0008-5472.CAN-08-4401
48. Knævelsrud H et al (2013) Membrane remodeling by the PX-BAR protein SNX18 promotes autophagosome formation. J Cell Biol 202(2):331–349. https://doi.org/10.1083/jcb.201205129
49. Komatsu M et al (2005) Impairment of starvation-induced and constitutive autophagy in Atg7-deficient mice. J Cell Biol 169(3):425–434. https://doi.org/10.1083/jcb.200412022
50. Kopan R (2012) Notch signaling. Cold Spring Harb Perspect Biol 4(10):a011213. https://doi.org/10.1101/cshperspect.a011213
51. Lampert MA, Gustafsson AB (2020) Mitochondria and autophagy in adult stem cells: proliferate or differentiate. J Muscle Res Cell Motil 41:355–362. https://doi.org/10.1007/s10974-019-09542-w
52. Lee IH et al (2012) Atg7 Modulates p53 activity to regulate cell cycle and survival during metabolic stress. Science 336(6078):225–228. https://doi.org/10.1126/science.1218395
53. Lee YK, Lee JA (2016) Role of the mammalian ATG8/LC3 family in autophagy: differential and compensatory roles in the spatiotemporal regulation of autophagy. BMB Rep 49:424–430. https://doi.org/10.5483/bmbrep.2016.49.8.081
54. Leedham SJ, Brittan M, McDonald SA, Wright NA (2005) Intestinal stem cells. Cell Mol Med 9(1):11–24. https://doi.org/10.1111/j.1582-4934.2005.tb00333.x
55. Levy A et al (2020) Innate immune receptor NOD2 mediates LGR5+ intestinal stem cell protection against ROS cytotoxicity via mitophagy stimulation. PNAS 117(4):1994–2003. https://doi.org/10.1073/pnas.1902788117
56. Li H, Jasper H (2016) Gastrointestinal stem cells in health and disease: from flies to humans. Dis Model Mech 9(5):487–499. https://doi.org/10.1242/dmm.024232
57. Liemburg-Apers DC et al (2015) Interactions between mitochondrial reactive oxygen species and cellular glucose metabolism. Arch Toxicol 89(8):1209–1226. https://doi.org/10.1007/s00204-015-1520-y
58. Liu S et al (2020) Metformin promotes Beclin1-dependent autophagy to inhibit the progression of gastric cancer. Onco Targets Ther 13:4445–4455. https://doi.org/10.2147/OTT.S242298
59. Lorzadeh S, Kohan L, Ghavami S, Azarpira N (2021) Autophagy and the Wnt signaling pathway: a focus on Wnt/β-catenin signaling. Biochimica et Biophysica Acta (BBA)-Mol Cell Res 1868(3):118926. https://doi.org/10.1016/j.bbamcr.2020.118926
60. Madison BB et al (2005) Epithelial hedgehog signals pattern the intestinal crypt-villus axis. Development 132:279–289. https://doi.org/10.1242/dev.01576
61. Marshman E, Booth C, Potten CS (2002) The intestinal epithelial stem cell. BioEssays 24(1):91–98. https://doi.org/10.1002/bies.10028
62. Martinez PS, Giuranno L, Vooijs M, Coppes RP (2021) The radiation-induced regenerative response of adult tissue-specific stem cells: models and signaling pathways. Cancers 13(4):855. https://doi.org/10.3390/cancers13040855
63. Matsuzawa-Ishimoto Y et al (2017) Autophagy protein ATG16L1 prevents necroptosis in the intestinal epithelium. J Exp Med 214(12):3687–3705. https://doi.org/10.1084/jem.20170558
64. Menshikov M, Zubkova E, Stafeev J, Parfyonova Y (2021) Autophagy, mesenchymal stem cell differentiation, and secretion. Biomedicines 9(9):1178. https://doi.org/10.3390/biomedicines9091178

65. Miguel-Aliaga I, Jasper H, Lemaitre B (2018) Anatomy and physiology of the digestive tract of Drosophila melanogaster. Genetics 210(2):357–396. https://doi.org/10.1534/genetics.118.300224
66. Mizushima N, Levine B (2010) Autophagy in mammalian development and differentiation. Nat Cell Biol 12:823–830. https://doi.org/10.1038/ncb0910-823
67. Mohammed I (2021) A critical review of the evidence that metformin is a putative anti-aging drug that enhances healthspan and extends lifespan. Front Endocrinol (Lausanne) 12:718942. https://doi.org/10.3389/fendo.2021.718942
68. Murthy A et al (2014) A Crohn's disease variant in Atg16l1 enhances its degradation by caspase 3. Nature 506(7489):456–462. https://doi.org/10.1038/nature13044
69. Na HJ et al (2018) Deficiency of Atg6 impairs beneficial effect of metformin on intestinal stem cell aging in *Drosophila*. Biochem Biophys Res Commun 498(1):18–24. https://doi.org/10.1016/j.bbrc.2018.02.191
70. Nagy P et al (2017) *Drosophila* Atg16 promotes enteroendocrine cell differentiation via regulation of intestinal Slit/Robo signaling. Development 144(21):3990–4001. https://doi.org/10.1242/dev.147033
71. Nagy P, Sándor GO, Juhász G (2018) Autophagy maintains stem cells and intestinal homeostasis in *Drosophila*. Sci Rep 8:4644. https://doi.org/10.1038/s41598-018-23065-3
72. Nagy P, Kovács L, Sándor GO, Juhász G (2016) Stem-cell-specific endocytic degradation defects lead to intestinal dysplasia in *Drosophila*. Dis Model Mech 9(5):501–512. https://doi.org/10.1242/dmm.023416
73. Nigro G et al (2014) The cytosolic bacterial peptidoglycan sensor Nod2 affords stem cell protection and links microbes to gut epithelial regeneration. Cell Host Microbe 15(6):792–798. https://doi.org/10.1016/j.chom.2014.05.003
74. Noda T, Suzuki K, Ohsumi Y (2002) Yeast autophagosomes: de novo formation of a membrane structure. Trends Cell Biol 12(5):231–235. https://doi.org/10.1016/s0962-8924(02)02278-x
75. Noda NN, Inagaki F (2015) Mechanisms of autophagy. Ann Rev Biophys 44:101–122. https://doi.org/10.1146/annurev-biophys-060414-034248
76. Ogawa C, Liu YJ, Kobayashi KS (2011) Muramyl dipeptide and its derivatives: peptide adjuvant in immunological disorders and cancer therapy. Curr Bioact Compd 7(3):180–197. https://doi.org/10.2174/157340711796817913
77. Ohlstein B, Spradling A (2006) Multipotent *Drosophila* intestinal stem cells specify daughter cell fates by differential notch signaling. Science 315(5814):988–992. https://doi.org/10.1126/science.1136606
78. Parzych KR, Klionsky DJ (2014) An overview of autophagy: morphology, mechanism, and regulation. Antioxid Redox Signal 20(3):460–473. https://doi.org/10.1089/ars.2013.5371
79. Pellegata NS, Antoniono RJ, Redpath JL, Stanbridge EJ (1996) DNA damage and p53-mediated cell cycle arrest: a reevaluation. Proc Natl Acad Sci USA 93(26):15209–15214. https://doi.org/10.1073/pnas.93.26.15209
80. Phadwal K, Watson AS, Simon AK (2013) Tightrope act: autophagy in stem cell renewal, differentiation, proliferation, and aging. Cell Mol Life Sci 70(1):89–103. https://doi.org/10.1007/s00018-012-1032-3
81. Richmond CA et al (2015) Dormant intestinal stem cells are regulated by PTEN and nutritional status. Cell Rep 13:2403–2411. https://doi.org/10.1016/j.celrep.2015.11.035
82. Rockenfeller P et al (2015) Phosphatidylethanolamine positively regulates autophagy and longevity. Cell Death Differ 22(3):499–508. https://doi.org/10.1038/cdd.2014.219
83. Rodríguez-Colman MJ et al (2017) Interplay between metabolic identities in the intestinal crypt supports stem cell function. Nature 543(7645):424–427. https://doi.org/10.1038/nature21673
84. Rodriguez-Fernandez IA, Tauc HM, Jasper H (2020) Hallmarks of aging *Drosophila* intestinal stem cells. Mech Ageing Dev 190:111285. https://doi.org/10.1016/j.mad.2020.111285
85. Rusten TE, Stenmark H (2010) p62, an autophagy hero or culprit? Nat Cell Biol 12(3):207–209. https://doi.org/10.1038/ncb0310-207

86. Sailaja BS, He XC, Li L (2016) The regulatory niche of intestinal stem cells. J Physiol 594(17):4827–4836. https://doi.org/10.1113/JP271931
87. Salzman NH et al (2010) Enteric defensins are essential regulators of intestinal microbial ecology. Nat Immunol 11(1):76–83. https://doi.org/10.1038/ni.1825
88. Sartor RB (2006) Mechanisms of disease: pathogenesis of Crohn's disease and ulcerative colitis nature clinical practice. Gastroenterology & Hepatology 3(7):390–407. https://doi.org/10.1038/ncpgasthep0528
89. Sato T, Nakashima A, Guo L, Tamanoi F (2009) Specific activation of mTORC1 by Rheb G-protein in vitro involves enhanced recruitment of its substrate protein. J Biol Chem 284(19):12783–12791. https://doi.org/10.1074/jbc.M809207200
90. Sato T et al (2011) Paneth cells constitute the niche for Lgr5 stem cells in intestinal crypts. Nature 469:415–418. https://doi.org/10.1038/nature09637
91. Strilbytska O et al (2017) Activation of TOR/MYC signaling axis in intestinal stem and progenitor cells affects longevity, stress resistance and metabolism in *Drosophila*. Comp Biochem Physiol B Biochem Mol Biol 203:92–99. https://doi.org/10.1016/j.cbpb.2016.09.008
92. Subramani S, Malhotra V (2013) Non-autophagic roles of autophagy-related proteins. EMBO Rep 14(2):143–151. https://doi.org/10.1038/embor.2012.220
93. Suzuki T, Yamamoto M (2015) Molecular basis of the Keap1-Nrf2 system. Free Radic Biol Med 88(Pt B):93–100. https://doi.org/10.1016/j.freeradbiomed.2015.06.006
94. Tong M et al (2019) The role of the Slit/Robo signaling pathway. J Cancer 10(12):2694–2705. https://doi.org/10.7150/jca.31877
95. Towers CG et al (2019) Cancer cells upregulate NRF2 signaling to adapt to autophagy inhibition. Dev Cell 50(6):690–703. https://doi.org/10.1016/j.devcel.2019.07.010
96. Trentesauxa C et al (2020) Essential role for autophagy protein ATG7 in the maintenance of intestinal stem cell integrity. PNAS 117(20):11136–11146. https://doi.org/10.1073/pnas.1917174117
97. Tsogtbaatar E, Landin C, Minter-Dykhouse K, Folmes C (2020) Energy metabolism regulates stem cell pluripotency. Front Cell Dev Biol 8:87. https://doi.org/10.3389/fcell.2020.00087
98. Umar S (2010) Intestinal stem cells. Curr Gastroenterol Rep 12(5):340–348. https://doi.org/10.1007/s11894-010-0130-3
99. Van der Heijden M, Vermeulen L (2019) Stem cells in homeostasis and cancer of the gut. Mol Cancer 18:66. https://doi.org/10.1186/s12943-019-0962-x
100. Vega-Rubín-de-Celis S (2020) The role of Beclin 1-dependent autophagy in cancer. Biology 9(1):4. https://doi.org/10.3390/biology9010004
101. Viswanathan VK (2014) Muramyl dipeptide. Gut Microbes 5(3):275–276. https://doi.org/10.4161/gmic.32181
102. Wang CW, Klionsky DJ (2003) The molecular mechanism of autophagy. Mol Med 9:65–76. https://doi.org/10.1007/BF03402040
103. Weeden CEE, Asselin-Labat ML (2018) Mechanisms of DNA damage repair in adult stem cells and implications for cancer formation. Biochimica et Biophysica Acta (BBA)—Mol Basis Dis 1864(1): 89–101. https://doi.org/10.1016/j.bbadis.2017.10.015
104. Wen JK et al (2017) Atg9 antagonizes TOR signaling to regulate intestinal cell growth and epithelial homeostasis in *Drosophila*. eLife 6:e29338. https://doi.org/10.7554/eLife.29338
105. Wittkopf N et al (2012) Lack of ıntestinal epithelial Atg7 affects paneth cell granule formation but does not compromise ımmune homeostasis in the Gut. Clin Dev Immunol 278059. https://doi.org/10.1155/2012/278059
106. Wu H, Wang MC, Bohmann D (2009) JNK protects *Drosophila* from oxidative stress by trancriptionally activating autophagy. Mech Dev 126(8–9):624–637. https://doi.org/10.1016/j.mod.2009.06.1082
107. Wu X et al (2016) Autophagy regulates Notch degradation and modulates stem cell development and neurogenesis. Nat Commun 7:10533. https://doi.org/10.1038/ncomms10533

108. Xie J (2020) Slit2/Robo1 mitigates DSS-induced ulcerative colitis by activating autophagy in intestinal stem cell. Int J Biol Sci 16(11):1876–1887. https://doi.org/10.7150/ijbs.42331
109. Xiong Q et al (2019) The role of ATG16 in autophagy and the ubiquitin proteasome system. Cells 8(1):2. https://doi.org/10.3390/cells8010002
110. Yang Y (2020) Autophagy promotes mammalian survival by suppressing oxidative stress and p53. Genes Dev 34(9–10):688–700. https://doi.org/10.1101/gad.335570.119
111. Yang Y, White E (2020) Autophagy suppresses TRP53/p53 and oxidative stress to enable mammalian survival. Autophagy 16(7):1355–1357. https://doi.org/10.1080/15548627.2020.1765522
112. Ye X, Zhou XJ, Zhang H (2018) Exploring the role of autophagy-related gene 5 (ATG5) yields important insights into autophagy in autoimmune/autoinflammatory diseases. Front Immunol 9:2334. https://doi.org/10.3389/fimmu.2018.02334
113. Yilmaz ÖH et al (2012) mTORC1 in the Paneth cell niche couples intestinal stem-cell function to calorie intake. Nature 486(7404):490–495. https://doi.org/10.1038/nature11163
114. Yin H et al (2018) The therapeutic and pathogenic role of autophagy in autoimmune diseases. Front Immunol 9:1512. https://doi.org/10.3389/fimmu.2018.01512
115. Yun CW, Lee SH (2018) The roles of autophagy in cancer. Int J Mol Sci 19(11):3466. https://doi.org/10.3390/ijms19113466
116. Zeng X et al (2015) Genome-wide RNAi screen identifies networks involved in intestinal stem cell regulation in *Drosophila*. Cell Rep 10(7):1226–1238. https://doi.org/10.1016/j.celrep.2015.01.051
117. Zhang P et al (2019) An SH3PX1-dependent endocytosis-autophagy network restrains intestinal stem cell proliferation by counteracting EGFR-ERK signaling. Dev Cell 49:574–589. https://doi.org/10.1016/j.devcel.2019.03.029
118. Zhang P, Holowatyj AN, Ulrich CM, Edgar BA (2019) Tumor suppressive autophagy in intestinal stem cells controls gut homeostasis. Autophagy 15(9):1668–1670. https://doi.org/10.1080/15548627.2019.1633863
119. Zhang J et al (2012) The fibroblast growth factor signaling axis controls cardiac stem cell differentiation through regulating autophagy. Autophagy 8(4):690–691. https://doi.org/10.4161/auto.19290
120. Zhou WJ, Geng ZH, Spence JR, Geng JG (2013) Induction of intestinal stem cells by R-spondin 1 and Slit2 augments chemoradioprotection. Nature 501(7465):107–111. https://doi.org/10.1038/nature12416
121. Zhu G, Hu J, Xi R (2021) The cellular niche for intestinal stem cells: a team effort. Cell Regen 10(1):1. https://doi.org/10.1186/s13619-020-00061-5
122. Zorov DB, Juhaszova M, Sollott SJ (2014) Mitochondrial reactive oxygen species (ROS) and ROS-Induced ROS release. Physiol Rev 94(3):909–950. https://doi.org/10.1152/physrev.00026.2013

Chapter 8
The Autophagy Lysosomal Pathway: Friend or Foe in Adult Neural Stem Cells?

Isabel Calatayud-Baselga and Helena Mira

Abstract Adult neural stem cells (NSCs) located in defined regions of the mammalian brain continuously produce new neurons throughout life. This process of stem cell-driven neurogenesis is considered as a form of structural plasticity that allows the mature brain to adapt to its environment. Preservation of a healthy proteome through autophagy is key for maintaining NSC function. The autophagy lysosomal pathway is directly linked to the activation of quiescent NSC reservoirs and has a cell-autonomous role in the maturation of newborn neurons. However, excessive autophagy can also lead to autophagic cell death of the adult NSCs. Since NSC homeostasis and neurogenesis decline during ageing and are altered in several pathological conditions, autophagy is emerging as a relevant pharmacological target. Interventions modulating autophagy may prevent the loss of NSCs and/or rescue their activity, helping perhaps to rejuvenate the brain neurogenic niches during ageing or recover their homeostasis in brain pathology.

Keywords Autophagy · Lysosome · Protein aggregates · Neural stem cell · Adult neurogenesis · Quiescence · Autophagic cell death

Abbreviations

ACD	Autophagic cell death
AMPK	5' AMP-activated protein kinase
Baf A	Bafilomycin A1
BMP	Bone morphogenetic protein

I. Calatayud-Baselga
Instituto de Biomedicina de Valencia, Consejo Superior de Investigaciones Científicas (IBV-CSIC), València, Spain

H. Mira (✉)
IBV-CSIC, Carrer Jaume Roig, 11, E-46010 València, Spain
e-mail: hmira@ibv.csic.es

B. V. Shravage and K. Turksen (eds.), *Autophagy in Stem Cell Maintenance and Differentiation*, Stem Cell Biology and Regenerative Medicine 73,
https://doi.org/10.1007/978-3-031-17362-2_8

CRS	Chronic restraint stress
CORT	Corticosterone
DG	Dentate gyrus
EGFP	Enhanced green fluorescent protein
EGFR	Epidermal growth factor receptor
FOXO	Forkhead Box O transcription factor
GCL	Granule cell layer
GLAST	Glutamate/Aspartate Transporter
GFAP	Glial Fibrillary Acidic Protein
Hcy	Homocysteine
3-MA	3-Methyladenine
ML	Molecular layer
NPCs	Neural progenitor cells
NSCs	Neural stem cells
qNSCs	Quiescent neural stem cells
aNSCs	Active neural stem cells
OB	Olfactory bulb
PI3K	Phosphatidylinositol 3-kinase
RFP	Red fluorescent protein
RMS	Rostral migratory stream
SGK	Serum/glucocorticoid regulated kinase
SGZ	Subgranular zone
TFEB	Transcription factor EB
mTOR	Mammalian target of Rapamycin
V-SVZ	Ventricular-subventricular zone

Introduction

Neurogenesis is defined as the process of generating new functional neurons from neural stem cells (NSCs). Although neurogenesis ceases in most brain areas after the developmental phase, stem cell-driven neurogenesis persists in restricted regions as a form of structural plasticity that allows the mature brain to adapt to environmental and age-related changes. Throughout most of the twentieth century, exciting work evidenced the existence of neurogenesis in two areas of the adult rodent brain, the ventricular–subventricular zone (V-SVZ) and the subgranular zone (SGZ) of the hippocampal dentate gyrus (Fig. 8.1a). Migrating long distances through the rostral migratory stream, the progeny of the V-SVZ NSCs ultimately differentiate into neurons in the olfactory bulb (OB) [32]. On the other hand, most newborn cells in the SGZ also differentiate into neurons and migrate to the adjacent granule cell layer, where they finally integrate [7]. In 1997, adult neurogenesis was described in the hippocampus of three primate species [17] and only 1 year later, it was reported in humans [14]. Currently, a long track of neuroscience research in multiple species such as birds, fishes, reptiles and mammals supports the seminal papers including those described above, and the existence of adult neurogenesis is now completely accepted [16]. Nevertheless, a controversy still exists between researchers as to how extensive this process is in the human brain [4, 25, 37, 44, 47–49].

It has been widely documented that preservation of a healthy proteome is key for maintaining cellular function and that proteostasis impairment is associated with neurodegeneration and ageing. A fundamental cornerstone of proteome-quality control that also removes and recycles other damaged cytosolic structures is autophagy. This well-characterised process relies on the degradation of cellular components by a vesicular structure termed the autophagolysosome that results from the fusion of autophagosomes, loaded with degradable cargoes and lysosomes. Previously, we reviewed the regulation of embryonic and adult neurogenesis by the autophagy lysosomal pathway [8]. For the purpose of the current chapter, we will only focus on the knowledge that deals with the role of this pathway in the regulation of mammalian adult NSCs and adult neurogenesis. First, we will briefly revisit findings on the discovery of adult neurogenesis and the process by which NSCs divide to generate neuronal progeny during adulthood in the two main neurogenic niches (V-SVZ and SGZ), highlighting new insight into the complex dynamics of adult NSCs. We will then focus on the consequences of blocking autophagy in postnatal and adult stem and progenitor cells by means of genetic and pharmacological approaches, summarising the role of autophagy along the different phases of the adult neurogenic cascade. We will emphasise studies describing the fascinating regulation of the adult NSC quiescent state by the autophagy lysosomal pathway. Finally, we will discuss the possibility of exploiting autophagy as a target to prevent the loss of NSCs and/or rescue their activity in order to rejuvenate the brain neurogenic niches during ageing.

Neural Stem Cells and Neurogenesis in the Adult Mammalian Brain

Adult Neural Stem Cell Niches and the Neurogenic Cascade

Located in the V-SVZ and SGZ, NSCs produce intermediate neural progenitor cells (NPCs) during the adult neurogenesis process (Fig. 8.1). Not only are these NPCs able to amplify through a few rounds of symmetric cell divisions but also give rise to the neuronally committed neuroblasts along a neurogenic cellular cascade. In turn, neuroblasts are still capable of restricted proliferation and finally differentiate into mature neurons, producing excitatory granule neurons in the hippocampus or inhibitory interneurons in the OB [38]. Both the V-SVZ and SGZ niches have a unique cytoarchitecture that allows cells to communicate with each other through intercellular contacts or through locally secreted molecules such as growth factors, cytokines, hormones or ligands of the principal signalling pathways [2]. To precisely

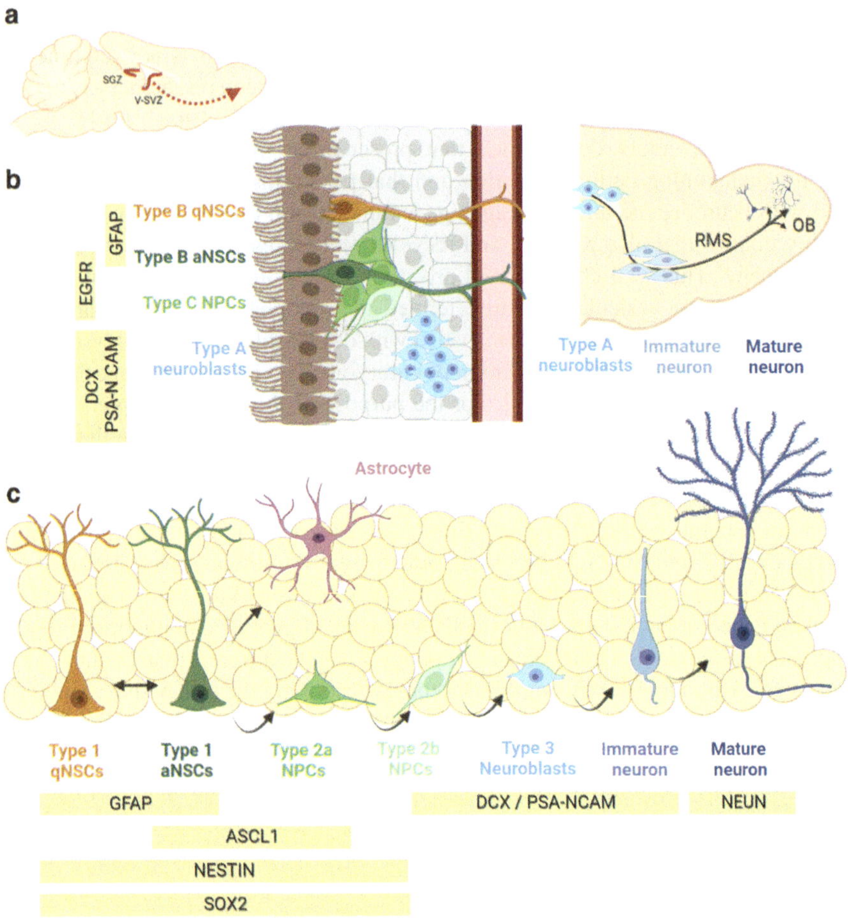

Fig. 8.1 Neurogenic niches in the adult mouse brain. a Schematic representation of the location of the two main adult neurogenic niches, the V- SVZ lining the lateral ventricles and the SGZ of the hippocampal dentate gyrus (red lines). The view corresponds to a sagittal section of the adult mouse brain. The arrow represents the rostral migratory stream (RMS) connecting the V-SVZ to the OB. **b** Detailed cellular composition of the V-SVZ niche. *Right*, the V-SVZ is located beneath a layer of multiciliated ependymal cells (E) lining the ventricles. Type B NSCs contact the ventricular lumen and the cerebrospinal fluid, as well as blood vessels (BV). Other cells belonging to the neurogenic V-SVZ cascade are also represented (Type C and A). Light orange boxes represent markers of each neurogenic cell type. *Left*, schematic representation of the tangential neuroblast migration through the RMS towards the OB. Note that once reaching the OB core, the cells switch to a radial migratory mode and terminally differentiate into several subtypes of mature interneurons. **c** Detailed cellular composition of SGZ niche showing the hippocampal neurogenic cascade. The soma of the NSCs (type 1 cells) is located within the subgranular zone, at the bottom of the granule cell layer (GCL). Their radial process crosses the GCL and branches on the adjacent

describe neurogenesis, we will first introduce the nomenclature used to reference different kinds of cells in the neurogenic cascade, which has slightly changed over the years and is not the same for each neurogenic niche due to historical reasons [34].

NSCs of the V-SVZ are designated as type B or type B1 cells and they are located in the wall of the lateral ventricles beneath the ependymal cells lining the ventricles (Fig. 8.1b). Their special morphology allows them to make contact with the cerebrospinal fluid and blood vessels, through an apical and basal process, respectively. These cells give rise to NPCs, also called type C cells, which proliferate and after some divisions generate mainly neuroblasts or so-called type A cells. Neuroblasts are located in the rostral migratory stream (RMS) where they undergo a long tangential migration from the V-SVZ up to the OB. Once they arrive to the OB, they switch to a radial migration mode in order to occupy their final location in the different OB layers. During this process, the cells initially express glial-related markers (GFAP, GLAST), stem cell markers (SOX2) and immature markers such as Nestin, but they progressively acquire more mature neuronal markers and finally differentiate into inhibitory granule and periglomerular interneurons [31]. At a functional level, adult neurogenesis in the V-SVZ/OB is related to the processing of the olfactory information [43].

Regarding NSCs in the hippocampus, they are referred to as radial glia-like cells or type 1 cells (Fig. 8.1c). The soma of the NSCs is located in the SGZ of the hippocampal dentate gyrus and their radial process spans through the densely packed granule cell layer and branches out in the molecular layer. NPCs generated from these cells are called firstly type 2a cells and after type 2b cells, when their neuronal fate is committed. Arising from type 2b cells, migratory neuroblasts or type 3 cells finally exit the cell cycle, migrate to the granule cell layer and become immature neurons that progressively differentiate giving rise to a fully mature granule neuron. The newly generated granule neurons integrate into the hippocampal circuit spreading dendrites across the molecular layer and sending their axon towards the hippocampal CA3 region [22, 26, 33]. The neurogenic process of the adult SGZ is associated to affective and cognitive behaviours, including learning and memory [18, 26].

The Quiescent State of Adult Neural Stem Cells

It is widely accepted that NSCs of the adult brain can be mainly found out of the cell cycle, in a dormant state known as quiescence, although a fraction of the NSCs abandon this resting state and actively engage in the neurogenic cascade. The predominant quiescent state that characterises adult NSCs is acquired differently by the two main adult NSC populations. The neurogenic V-SVZ niche is set up during development from embryonic radial glia NSCs. Although the majority of the embryonic NSCs continue dividing during this period, a subpopulation is already set aside as a quiescent reservoir that will lead to the adult type B NSC pool [15]. However, the SGZ hippocampal niche is established after birth, during the second postnatal week. At this stage, type 1 NSCs acquire their radial glia-like morphology and their definitive location, and finally, they also enter into quiescence [3].

Interestingly, not only can active and quiescent NSCs be classified by the expression of cell cycle markers but also can be distinguished by their transcriptomic signature and their metabolic state. While active NSCs are enriched in genes involved in transcription, translation and DNA repair, quiescent NSCs increase the expression of genes related to cellular adhesion, extracellular matrix, cell–cell communication and autophagy [36]. Regarding the metabolic state, quiescent cells show, in general, a lower metabolic rate, a drop in protein and RNA synthesis and high levels of fatty acid β-oxidation [27]. This quiescent NSC state is thought to prevent DNA damage, cellular deterioration and senescence while maintaining the germinal cells' ability to re-enter the cell cycle and respond adequately to a variety of stimuli [51]. Nevertheless, ageing causes a marked decrease in the NSC pool and the neurogenic process is drastically reduced throughout life [29]. As we will see throughout the chapter, there is extensive evidence supporting a role for the autophagy lysosomal pathway in the maintenance of adult NSC populations and in regulating the switch from NSC quiescence to activation.

Multiple niche molecules are involved in the regulation of the transition between the NSC quiescent and proliferation states and, in general, in the regulation of adult neurogenesis, including BMPs, Wnts, Notch ligands or Sonic hedgehog. The BMP and Notch signalling pathways are considered the main regulators of the adult NSC quiescent state and their receptors are expressed in quiescent NSCs [9], while EGFR (epidermal growth factor receptor) accumulates in active NSCs and early NPCs, boosting the cells for a rapid activation [51]. Consequently, NSCs can be isolated and expanded in vitro in the presence of EGF and FGF2 as mitogens, either as adherent cultures or as suspension cultures of floating clonal aggregates termed neurospheres. Moreover, NSC quiescence can be induced in vitro by removing EGF and supplementing the medium with the BMP4 ligand [35]. These in vitro systems have been widely employed to assess the role of multiple cellular pathways and genes, including those related to autophagy, in the regulation of stem cell properties. Although the primary cultures typically contain a mix of NSCs and NPCs, and are often referred to as NSPC cultures in the literature, for the sake of simplicity, we will refer to them as NSC cultures.

One of the most controversial issues in the continuous debate is related to the activation-quiescence dynamics of the adult NSCs, their ability to self-renew and their loss during ageing. A recent work suggests that NSCs of the V-SVZ exploit two modes of division to ensure their life-long maintenance: 20% of the NSCs divide by symmetric self-renewing divisions to maintain the NSC pool, whereas the other 80% divide by symmetric consuming divisions to produce type C progenitors [39]. In 2011, two opposite models were proposed to explain the dynamics of NSCs in the hippocampal SGZ of the adult brain. On one hand, the 'disposable stem cell model' proposed that when a stem cell exits from the quiescence state, the NSC undergoes a few rounds of asymmetric divisions to generate neuronal progeny and is depleted afterwards, most likely due to terminal differentiation into astrocytes (Fig. 8.1c). Therefore, NSC divisions result in the depletion of the active NSCs [13].

On the other hand, the ‘long-term self-renewal model’ suggested that NSCs are able to divide in both symmetric and asymmetric ways, generating not only differentiated cell clones but also multipotent NSCs with the ability to return to the quiescence state. In this model, the homeostatic balance between the two modes of division would prevent the exhaustion of NSCs [5].

Recently, a new ‘sequential model’ has been suggested to explain the quick fall of SGZ type 1 NSCs during the postnatal period and the slowing down of their reduction with age. This new model proposes a change in the behaviour of NSCs. While NSCs activate, divide and exhaust as ‘disposable’ stem cells in postnatal ages, a residual heterogeneous population of NSCs persists during adulthood [6, 23]. This population can be further classified into deeply dormant and resting states, according to their cellular properties. Dormant cells would correspond to the reservoir of long-term quiescent NSCs that have never proliferated while resting cells would be NSCs that return to a shallow quiescent state after their activation [21]. Furthermore, NSCs seem to lengthen their resting phase with each round of division and progressively switch to a symmetric cell fate choice, two events that together contribute to the SGZ NSC dynamics and SGZ neurogenesis loss during ageing [23].

Autophagy, Neural Stem Cells and Adult Neurogenesis

Accumulating evidence points to a fundamental role of the autophagy lysosomal pathway in the regulation of the maintenance of the NSC reservoirs throughout adulthood, their activation and their survival, as well as on the maturation of the newly generated neurons. In this section, we will attempt to review the main findings described in the literature so far. It is important to note that research in which autophagy is blocked by selective deletion of autophagy genes in NSC or NPC after the developmental period remains relatively sparse [8]. Some studies employ mice with germline deleted genes or conditionally ablate them at embryonic stages, yet they analyse the phenotype in the niches during the postnatal and adult ages. Thus, it is sometimes complicated to firmly drive conclusions regarding the cell-autonomous role of autophagy in adult NSCs in vivo, since the observed effects could be indirect and/or result from changes in development. Other studies employ pharmacological strategies and mainly perform in vitro experiments that can be occasionally extrapolated in vivo, as explained below.

The Autophagy Lysosomal Pathway in the Adult Neurogenic Niches

Autophagy genes play a fundamental role in the maintenance of adult and postnatal NSCs. In vivo work demonstrated that *Rb1cc1*/FIP200 (FAK-family Interacting

Protein of 200 kDa) that is part of the autophagy induction complex ULK1-Atg13-FIP200-Atg101, is required to maintain proper functioning of adult and postnatal NSCs through the regulation of their oxidative state [52]. Conditional deletion of *FIP200* in hGFAP-Cre transgenic mice expressing Cre recombinase in radial glia during development ($FIP200^{hGFAP}$cKO animals) influences the SGZ and V-SVZ neurogenic niches. Despite the fact that the niches appear normal at postnatal day 0 (P0) in $FIP200^{hGFAP}$ cKOs, at P28, the dentate gyrus (DG) is reduced in size and the number of type 1 NSCs ($GFAP^+Nestin^+$ cells with radial morphology) and PSA-$NCAM^+$ neuroblasts diminishes. Concomitantly, astrocytes increase in the SGZ and form a thick band. At the same age (P28), the V-SVZ appears to be thinner. Type B cells ($GFAP^+Nestin^+$ or $GFAP^+SOX2^+$ cells) adjacent to the lateral ventricles and PSA-$NCAM^+$ neuroblasts in the niche and RMS are exhausted. Moreover, the OBs of 8-week-old cKO mice show a smaller size, suggesting that V-SVZ neurogenesis is reduced in these animals. Thus, the depletion of NSCs/NPCs and the decrease in neurogenesis occur in both brain niches postnatally in the absence of FIP200.

In addition, in the $FIP200^{hGFAP}$-deficient animals, NSC proliferation is compromised and apoptosis is increased (although the study does not demonstrate that apoptosis affects specifically NSCs). Therefore, the loss of NSCs in the neurogenic niches at P28 could be due to a decrease in proliferation that would impair the maintenance of the NSC pool and/or to the (possible) increase of programmed cell death. On the other hand, in vitro assays with neurospheres from $FIP200^{hGFAP}$ cKOs show a decrease in the ability of the NSC cultures to survive, differentiate and self-renew. Interestingly, in older animals (P56 adult cKOs) an increase in the number, size and heterogeneity of mitochondria was found by transmission electron microscopy not only in the neurospheres but also in V-SVZ cells. Given mitochondria are the main source of reactive oxygen species (ROS) and since autophagy is required to eliminate damaged and/or excess mitochondria, oxidative stress was evaluated in the V-SVZ and SGZ of $FIP200^{hGFAP}$ cKO. An increase in both ROS levels and tumour suppressor p53, that is activated in response to many stress stimuli, was detected. Consistently, the effects in proliferation and apoptosis of the $FIP200^{hGFAP}$ cKOs were rescued in the double-cKO $FIP200^{hGFAP}p53^{hGFAP}$. The defects in differentiation were not rescued, suggesting that autophagy plays a role in the regulation of adult NSCs through the p53 axis, but that its function in the newborn neurons occurs in a p53-independent manner. Lastly, this work shows that scavenging the unusual elevated ROS levels in the cKOs with N-acetyl-cysteine, an antioxidant treatment, maintains the pool of NSCs in the V-SVZ as well as in the SGZ and rescues the neuronal differentiation defects of the $FIP200^{hGFAP}$ cKO mice [52].

Furthermore, $FIP200^{hGFAP}$ cKOs show an accumulation of p62/SQSTM1 aggregates in the V-SVZ and SGZ, as well as in neurospheres derived from the V-SVZ. p62 is a well-known autophagy receptor that recruits ubiquitinated cargos to the forming autophagosome. The abnormal accumulation of p62 can be due to a failure in its degradation when the autophagic flux is blocked. Interestingly, the p62 aggregates observed in V-SVZ neurospheres of the cKOs reduced superoxide dismutase 1 (SOD1) activity, and in turn, this resulted in the accumulation of the superoxide ion and in oxidative stress. In $FIP200^{hGFAP}$ cKO mice with a p62 deletion, or treated with compounds that mimic SOD activity, superoxide levels were rescued and so were the defects in NSCs of both niches. Thus, the depletion of NSCs and their reduction in proliferation could be due to the abnormal accumulation of p62 aggregates and ROS. In the same work, mice lacking genes for the ubiquitin conjugation system and autophagosome formation were also analysed (cKOs for autophagy-related genes; *Atg5*^{hGFAP} and *Atg16L1*^{hGFAP}). The animals showed the same phenotype as the *FIP200*^{hGFAP} cKO, displaying defects in autophagy and increases in mitochondrial mass in postnatal NSCs. However, contrary to *FIP200*^{hGFAP} cKOs, no defects were found in the maintenance, self-renewal or differentiation of the NSCs. Accumulation of p62 aggregates and superoxide ions in the V-SVZ and SGZ was not detected. Therefore, this work highlights the key role of p62/SQSTM1 in the regulation of NSCs [53]. In addition, FIP200 and p62 also regulate neurogenesis in an indirect way. The p62 aggregates accumulated in NSCs of *FIP200*^{hGFAP} cKO activate NFκB and promote the production of CCL5 and CXCL10 chemokines in the postnatal V-SVZ. As a result, microglia is activated and infiltrates into the niche, thereby interfering in the differentiation of the neural progenitors [54].

Another study analysed the levels of the autophagy marker microtubule-associated protein 1 light chain (*Map1lc3b*/LC3) in the V-SVZ niche taking advantage of transgenic mice expressing the green fluorescent protein fused to LC3 (GFP-LC3 mice). High levels of GFP-LC3 were observed in the adult V-SVZ and RMS [57]. The work also shows that the autophagy regulators Ambra 1 and Beclin 1 are expressed both in Nestin$^+$ NSCs and in DCX$^+$ neuroblasts of the V-SVZ. Thus, adult V-SVZ stem and progenitor cells seemingly have a high level of autophagy. Heterozygote Beclin 1$^{\pm}$ mice show a reduction in proliferation and an increase in apoptosis in the V-SVZ. In vitro assays employing neurospheres isolated from V-SVZ of Beclin 1$^{\pm}$ mice show a reduction in the number of neurospheres, less proliferation and an increase in apoptotic cells. Differentiation assays also show defects in the generation of neurons and increased apoptosis. Similar phenotypes were also observed in Ambra 1$^{+/gt}$ heterozygotes; thus, the combined in vivo and in vitro findings in the Ambra 1 and Beclin 1-deficient animals already illustrates an important role of autophagy in adult V-SVZ neurogenesis [57].

Autophagy is also required for the adequate radial migration of the newly born neurons from the adult V-SVZ once they reach the OB core. These new neurons show an enrichment in the expression of the *let-7* family of microRNAs, which is involved in the regulation of the autophagy pathway through the modulation of amino acid-sensing pathway genes [12]. *Let-7* knockdown using a lentivirus reduced autophagy levels in the new OB neurons, prevented their radial migration and decreased their morphological maturation. These effects were partially recovered when the autophagy inducers Beclin 1 and TFEB (the master transcription factor that activates autophagy and lysosome biogenesis [46]) were simultaneously overexpressed. Despite the fact that migratory defects were restored, morphological changes still persisted, probably due to autophagy-independent effects of *let-7* knockdown [41].

Another master transcription factor that directly regulates the autophagy gene network and the induction of autophagy in adult NSCs is the Forkhead Box O family member FOXO3 [1]. *FoxO* genes are required for the long-term maintenance of adult NSCs since their developmental deletion leads to the precocious depletion of stem and progenitor cells in the adult neurogenic niches [40, 42]. More recently, the adult NSC-specific ablation of FOXO function has been achieved employing the Tamoxifen-inducible *GLAST-CreERT2 FoxO1/3/4*$^{flox/flox}$ cKO mouse model [45]. At early time points after gene deletion, loss of FOXOs led to the activation of type 1 radial glia-like NSCs in the SGZ. At later time points, there was a decrease in type 1 NSCs in the cKO (perhaps due to their exhaustion according to the 'disposable' model) and a marked reduction in hippocampal neurogenesis, supporting the previously reported role of FOXOs in NSC maintenance. In accordance with this, FOXO cKO NSCs showed increased proliferation in vitro and displayed higher levels of LC3-II, p62 and more autophagic vesicles as assessed by transmission electron microscopy. Indeed, LC3 turnover assays indicated that the loss of FOXOs impaired the autophagy flux.

In the hippocampal niche, a strategy based on the injection of retroviruses has been also employed to both determine autophagy levels and delete the autophagy-related gene *Atg5* specifically in dividing NPCs during adulthood. This approach has allowed to clearly assign a cell-autonomous function of autophagy in SGZ neurogenesis. A tandem-tagged mCherry-EGFP-LC3 sensor retrovirus was injected in the DG of young mice to analyse the autophagy flux of transduced cells and their progeny along the neurogenic cascade, in particular at 3, 7, 14 and 30 days after the injection. This is a widely used reporter that takes advantage of the differential sensitivity of mCherry and GFP to low pH. In the acidic milieu of the lysosome, GFP loses fluorescence while mCherry is stable, allowing to determine simultaneously the overall level of autophagy and the autophagy flux. Autolysosomes (mCherry$^+$ puncta) were found at all neurogenic differentiation stages suggesting a quick autophagy flux. In addition, at 30 days post-injection, a decrease in autolysosome content was shown in the processes of the new neurons. However, autolysosomes seemed to slightly accumulate at 7 and 14 days post-injection, although this increase was not significant. Collectively, these results suggest that the autophagy flux is faster in the processes of young neurons still maturing (less than 30 days since their birth). To study the function of autophagy in this process, the genetic *Atg5*$^{flox/flox}$ cKO model was co-injected with a GFP-Cre retrovirus and either the mCherry-EGFP-LC3 sensor or RFP retroviruses were used as a control (depending on the experiment). *Atg5* deletion resulted in a decrease in the autophagy flux, autolysosome numbers and progeny survival at 3 or 7 days after injection, without showing neither changes in proliferation at 3 days post-injection (the first timepoint analysed) nor defects in the neuronal fate of the progeny. Moreover, neurons lacking *Atg5* presented a maturation delay, a transient reduction in dendritic spine density and a prolonged expression of the neuroblast/immature neuron marker DCX 30 days after injection. The phenotype was rescued in the double *Atg5* mutant and proapoptotic *Bax* mutant, suggesting that *Atg5* functions upstream of the Bax apoptotic pathway in immature neurons [56]. The immature phenotype displayed by the new neurons lacking *Atg5* in the DG is reminiscent of that of the new neurons generated in old mice, which require more time to mature completely compared to those from young mice. This data suggest that autophagy defects may partly underlie the alterations in neuronal maturation found in the aged hippocampal niche [50].

Interestingly enough, in another set of similar in vivo experiments, conditional deletion of *FoxO1/3/4*$^{flox/flox}$ achieved employing Cre-expressing retroviruses showed that the FoxO genes are required for the survival of the new hippocampal neurons and for their proper maturation [45]. FOXO-deficient neurons showed a transient decrease in dendritic length, a shortening of their apical dendrite and an aberrant spine development that compromised their integration into the hippocampal circuit. The autophagy flux of the FOXO-deficient neurons was markedly impaired and the pharmacological treatment with autophagy inducers (Rapamycin and Trehalose) partially corrected the phenotype, indicating that the neurogenic defects found in the absence of FOXOs are caused by deficient autophagic activity.

Besides the importance in neurogenesis, autophagy is also involved in astrocyte differentiation, at least in vitro [20]. When the mRFP-GFP-LC3 sensor was used to analyse hippocampal NSC cultures from adult rats undergoing astrogliogenesis, a natural increase in autophagy was observed during the first days of glial differentiation. Several strategies employed to block autophagy (*Atg7* and *Map1lc3b* knockdown with shRNA, p62 deletion with the CRISPR/Cas9 system, pharmacological inhibition with Bafilomycin A1 (Baf A, an inhibitor of the autophagic flux that prevents the acidification of lysosomes) impeded the optimal differentiation of astrocytes, evidencing the important role of autophagy during this process [20]. This interesting observation is awaiting validation in vivo.

The Autophagy Lysosomal Pathway in Adult NSC Quiescence

Recently, new works have been published directly linking the autophagy lysosomal pathway to the regulation of adult NSC quiescence (Fig. 8.2). The first study by Leeman and collaborators [30] showed that quiescent NSCs (qNSCs) rely on the autophagy–lysosome proteolytic system for proteostasis, while active NSCs (aNSCs) mainly use the proteasome proteolytic branch for the same purpose. They demonstrated that, compared to aNSCs, freshly isolated qNSC from the V-SVZ of young GFAP-GFP adult mice accumulated protein aggregates stored in large lysosomes and had a reduced autophagy flux (slower degradation of autophagosomes, less accumulation of LC3 upon Baf A treatment). Consistently, the qNSCs also exhibited an increased expression of lysosome-related genes that are enriched in a binding motif for the TFEB transcription factor (TFEB target genes, Fig. 8.2a). Adult NSC cultures isolated from the V-SVZ behaved similarly in vitro when the quiescent state was induced with BMP4. The use of the mCherry-GFP-LC3 sensor further demonstrated that qNSCs are endowed with more and larger autolysosomes compared to aNSCs. Experiments with the BMP4-induced qNSCs showed that nutrient deprivation in Hank's balanced salt solution (HBSS) boosted the clearance of the protein aggregates and the activation of the stem cells in response to growth factors (EGF). The expression of a constitutively active form of TFEB (CA-TFEB) in qNSCs also cleared the aggregates and enhanced qNSCs activation. Consistently, the activation was completely blocked with Baf A treatment. Together these results suggest that the mobilisation of aggregates through the autophagy lysosomal pathway may play a physiological role, perhaps constituting an energy source for the exit from the quiescent state. During ageing, quiescent NSCs accumulate more aggregates and higher amounts of GFP-LC3, perhaps as a result of a defective autophagy flux, while proteasome activity is not affected (Fig. 8.3). Notwithstanding, overexpression of CA-TFEB in old qNSCs or fasting was sufficient to reduce the accumulation of aggregates and GFP-LC3. Furthermore, old qNSCs in culture were activated less efficiently by growth factors compared to young NSCs, yet their activation could be improved by CA-TFEB expression. In old mice, the systemic administration over 3 months of Rapamycin (an mTOR inhibitor that enhances the autophagy lysosomal pathway) increased the number of aNSCs/NPCs ($EGFR^+$ cells) in the V-SVZ [30]. Thus, although Rapamycin could also have indirect effects, the data suggest that enhancing autophagy in the old V-SVZ niche counteracts the age-related decline in NSC proliferation.

Another work published one year later by Kobayashi and colleagues [28] also shows that qNSCs are endowed with high lysosomal content but contradicts the view that lysosomal function enhances qNSC activation (Fig. 8.2). On the one hand, it was verified that lysosomes are more abundant in quiescent NSCs than in active NSCs in both the adult V-SVZ and adult SGZ. They further showed that radial glia-like type 1 NSCs of the SGZ accumulated lysosomes in the basal region of their radial process. When quiescence was induced in NSC cultures from the adult V-SVZ and SGZ through the addition of BMP4, an increase in lysosomal content and lysosomal activity was detected compared to active NSCs. LC3 turnover assays showed higher LC3-II levels in qNSCs, demonstrating an increased endogenous autophagy in the quiescent state. On the other hand, overexpression of CA-TFEB mutants that enhance autophagy and lysosome biogenesis reduced the proliferation of NSCs. Along the same line, lysosomal inhibition with a low dose of Baf A, that does not affect cell survival, induced proliferation of qNSCs in striking contrast to Leeman and colleagues' results. The increase in proliferation was also detected in organotypic hippocampal slice cultures treated with Baf A, while a decrease was found in the DG of mice injected with a lentivirus to overexpress CA-TFEB in NSCs under the control of the *Hes5* or *GFAP* promoter. At a mechanistic level, Kobayashi and colleagues showed that Baf A caused the accumulation of active phosphorylated EGF receptor (P-EGFR) and active Notch (NICD) in qNSCs while inducing cell cycle genes such as Cyclin D1. This effect was not observed when the Class III PI3K VPS34 that plays essential roles in autophagy was specifically inhibited by SAR405. These results collectively show that the effect of Baf A in these assays depends on the decrease in lysosome activity (not autophagy) in quiescent NSCs. Lysosomal activity inhibition would cause the accumulation of P-EGFR that in turn would lead to the increased activation of hippocampal quiescent NSCs in vitro and in vivo. Furthermore, when *Tfeb* is deleted in NSCs isolated from *Tfeb*$^{flox/flox}$ mice by means of transient Cre expression, a delay in the entrance into quiescence is observed and higher levels of P-EGFR and NICD are detected compared to control NSCs.

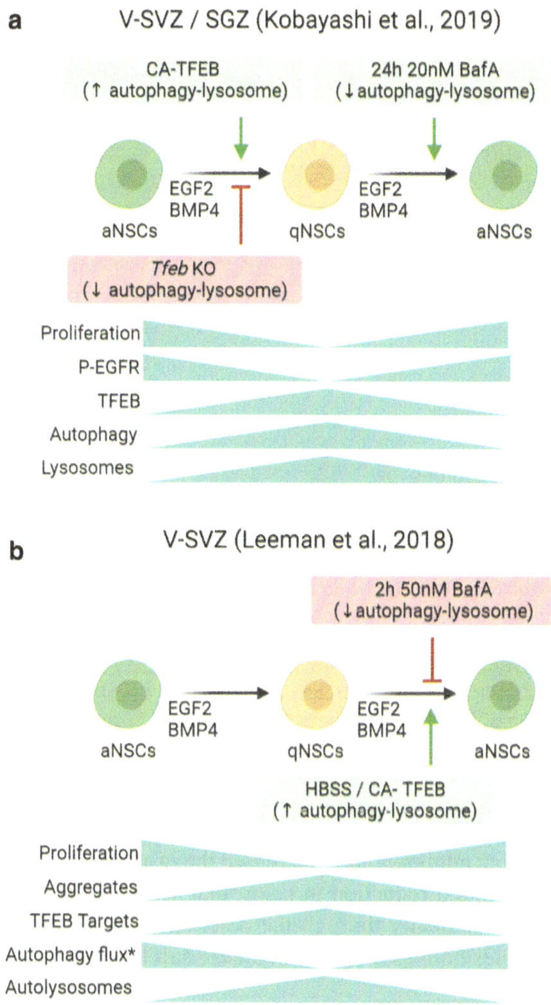

Fig. 8.2 Effects of autophagic manipulation on the dynamics of quiescent and active adult NSCs. Schematic representation illustrating the results from the two main works addressing the role of the autophagy lysosomal pathway in NSC quiescence [28, 30]. Similar in vitro assays were employed in both studies. Briefly, NSC cultures were derived from the V-SVZ and/or the SGZ of the dentate gyrus of adult mice. Cells were activated in the presence of mitogens. EGF withdrawal and BMP4 supplementation induced a reversible quiescent sate. Both studies show an increase in the lysosomal pathway in qNSCs. However, functional in vitro assays differ in their results after treatment with activators or inhibitors of the autophagy lysosomal pathway. *the autophagy flux was measured in cells freshly isolated from the V-SVZ

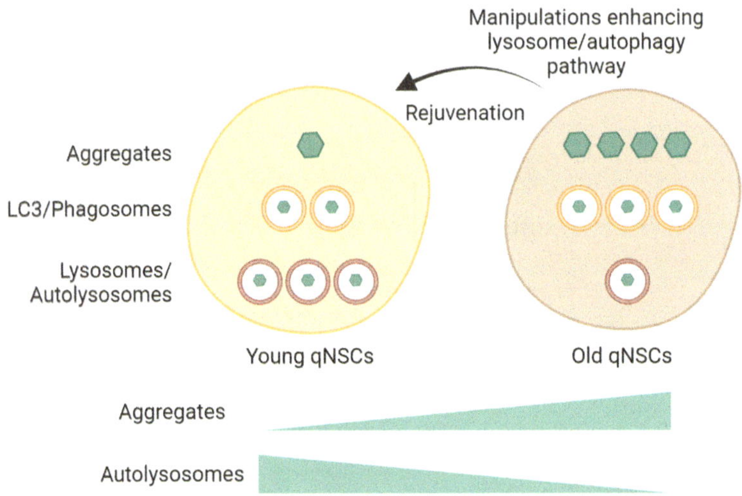

Fig. 8.3 Aggregates and the autophagy lysosomal pathway in V-SVZ quiescent NSCs during aging. Model depicting the dynamics of protein aggregates, autophagosomes and autolysosomes in quiescent NSCs (qNSCs) from the V-SVZ niche of young and old mice. While young adult qNSCs degrade aggregates through an enhanced lysosomal pathway, old qNSCs accumulate aggregates and autophagosomes that have not fused to lysosomes. Consequently, there is a decrease in autolysosomes in aged qNSCs. Genetic, pharmacological and dietary interventions that enhance the autophagy lysosomal pathway may favour the maintenance of a healthy proteome in old qNSCs, leading to rejuvenation of the stem cell compartment and to the recovery of the capacity of qNSC to respond to activation cues. Adapted from Leeman et al. [30]

Conditional deletion of *Tfeb* in vivo in NSCs of *Tfeb*$^{flox/flox}$ *GLAST-CreERT2* 8 weeks old mice by Tamoxifen administration increased the proportion of active NSCs in the SGZ. This indicates that a reduction in lysosomal activity and likely in autophagy in NSCs, achieved specifically through the deletion of *Tfeb* in the adult stem cell compartment, activates both EGF signalling and adult hippocampal NSCs [28].

The Autophagy Lysosomal Pathway in the Autophagic Cell Death of Adult NSCs

Even though autophagy generally acts as a cytoprotective mechanism, under certain conditions, prolonged or excessive autophagy can lead to a type of cell death that occurs without signs of apoptosis. Insulin withdrawal induces this kind of autophagic cell death (ACD) in vitro in hippocampal NSC cultures from adult rats (Fig. 8.4). In NSC cultures, mitogens such as FGF2 and EGF are required for proliferation and self-renewal, while insulin is essential for survival. When insulin is removed from the medium, NSC viability markedly decreases with no concurrent caspase 3 activation or DNA fragmentation (apoptotic signals). However, Beclin 1 and LC3-II are upregulated leading to autophagosome accumulation and ACD [58]. In addition, insulin withdrawal activates AMP-activated protein kinase (AMPK) through the Ca^{2+} CaMKK pathway, resulting in the direct phosphorylation of p62 by AMPK in a well-conserved Serine residue (Ser293 in rat). Phosphorylation at this novel p62 site is required for ACD to occur. Next, p62 translocates to the mitochondria producing mitochondrial fragmentation, mitophagy and finally cell death [19]. The ACD phenotype of NSCs in the absence of insulin is partially rescued by inhibiting autophagy through the silencing of the *Atg7* gene or with 3-methyladenine (3-MA), a widely used compound that blocks autophagy via its inhibitory effect on class III PI3K. Conversely, the mTOR inhibitor Rapamycin promotes autophagy and accentuates cell death [58].

In insulin withdrawal conditions, ACD is the cell death pathway by default. However, hippocampal adult NSCs in vitro can change the type of cell death from ACD to apoptosis, according to the activity levels of calpain. Calpains (cytosolic calcium ($Ca2+$)-activated cysteine) are ubiquitously expressed proteases in mammals and their activity is calcium-dependent. Calpain 1 levels are almost undetectable in NSCs; however, Calpain 2 is present in the NSC cultures in vitro. Under conditions of insulin deprivation, Calpain 2 is mainly degraded by the proteasome, and its activity levels remain low. When this degradation is blocked or when calpain activity is increased by the overexpression of Calpain 1, a switch occurs and the NSCs die by apoptosis instead of dying through ACD. These results suggest that in culture, NSCs show a preference for autophagic over apoptotic death, although both processes are interconnected through calpain [11].

Other in vitro experimental models have been used to study autophagic cell death in adult NSCs, such as oxygen–glucose deprivation (OGD) that simulates cerebral ischemia, and corticosterone (CORT) treatment that simulates glucocorticoid exposure during stress. Both treatments increase the ACD of NSCs (Fig. 8.4). A hypoxic condition with glucose deprivation for a short time (1–3 h) to mimic ischemia, followed by the restoration of normoxia and glucose supplementation for 24 h to simulate reperfusion (OGD/R treatment) decreases cell viability in NSC cultures. At the same time, during the reoxygenation phase, autophagy increases as measured by the rise in the number of autophagosomes, as well as in the levels of Beclin 1 and LC3-II, and the decrease in p62 [10, 10]. Exposure to homocysteine (Hcy, a neurotoxic metabolite whose elevated levels are considered a risk factor for ischemic stroke) during OGD/R accentuates autophagy and the loss of viability in NSCs isolated from the rat V-SVZ [56]. Exposure to Hcy inactivates mTOR, potentially through the ERK and PI3K-AKT signalling pathways, leading to the activation of autophagy. This produces an increase in autophagic markers and autophagosome formation, which is reversed by treating the cultures with an activator of mTOR (MHY1485) or with the autophagy inhibitor 3-MA.

Treatment of NSCs with corticosterone (CORT) also induces autophagic death *in vitro* in a time- and dose-dependent manner [24]. At concentrations of 200 μM CORT, there is an increase in cell death without signs of apoptosis, which is accompanied by the accumulation of autophagosomes and an increase in the autophagic flux, measured by the mRFP-EGFP-LC3 sensor. Blocking autophagy

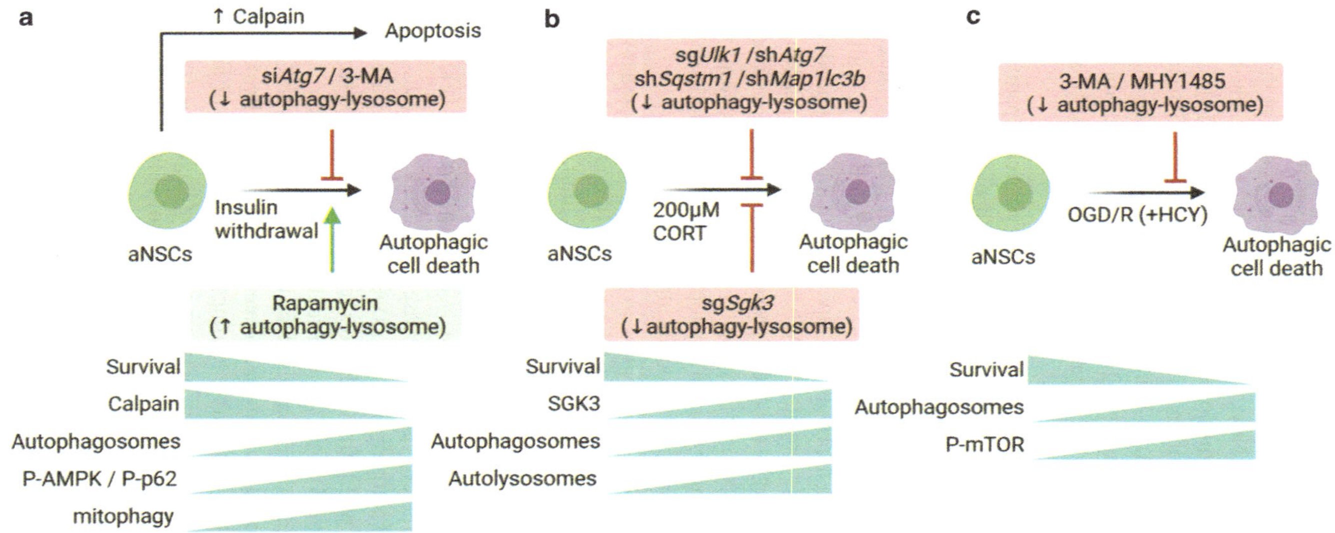

Fig. 8.4 Experimental models triggering autophagic cell death of adult NSCs. Summary of the three in vitro conditions reported to induce autophagic cell death (ACD) of adult NSCs. Schematic representation of the effects of pharmacological and genetic interventions modulating autophagy on survival and autophagy markers in NSCs

by knockdown or CRISPR/Cas9 deletion of several genes involved in different phases of autophagosome formation reduces CORT-induced death (Fig. 8.4). On the other hand, it is known that high CORT concentrations activate the glucocorticoid receptor GR/NR3C1, which in turn transcriptionally upregulates the SGK kinases (serum/glucocorticoid-regulated kinases). Specifically, SGK3 is increased in NSCs treated with CORT and co-localises with autophagosomes. Deletion of this gene by CRISPR/Cas9 or its mutation at a critical residue (Arg90) required for the binding to phosphatidylinositol 3-phosphate blocks the autophagic flux and also reduces the CORT-induced ACD. In SGK3-deficient NSCs, the formation of EGFP-ZFYVE1/DFCP1 puncta induced by CORT is also reduced. This data points to a regulation of autophagosome nucleation by SGK3 and highlights the role of this kinase in the signal transduction pathway that leads to the initiation of autophagy in NSCs.

In vivo, both psychological stress and exposure to the stress hormone CORT possibly induce the autophagic cell death of adult hippocampal NSCs through SGK3 activation. Under specific conditions of chronic restraint stress (CRS), a decrease in the $SOX2^+$ cell pool and a reduction in the expression of other NSCs markers such as Nestin is observed in the SGZ. However, the proportion of $SOX2^+$ cells that are proliferating is only slightly reduced and apoptotic markers are not altered. At the same time, an increase in autophagosomes is also observed in $SOX2^+$ cells by correlative light and electron microscopy. This correlates with the increase in autophagosomes and autolysosomes detected in the SGZ of the animals exposed to CRS following the injection of a lentivirus expressing the mRFP-EGFP-LC3 sensor under the regulation of the Nestin promoter. A more detailed analysis shows a decrease in the total number of type 1 NSCs ($GFAP^+$ $SOX2^+$), type 1 active NSCs ($GFAP^+$ $SOX2^+$ $Ki67^+$), type 2a ($SOX2^+$ $ASCL1^+$), type 2b ($SOX2^+$ DCX^+) and type 3 (DCX^+) cells; summarising, a marked reduction in the NSCs and the entire cascade of neurogenic progenitors derived from them.

The deletion of *Atg7* specifically in the NSCs of adult age using the *Nestin-CreERT2 Atg7$^{flox/flox}$* cKO rescues the defects caused by CRS. Considering that, the loss of the NSCs could be due to autophagic death. Furthermore, gene editing of *Sgk3* in vivo using adeno-associated vectors injected into the DG attenuates the CRS-induced loss of NSCs. Therefore, it is possible to prevent autophagic death and to keep an intact pool of NSCs, counteracting the effect of chronic stress on adult neurogenesis [24].

A blockade of autophagy could also have a protective effect in a biological context where there is oxygen–glucose deprivation, such as the ischemic brain. As previously noted, Hcy produces neurotoxic effects and reduces the viability of NSCs in OGD/R in vitro, so potentially Hcy could be enhancing autophagic death in vivo in ischemic brains. In a rat model of transient focal ischemia reperfusion by MCAO (Middle Cerebral Artery occlusion), LC3 puncta levels increased in $Nestin^+$ NSCs/NPCs of the V-SVZ. This was accentuated when Hcy was administered systemically, while suppression of autophagy with 3-MA attenuated the effects produced by Hcy [56].

Concluding Remarks

As reviewed in this chapter, in recent years we have gained insight into the role of the autophagy lysosomal pathway in the regulation of adult neural stem cells and adult neurogenesis, in both the V-SVZ and SGZ brain niches (see Table 8.1). Emerging results point to a cell-autonomous role of autophagy in the regulation of the survival and maturation of the new neurons. There is also considerable consensus regarding the absolute requirement of adequate levels of autophagy for the proper maintenance and homeostasis of adult NSC reservoirs. This seems to be due to a combination of factors: (i) the dependency of the quiescent NSCs on autophagy as their main proteostatic pathway in order to keep a healthy proteome, (ii) the dependency of NSCs on adequate levels of autophagy to restrain their overactivation (note the overproliferation of the NSCs at young ages in certain autophagy cKO models, that would lead in the long run to NSC depletion), (iii) the mechanistic links uncovered by some authors suggesting that blocking lysosomal function allows for a quick re-activation of the qNSCs and/or delays the return of the aNSC pool into quiescence and (iv) the possible loss of NSCs due to autophagic cell death.

An interesting observation is also the accumulation of insoluble protein aggregates in qNSCs and the identification of enlarged lysosomes filled with these aggregates. The origin of the aggregates is currently unknown and is quite puzzling given qNSCs have reduced protein synthesis compared to aNSCs. It also is currently unclear if the activation of the autophagy lysosomal pathway in qNSCs simply represents a protective mechanism in order to clear potentially toxic aggregates and prevent their distribution to the progeny, or if it truly represents a type of store that may be of use when a burst of energy is required, for instance during NSC activation. Another interesting conclusion based on the available data is that the improvement of the autophagy flux during ageing may allow to rejuvenate the old qNSC reservoirs. We should nevertheless bear in mind that an exacerbation of autophagy can be also detrimental and cause autophagic cell death of NSCs. In pathological scenarios where autophagic cell death has been reported, such as during oxygen–glucose deprivation, stress-related models or insulin deficiency, counteracting excessive autophagy is emerging as a neuroprotective strategy aimed at preserving adult NSCs. This approach may be of use in conditions such as stroke, chronic stress or Alzheimer's Disease. Indeed, supporting the survival of NSCs during injury or in neurodegeneration may allow to partly recover normal levels of adult neurogenesis. Nevertheless, studies are still needed to shed more light on the possible utility of adequately modulating autophagy to restore neurogenic niche homeostasis, both under pathological conditions and during ageing.

Table 8.1 Role of autophagy in adult neurogenesis

	Effect	Model	Species/Experiment (Adult niche)	References
Activation	↓ NSC maintenance, ↓ self-renewal, ↓ proliferation	*FIP200 (hGFAP* -Cre) cKO	*Mus musculus*. In vitro and In vivo (V-SVZ, SGZ)	Wang et al. [52, 54]
	Rescue	*FIP200 (hGFAP* -Cre) cKO + N-acetyl-cystelne antioxidant		Wang et al. [52]
	Rescue	*FIP200/Trp53 (hGFAP -Cre)* dcKO	*Mus musculus*. In vivo (V-SVZ, SGZ)	
	↑ qNSC/NPC activation	Nutrient deprivation/CA-TFEB	*Mus musculus*. In vitro (V-SVZ, young/old animals)	Leeman et al. [30]
		Rapamycin	*Mus* musculus. In vivo (V-SVZ, old animals)	
	↑ qNSC/NPC activation	Bafilomycin A	*Mus musculus*. In vitro *(V-SVZ and SGZ) and organotypic hlppocompol slice cultures*	Kobayashi et al. [28]
		Tfeb$^{flox/flax}$ *GLAST-CreERT2*	*Mus musculus*. In vitro and in vivo (SGZ)	
	↑ NSC proliferation ↓ NSC maintenance	*FoxOl.3.4, cKO*	Mus musculus. in vitro and In vivo (SGZ)	Schäffner et al. [47]
Cell death	↑ Apoptosis	*FIP200 (hGFAP* -Cre) CKO	*Mus musculus*. In vitro and in vivo (V-SVZ, SGZ)	Wang et al. [52]
	Rescue	*FIP200/Trp53* (hGFAP-Cre) dcKO		
	↑ Apoptosis	*Beclin 1*$^{+/-}$	*Mus musculus*. In vitro and in vivo (V-SVZ)	Yazdankhah et al. [57]
		Ambra 1$^{+/gt}$		
	↑ ACD	Insulin withdrawal	*Rattus norverglcus*. In vitro (SGZ)	Chung et al. [11], Ha et al. [19], Yu et al. [58]

(continued)

Table 8.1 (continued)

	Effect	Model	Species/Experiment (Adult niche)	References
	Rescue	Insulin withdrawal Atg7 siRNA		Chung et al. [11]
	Enhanced ACD	Insulin withdrawal + rapamycin		Yu et al. [58]
	↑ ACD	OGD	*Rottus norvergicus*. In vitro (SGZ)	Chung et al. [10]
	Enhanced ACD	OGD + Hcy	*Rattus norvergicus*. In vitro (V-SVZ)	
	Partial rescue	OGD + Hcy + 3-MA	*Mus musculus*. In vivo (V-SVZ)	Wang et al. [55]
	↑ ACD	CORT/Stress	*Mus musculus*. In vitro and in vivo (SGZ)	Jung et al. [24]
	Rescue	CORT + *SGK3* deletion (CRISPR/Cas9)		
		Nestm-CreERT2 Atg7$^{flox/flox}$, *Sgk3* (adeno-associated vectors)		
Differentiation	↓ Neuronal differentiation (infiltration and activation of microglia), ↑ astrogenesis	*FIP200 (hGFAP* -Cre) cKO	*Mus musculus*. In vitro and in vivo (V-SVZ, SGZ)	Wang et al. [52, 54]
	Rescue neuronal differentiation	*FIP200 (hGFAP* -Cre) cKO + N-acetyl-cysteine antioxidant	*Mus musculus*. In vivo (V-SVZ, SGZ)	Wang et al. [52]
	Defects in neuronal differentitation	*Beclin 1$^{+/-}$*	Mus musculus. In vitro and in vivo (V-VSVZ)	Yazdankhah et al. [57]
		Ambra 1$^{+/gt}$		
	↓ Radial migration, ↓ morophologlcal maturation	*Let-7* (LV)	*Mus musculus*. In vivo (OB)	Petri et al. [41]
	Rescue migration	*Let-7/Beclin o.e.* (LV), let-7/TFEB o.e. LV)		
	↓ Maturation delay	*Atg5$^{flox/flox}$* (RV-Cre) cKO	*Mus musculus*. In vivo (SGZ)	Xi et al. [56]

(continued)

Table 8.1 (continued)

	Effect	Model	Species/Experiment (Adult niche)	References
	Rescue	*Atg5*$^{flox/flox}$ (RV-Cre) *cKO/Bax*$^{-/-}$		
	↓ Astrogenesls	*Atg7 and LC3 knockdown with shRNA, p62 deletion with CRISPR/Cas9 scystem, Bafilomycin A1*	*Rattus norvergicus*. In vitro (SGZ)	Ha et al. [20]
	Defects in neuronal differentitation	*FoxOl.3,4, cKO*	Mus musculus. In vitro and In vivo (SGZ)	Schäffner et al. [47]
	Rescue	*FoxOl.3,4, cKO + Rapamycin/trehalose*		

ACD, autophagy cell death; cKO, conditional knockout; dcKO, double conditional knockout, LV, lentivirus; OGD, oxygen-glucose deprivation; RV, retrovirus

Acknowledgements We thank Lucía Casares-Crespo and Arancha Baselga for critically reading the manuscript. We apologise to those colleagues whose work has not been cited due to space limitations. All figures were created with BioRender.

Compliance with Ethical Standards This study was funded by the Spanish MICINN (grant number PID2019-111225RB-I00 to HM; fellowship number BES-2016–077156 to IC-B) and by Generalitat Valenciana (grant number PROMETEO/2018/055 to HM). All authors declare they have no conflict of interest. This review article does not contain any studies with human participants performed by any of the authors. This review article does not contain any studies with animals performed by any of the authors.

References

1. Audesse AJ et al (2019) FOXO3 directly regulates an autophagy network to functionally regulate proteostasis in adult neural stem cells. PLoS Genet 15(4):e1008097. https://doi.org/10.1371/journal.pgen.1008097
2. Bátiz LF et al (2015) Exosomes as novel regulators of adult neurogenic niches. Front Cell Neurosci 9:501. https://doi.org/10.3389/fncel.2015.00501
3. Berg DA et al (2019) A common embryonic origin of stem cells drives developmental and adult neurogenesis. Cell 177(3):654–668.e15. https://doi.org/10.1016/j.cell.2019.02.010. Elsevier Inc
4. Boldrini M et al (2018) Human hippocampal neurogenesis persists throughout aging. Cell Stem Cell 22(4):589-599.e5. https://doi.org/10.1016/j.stem.2018.03.015
5. Bonaguidi MA et al (2011) In vivo clonal analysis reveals self-renewing and multipotent adult neural stem cell characteristics. Cell 145(7):1142–1155. https://doi.org/10.1016/j.cell.2011.05.024. Elsevier Inc
6. Bottes S et al (2021) Long-term self-renewing stem cells in the adult mouse hippocampus identified by intravital imaging. Nat Neurosci 24(2):225–233. https://doi.org/10.1038/s41593-020-00759-4.
7. Cameron HA et al (1993) Differentiation of newly born neurons and glia in the dentate gyrus of the adult rat. Neuroscience 56(2):337–344. https://doi.org/10.1016/0306-4522(93)90335-d. United States
8. Casares-Crespo L et al (2018) On the role of basal autophagy in adult neural stem cells and neurogenesis. Front Cell Neurosci 12:339. https://doi.org/10.3389/fncel.2018.00339
9. Choe Y, Pleasure SJ, Mira H (2015) Control of adult neurogenesis by short-range morphogenic-signaling molecules. Cold Spring Harbor Perspect Biol 8(3):a018887. https://doi.org/10.1101/cshperspect.a018887
10. Chung H, Choi J, Park S (2018) Ghrelin protects adult rat hippocampal neural stem cells from excessive autophagy during oxygen-glucose deprivation. Endocrine J 65(1):63–73. https://doi.org/10.1507/endocrj.EJ17-0281. Japan
11. Chung KM et al (2015) Calpain determines the propensity of adult hippocampal neural stem cells to autophagic cell death following insulin withdrawal. Stem Cells (Dayton, Ohio) 33(10):3052–3064. https://doi.org/10.1002/stem.2082. United States
12. Dubinsky AN et al (2014) Let-7 coordinately suppresses components of the amino acid sensing pathway to repress mTORC1 and induce autophagy. Cell Metab 20(4):626–638. https://doi.org/10.1016/j.cmet.2014.09.001
13. Encinas JM et al (2011) Division-coupled astrocytic differentiation and age-related depletion of neural stem cells in the adult hippocampus. Cell Stem Cell 8(5):566–579. https://doi.org/10.1016/j.stem.2011.03.010
14. Eriksson PS et al (1998) Neurogenesis in the adult human hippocampus. Nat Med 4(11):1313–1317. https://doi.org/10.1038/3305. United States

15. Fuentealba LC et al (2015) Embryonic origin of postnatal neural stem cells. Cell 161(7):1644–1655. https://doi.org/10.1016/j.cell.2015.05.041
16. Ghosh HS (2019) Adult neurogenesis and the promise of adult neural stem cells. J Experiment Neurosci 13:1179069519856876. https://doi.org/10.1177/1179069519856876
17. Gould E et al (1997) Neurogenesis in the dentate gyrus of the adult tree shrew is regulated by psychosocial stress and NMDA receptor activation. J Neurosci Official J Soc Neurosci 17(7):2492–2498. https://doi.org/10.1523/JNEUROSCI.17-07-02492.1997
18. Gould E et al (1999) Learning enhances adult neurogenesis in the hippocampal formation. Nat Neurosci 2(3):260–265. https://doi.org/10.1038/6365
19. Ha S et al (2017) Phosphorylation of p62 by AMP-activated protein kinase mediates autophagic cell death in adult hippocampal neural stem cells. J Biol Chem 292(33):13795–13808. https://doi.org/10.1074/jbc.M117.780874. United States
20. Ha S et al (2019) Autophagy mediates astrogenesis in adult hippocampal neural stem cells. Exp Neurobiol 28(2):229–246. https://doi.org/10.5607/en.2019.28.2.229
21. Harris L et al (2021) Coordinated changes in cellular behavior ensure the lifelong maintenance of the hippocampal stem cell population. Cell Stem Cell 28(5):863-876.e6. https://doi.org/10.1016/j.stem.2021.01.003
22. Hastings NB, Gould E (1999) Rapid extension of axons into the CA3 region by adult-generated granule cells. J Compar Neurol 413(1):146–154. https://doi.org/10.1002/(sici)1096-9861(19991011)413:1<146::aid-cne10>3.0.co;2-b. United States
23. Ibrayeva A et al (2021) Early stem cell aging in the mature brain. Cell Stem Cell 28(5):955–966.e7. https://doi.org/10.1016/j.stem.2021.03.018. United States
24. Jung S et al (2020) Autophagic death of neural stem cells mediates chronic stress-induced decline of adult hippocampal neurogenesis and cognitive deficits. Autophagy 16(3):512–530. https://doi.org/10.1080/15548627.2019.1630222
25. Kempermann G et al (2018) Human adult neurogenesis: evidence and remaining questions. Cell Stem Cell 23(1):25–30. https://doi.org/10.1016/j.stem.2018.04.004. United States
26. Kempermann G, Song H, Gage FH (2015) Neurogenesis in the adult hippocampus. Cold Spring Harbor Perspect Biol 7(9):a018812. https://doi.org/10.1101/cshperspect.a018812
27. Knobloch M et al (2017) A fatty acid oxidation-dependent metabolic shift regulates adult neural stem cell activity. Cell Rep 20(9):2144–2155. https://doi.org/10.1016/j.celrep.2017.08.029
28. Kobayashi T et al (2019) Enhanced lysosomal degradation maintains the quiescent state of neural stem cells. Nat Commun 10(1):2–4.https://doi.org/10.1038/s41467-019-13203-4. Springer US
29. Kuhn HG, Dickinson-Anson H, Gage FH (1996) Neurogenesis in the dentate gyrus of the adult rat: age-related decrease of neuronal progenitor proliferation. J Neurosci Off J Soc Neurosci 16(6):2027–2033. https://doi.org/10.1523/JNEUROSCI.16-06-02027.1996
30. Leeman DS et al (2018) Lysosome activation clears aggregates and enhances quiescent neural stem cell activation during aging. Science 359(6381):1277–1283. https://doi.org/10.1126/science.aag3048. (New York, N.Y.)
31. Lim DA, Alvarez-Buylla A (2014) Adult neural stem cells stake their ground. Trends Neurosci 37(10):563–571. https://doi.org/10.1016/j.tins.2014.08.006
32. Lois C, Alvarez-Buylla A (1994) Long-distance neuronal migration in the adult mammalian brain. Science (New York, N.Y.) 264(5162):1145–1148. https://doi.org/10.1126/science.8178174. United States
33. Markakis EA, Gage FH (1999) Adult-generated neurons in the dentate gyrus send axonal projections to field CA3 and are surrounded by synaptic vesicles. J Comp Neurol 406(4):449–460. United States
34. Martínez-Cerdeño V, Noctor SC (2018) Neural progenitor cell terminology. Front Neuroanat 12:104. https://doi.org/10.3389/fnana.2018.00104
35. Mira H et al (2010) Signaling through BMPR-IA regulates quiescence and long-term activity of neural stem cells in the adult hippocampus. Cell Stem Cell 7(1):78–89. https://doi.org/10.1016/j.stem.2010.04.016

36. Morales AV, Mira H (2019) Adult neural stem cells: born to last. Front Cell Dev Biol 7:96. https://doi.org/10.3389/fcell.2019.00096
37. Moreno-Jiménez EP et al (2019) Adult hippocampal neurogenesis is abundant in neurologically healthy subjects and drops sharply in patients with Alzheimer's disease. Nat Med 25(4):554–560. https://doi.org/10.1038/s41591-019-0375-9. United States
38. Morrens J, Van Den Broeck W, Kempermann G (2012) Glial cells in adult neurogenesis. Glia 60(2):159–174. https://doi.org/10.1002/glia.21247. United States
39. Obernier K et al (2018) Adult neurogenesis is sustained by symmetric self-renewal and differentiation. Cell Stem Cell 221–234.e8. https://doi.org/10.1016/j.stem.2018.01.003
40. Paik J et al (2009) FoxOs cooperatively regulate diverse pathways governing neural stem cell homeostasis. Cell Stem Cell 5(5):540–553. https://doi.org/10.1016/j.stem.2009.09.013
41. Petri R et al (2017) let-7 regulates radial migration of new-born neurons through positive regulation of autophagy. EMBO J 36(10):1379–1391. https://doi.org/10.15252/embj.201695235
42. Renault VM et al (2009) FoxO3 regulates neural stem cell homeostasis. Cell Stem Cell 5(5):527–539. https://doi.org/10.1016/j.stem.2009.09.014
43. Sakamoto M et al (2014) Continuous postnatal neurogenesis contributes to formation of the olfactory bulb neural circuits and flexible olfactory associative learning. J Neurosci Official J Soc Neurosci 34(17):5788–5799. https://doi.org/10.1523/JNEUROSCI.0674-14.2014
44. Sanai N et al (2011) Corridors of migrating neurons in the human brain and their decline during infancy. Nature 478(7369):382–386. https://doi.org/10.1038/nature10487
45. Schäffner I et al (2018) FoxO function is essential for maintenance of autophagic flux and neuronal morphogenesis in adult neurogenesis. Neuron 99(6):1188–1203.e6. https://doi.org/10.1016/j.neuron.2018.08.017
46. Settembre C et al (2011) TFEB links autophagy to lysosomal biogenesis. Science (New York, N.Y.) 332(6036):1429–1433. https://doi.org/10.1126/science.1204592
47. Sorrells SF et al (2018) Human hippocampal neurogenesis drops sharply in children to undetectable levels in adults. Nature 555(7696):377–381. https://doi.org/10.1038/nature25975
48. Spalding KL et al (2013) Dynamics of hippocampal neurogenesis in adult humans. Cell 153(6):1219–1227. https://doi.org/10.1016/j.cell.2013.05.002
49. Tobin MK et al (2019) Human hippocampal neurogenesis persists in aged adults and alzheimer's disease patients. Cell Stem Cell 24(6):974–982.e3. https://doi.org/10.1016/j.stem.2019.05.003
50. Trinchero MF et al (2017) High plasticity of new granule cells in the aging hippocampus. Cell Rep 21(5):1129–1139. https://doi.org/10.1016/j.celrep.2017.09.064
51. Urbán N, Blomfield IM, Guillemot F (2019) Quiescence of adult mammalian neural stem cells: a highly regulated rest. Neuron 104(5):834–848. https://doi.org/10.1016/j.neuron.2019.09.026. United States
52. Wang C et al (2013) FIP200 is required for maintenance and differentiation of postnatal neural stem cells. Nat Neurosci 16(5):532–542. https://doi.org/10.1038/nn.3365
53. Wang C et al (2016) Elevated p62/SQSTM1 determines the fate of autophagy-deficient neural stem cells by increasing superoxide. J Cell Biol 212(5):545–560. https://doi.org/10.1083/jcb.201507023
54. Wang C et al (2017) Autophagy gene FIP200 in neural progenitors non-cell autonomously controls differentiation by regulating microglia. J Cell Biol 216(8):2581–2596. https://doi.org/10.1083/jcb.201609093
55. Wang M et al (2019) Homocysteine enhances neural stem cell autophagy in in vivo and in vitro model of ischemic stroke. Cell Death Dis 10(8):561. https://doi.org/10.1038/s41419-019-1798-4
56. Xi Y et al (2016) Knockout of Atg5 delays the maturation and reduces the survival of adult-generated neurons in the hippocampus. Cell Death Disease 7(3):e2127. https://doi.org/10.1038/cddis.2015.406

57. Yazdankhah M et al (2014) The autophagy regulators Ambra1 and Beclin 1 are required for adult neurogenesis in the brain subventricular zone. Cell Death Disease 5(9):e1403. https://doi.org/10.1038/cddis.2014.358
58. Yu S-W et al (2008) Autophagic death of adult hippocampal neural stem cells following insulin withdrawal. Stem Cells 26(10):2602–2610. https://doi.org/10.1634/stemcells.2008-0153

Chapter 9
Autophagy in Mesenchymal Stem Cell-Based Therapy

Carl Randall Harrell, Dragica Pavlovic, and Vladislav Volarevic

Abstract Mesenchymal stem cells (MSCs) are adult stem cells which are, due to their huge differentiation potential, potent immunomodulatory and pro-angiogenic properties, considered as new therapeutic agents in regenerative medicine. Although MSC-based therapy holds a great potential in the treatment of inflammatory and degenerative diseases, there are several issues that limit therapeutic efficacy of MSCs. Due to the low survival of engrafted cells, high number of MSCs has to be transplanted to achieve optimal therapeutic benefits. A large number of evidence demonstrated that modulation of autophagy-related pathways in engrafted MSCs may increase viability and survival of transplanted MSCs, enhancing their potential for differentiation. immunomodulatory and pro-angiogenic properties. In this chapter, we summarized current knowledge about the role of autophagy in MSC-based therapy of inflammatory, ischemic, and degenerative diseases.

Keywords Mesenchymal stem cells · Autophagy · Immunomodualtion · Neo-angiogenesis · Regeneration.

Abbreviations

3-MA	3-methyladenine
aGVHD	Acute Graft-versus-Host Disease
AT-MSCs	Adipose tissue-derived mesenchymal stem cells
AD	Alzheimer's disease

C. R. Harrell
Regenerative Processing Plant, LLC, 34176 US Highway 19 N Palm Harbor, Palm Harbor FL 34684, USA

D. Pavlovic · V. Volarevic (✉)
Department of Genetics and Microbiology and Immunology, Faculty of Medical Sciences, University of Kragujevac, 69 Svetozar Markovic Street, 34000 Kragujevac, Serbia
e-mail: drvolarevic@yahoo.com

B. V. Shravage and K. Turksen (eds.), *Autophagy in Stem Cell Maintenance and Differentiation*, Stem Cell Biology and Regenerative Medicine 73,
https://doi.org/10.1007/978-3-031-17362-2_9

ATG7	Autophagy Related gene 7
bFGF	Basic fibroblast growth factor
Bcl-2	B-cell leukemia/lymphoma-2
CNS	Central nervous system
CMA	Chaperone-mediated autophagy
DCs	Dendritic cells
EAE	Experimental autoimmune encephalomyelitis
iPSC-MSC-EVs	Extracellular vesicles isolated from MSCs previously derived from human induced pluripotent stem cells
GIOP	Glucocorticoid-induced osteoporosis
HLA	Human leukocyte antigen
HD	Huntington disease
IDO	Indoleamine 2,3-dioxygenase
IFN-γ	Interferon gamma
JAK-STAT	Janus kinase-Signal transducer and activator of transcription
JNK	Jun N-terminal kinases
mTORC1	Mechanistic target of rapamycin complex 1
MSCs	Mesenchymal stem cells
MAPKs	Mitogen-activated protein kinases
MAIT	Mucosal-associated invariant T
NF-κB	Nuclear factor kappa-light-chain-enhancer of activated B cells
PD	Parkinson's disease
PE	phosphatidyl ethanol amine
PI3K	Phosphatidylinositol-3-kinase complex
PGE2	Prostaglandin E2
PKB/AKT	protein kinase B activation
ROS	reactive oxygen species
SDF-1	stromal cell derived factor 1
SLE	ystemic lupus erythematosus
(T-MSCs)	Tonsil-derived MSCs
(TGF-β)	transforming growth factor beta
TNF-α	Tumor necrosis factor alpha
UVRAG	UV irradiation resistance-associated tumor suppressor gene
Vps34	Vacuolar protein sorting 34

Introduction

Mesenchymal stem cells (MSCs) are non-hematopoietic cells that were first found in the bone marrow by Friedenstein and colleagues in the late 1960s [1, 2]. The International Society for Cellular Therapy (ISCT) established three minimal criteria for the definition of human MSCs: (i) plastic adherence, (ii) membrane expression of cluster of differentiation (CD)73, CD90, and CD105, lack of CD34, CD45, CD14,

CD19, CD79a, CD11b, human leukocyte antigen (HLA)-DR expression, and (iii) in vitro differentiation into the cells of mesodermal lineage: osteoblasts, adipocytes, and chondrocytes (Fig. 9.1) [3]. Following the initial isolation of MSCs from the BM, a number of studies suggested that cells meeting the aforementioned criteria and sharing similar properties can be harvested from a wide variety of adult human tissues, including adipose tissue, dental pulp, peripheral blood, menstrual blood, endometrium, and fetal tissues, such as amniotic fluid, placenta, umbilical cord, Wharton jelly, and umbilical cord blood [4–10].

Due to their huge differentiation potential, MSCs are considered as new therapeutic agents in regenerative medicine. In vitro, under specific culture conditions, MSCs can differentiate into endothelial cells, hepatocytes, pancreatic beta cells, neurons, and glial cells [11–14]. Additionally, MSCs produce large number of antimicrobial, immunoregulatory, angiomodulatory and growth factors which affect cell proliferation, differentiation, angiogenesis, phenotype, and function of immune cells, crucially contributing in the MSC-dependent enhanced repair and regeneration of injured tissues [14, 15]. The low expression of human leukocyte antigen (HLA) class I and the absence of expression of HLA class II and co-stimulatory molecules (CD40, CD80, CD83, CD86, and CD154) indicate low immunogenicity of MSCs enabling their transplantation in HLA-miss-matched recipients [16].

Although MSC-based therapy holds a great potential in the treatment of inflammatory and degenerative diseases, there are several issues that limit therapeutic efficacy of MSCs [17, 18]. Due to the low survival of engrafted cells, high number of MSCs has to be transplanted to achieve optimal therapeutic benefits [18]. A large number of evidence demonstrated that modulation of autophagy-related pathways in engrafted MSCs may increase viability and survival of transplanted MSCs, enhancing their therapeutic efficacy [19–23]. Accordingly, in this chapter we emphasized current knowledge about the role of autophagy in MSC-based therapy of inflammatory and degenerative diseases.

Autophagy-An Intracellular Mechanism for the Restoration of Cell Energy

Autophagy is a highly conserved process that functions as an intracellular recycling system for the restoration of cell energy [24]. Autophagy maintains cellular homeostasis [24]. During autophagy, an energy for vital intracellular metabolic pathways is generated from the dysfunctional organelles, altered, and misfolded proteins [24]. Autophagy is a survival process which is activated in response to stress elicited by starvation, hypoxia, microbial infection, or irradiation [24]. The coordinated activity of more than 30 autophagy-related (Atg) proteins regulates the three types of autophagy: macroautophagy, microautophagy, and chaperone-mediated autophagy (CMA) [25].

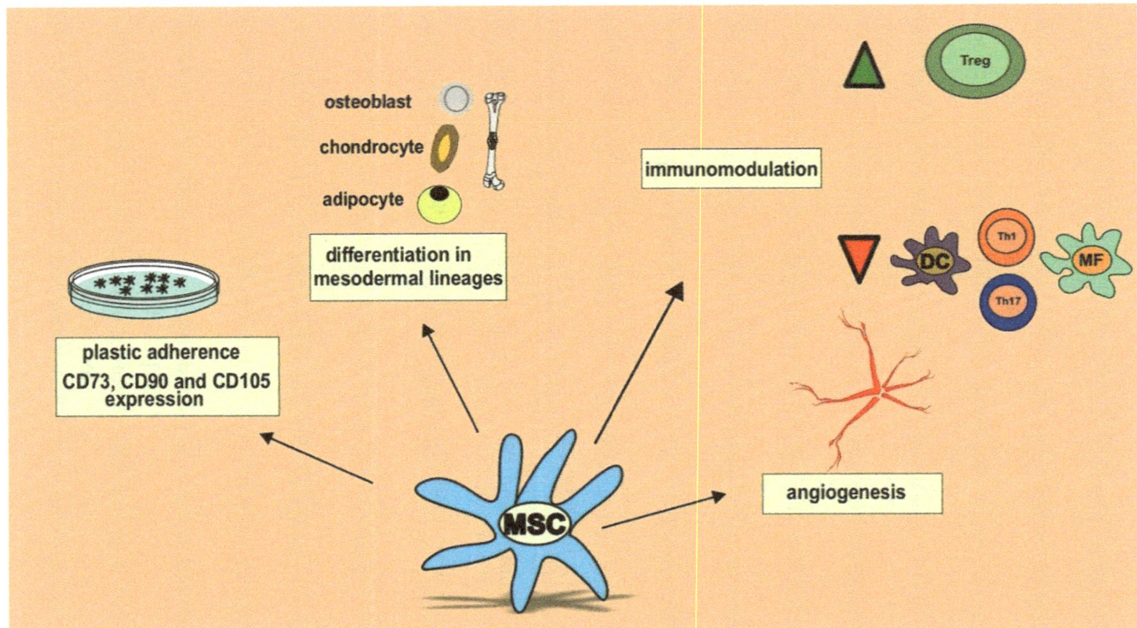

Fig. 9.1 Phenotype and function of mesenchymal stem cells (MSCs). MSCs are spindle-shape, plastic adherence cells that express CD73, CD90 and CD105, spontaneously differentiate in the cells of mesodermal origin (osteoblasts, adipocytes and chondrocytes) and possess angiomodulatory and immunoregulatory properties. MSCs promote expansion of immunosuppressive T regulatory cells (Tregs) and suppress proliferation and effector function of inflammatory dendritic cells (DCs), macrophages (MF), Th1 and Th17 lymphocytes

Macroautophagy involves several steps: initiation, vesicle nucleation (creation of a cup-shaped double-membrane structure-isolation membrane), vesicle elongation, fusion, and degradation [26]. The ULK protein complex, which consists of ULK1, Atg13, FIP200, and Atg101, is linked with mTOR complex 1 (mTORC1), which phosphorylates and inactivates ULK1/2 and Atg13 proteins. Cellular stress induces dissociation of mTORC1 from the ULK protein complex, resulting in the induction of macroautophagy [27]. Initiation of macroautophagy is followed by the vesicle nucleation, the process that requires the formation of Beclin 1/Class III phosphatidylinositol-3-kinase (PI3K) complex, coordinated by the interactions of Beclin 1, UV irradiation resistance-associated tumor suppressor gene (UVRAG), Atg14, B-cell leukemia/lymphoma-2 (Bcl-2), p150, ambra1, endophilin B1, and Vacuolar protein sorting 34 (Vps34). After nucleation, several Atg proteins (assembled into two ubiquitin-like conjugation systems, Atg12–Atg5–Atg16 L and Atg8 (LC3)–phosphatidylethanolamine (PE)) are being attracted to the membrane of pre-autophagosomes to promote vesicle elongation. As a final step, the expanding vesicle mature and close to form a completed autophagosome which fuses with an endosome and/or lysosome, becoming an autophagolysosome, the structure where damaged organelles or proteins are degraded [27]. The reactivation of mTOR (an essential component of the mTORC1 complex) by nutrients created during macroautophagy can be used to stop the autophagy. This is an example of a feedback mechanism that prevents excessive autophagy activation during periods of starvation [26, 27].

The process of microautophagy is characterized by the degradation of damaged cytoplasmic molecules which enter the lysosome by an invagination or distortion of the lysosomal membrane [28]. Microautophagy cooperates with macroautophagy and CMA in maintaining structure and function of cellular membrane and intracellular organelles [28].

CMA is a unique, selective form of autophagy in which various cytosolic proteins (glycolytic enzymes, transcription factors and their inhibitors, calcium and lipid binding proteins, proteasome subunits) are carried one by one across the lysosomal membrane for consequent degradation [29].

Autophagy-Dependent Regulation of Stemness and Differentiation Potential of MSCs

Stemness and differentiation potential of MSCs are regulated by autophagy [30, 31]. Low level of autophagy maintains stemness of MSCs, whereas deletion of autophagy-related genes causes genomic instability and telomere alterations in MSCs, hastening their senescence [30]. MSCs with extensive autophagy activation, as evidenced by over expression of Beclin-1, Atg5, Atg7, and enhanced LC3-II conversion, experienced premature senescence, as evidenced by expanded and flat morphology and lower proliferative potential [32]. Suppression of autophagy (induced by 3-methyladenine (3-MA)) preserved stemness of MSCs which were exposed to hyperglycemic stress [32]. These seemingly contradictory data suggest that autophagy is a stress adaption response that must be carefully managed [19–23]. Reduced autophagy has been found to cause senescence, but significantly increased autophagy may provoke development of irreversible functional changes in MSCs [33]. Autophagy is essential at the housekeeping level to avoid MSC senescence; however, excessive autophagic activation reduces MSC lifetime and stem cell properties [33]. Under standard culture conditions, activation of autophagy increased proliferation, while suppression of autophagy reduced survival of MSCs [34]. Given the need of a proper cellular stress response for maintaining homeostasis, autophagy appears to be a mechanism for protecting transplanted MSCs from external and internal stressors [19–21]. In MSCs exposed to oxidative or irradiation-induced stress conditions, autophagy is elicited to counteract deteriorative processes and to prevent senescence or cell death [34].

Autophagy attenuates differentiation capacity of MSCs by controlling their commitment to the adipogenic and osteoblastic lineages [35]. Undifferentiated MSCs appear to be in a condition of stalled autophagy, as evidenced by the large number of undegraded autophagic vacuoles found in MSCs [35]. Autophagy induction inhibited adipogenic and increased osteogenic differentiation in MSCs [35]. Rapamycin inhibits MSCs' ability to differentiate into adipocytes by decreasing activity of mammalian target of rapamycin (mTOR), which promotes the formation of white and brown adipocytes [36]. In contrast to adipogenic differentiation, autophagy activation (either by mTOR inhibition or by protein kinase B (PKB/AKT) activation) stimulated osteoblast differentiation in dental pulp-derived MSCs [35]. Similarly, activation of autophagy-related genes increased bone repair in MSC-treated rats [37]. Therapeutic potential of autophagy-dependent enhanced differentiation of MSCs towards osteoblast lineage has been demonstrated in glucocorticoid-induced osteoporosis (GIOP), the most frequent type of secondary osteoporosis [38]. Oral glucocorticoids inhibit osteoblast proliferation and apoptosis, prolong osteoclast survival, and promote bone resorption, resulting in a significantly increased incidence of bone fractures in patients who receive glucocorticoid therapy [39]. Because of its effects on the survival of transplanted MSCs, autophagy plays a critical role in the preservation of bone tissue homeostasis in GIOP [38]. GIOP protected engrafted MSCs from starvation-induced apoptosis, while autophagy inhibition reduced survival and proliferation of MSCs, diminishing their therapeutic effects in GIOP treatment [38]. Since autophagy protects engrafted MSCs from apoptosis and promotes their differentiation into osteoblasts, inducing autophagy could be considered as a new approach for improving MSC therapeutic effects in the treatment of GIOP patients [19–21, 38].

Autophagy regulation is crucial for MSC differentiation into myocytes, hepatocytes, and neural cells [19–23]. Tonsil-derived MSCs (T-MSCs) require autophagy induction to differentiate into myoblasts and skeletal myocytes [40]. Likewise, activating autophagy in T-MSCs increased their ability to differentiate into hepatocyte-like and neuron-like cells and significantly improved beneficial effects of MSCs in the treatment of liver fibrosis and neurodegenerative diseases [41–43]. Importantly, inhibiting autophagy in T-MSCs completely diminished these effects [43], indicating that autophagy induction can be used for enhancement of MSCs' differentiation and therapeutic potential [19–23].

Activation of Autophagy as New Approach for Attenuating Apoptosis of Transplanted MSCs

Numerous lines of evidence imply that autophagy is crucially important for the survival of transplanted MSC [35]. MSCs cultivated in serum-free media use autophagy to recycle macromolecules and to induce synthesis of anti-apoptotic factors in order to survive chronic serum deprivation [44]. Autophagy protects MSCs from injury caused by irradiation-induced reactive oxygen species (ROS) [45]. Starvation or rapamycin-induced autophagy can considerably reduce ROS accumulation-related DNA damage [45]. Additionally, autophagy suppression leads to increased ROS accumulation and DNA damage, resulting in genomic rearrangements and decreased viability of irradiated MSCs [45]. ROS stimulates autophagy in MSCs, as evidenced by increased LC3-II expression and decreased p62 expression, and has a substantial impact on the interaction between autophagy and apoptosis [46]. Bcl-2 plays an important role in signaling crosstalk between these two cell death pathways [47]. Bcl-2 interacts with Beclin 1 and influences the development of the Beclin 1/Vsp34 complex [47]. Bcl-2 phosphorylation and disruption of the Bcl-2/Beclin1 complex are implicated in the induction of autophagy by mitogen-activated protein kinases (MAPKs), such as Jun N-terminal kinases (JNK) [46]. JNK-mediated Bcl-2 degradation activates Beclin1-mediated autophagy in irradiated MSCs, since irradiated-induced ROS activates JNK [45, 46]. Other mechanisms for autophagy induction include Beclin-1 dissociation from Bcl-2 by pro-apoptotic BH3 proteins (such as Bad) and Beclin-1 phosphorylation by DAP kinase (DAPK) [48, 49]. Furthermore, activation of protein kinase D (PKD), which phosphorylates and activates Vps34, initiates autophagy [49]. Although ROS-activated autophagy protects cells from apoptosis, apoptosis is overwhelming, and autophagy cannot prevent apoptosis without significant induction. As a result, increasing autophagy in MSCs prior to transplantation is required to minimize apoptosis and prolong survival of transplanted MSCs [45].

Activation of autophagy in MSCs prior to their transplantation significantly increased therapeutic effects of MSCs in the treatment of myocardial infarction and diabetic limb ischemia [50]. Similarly, rapamycin-induced activation of autophagy significantly increased survival and hepatoprotective properties of adipose tissue-derived mesenchymal stem cells (AT-MSCs) [51]. Results obtained in animal model of cisplatin-induced liver injury showed that rapamycin-pretreated AT-MSCs promoted liver regeneration and completely restored liver function in experimental animals [51]. Up-regulated expression of *BCL2* gene and increased synthesis of anti-apoptotic Bcl2 protein were responsible for improved survival of rapamycin-pretreated AT-MSCs. Rapamycin-pretreated AT-MSCs modulated transforming growth factor beta (TGF-β)/Smad and Phosphatidylinositol-3-Kinase (PI3K)-AKT signaling pathways which down-regulated secretion of pro-fibrotic TGF-β, inhibited nuclear factor kappa-light-chain-enhancer of activated B cells (NF-κB)-driven synthesis of inflammatory cytokines and increased stromal cell derived factor 1 (SDF-1)-dependent hepatocyte proliferation and liver regeneration in experimental animals [51].

MSCs Modulate Autophagy in Parenchymal and Immune Cells Affecting Progression of Degenerative and Inflammatory Diseases

Autophagy plays an important role in the pathogenesis of many degenerative and inflammatory diseases [52–55]. Lysosomal dysfunction, altered autophagosome-lysosome fusion and autophagosome accumulation have been observed in patients suffering from Alzheimer's disease (AD), Parkinson's disease (PD), and Huntington disease (HD) [56]. The aggregation of abnormal and misfolded amyloid beta, alpha-synuclein, and huntingtin proteins are considered the main cause for neuronal toxicity and altered axonal transport in AD, PD, and HD patients [56]. MSCs increased neuronal survival by stimulating autophagy in damaged neurons exposed to toxic protein aggregates [57]. MSC-dependent modulation of autophagy in neural cells resulted in increased degradation of misfolded proteins that lead to the improved survival of neural cells of MSC-treated AD rats [57]. MSCs significantly increased the amount of autophagosomes fused to lysosomes which lead to the increased degradation of intracellular amyloid beta clumps and resulted in an improved survival of neural cells [57].

Dysfunction of autophagy in hepatocytes is responsible for massive accumulation of abnormal mitochondria which is followed by increased synthesis and release of ROS, resulting in the development of oxidative stress and inflammation in the liver [54]. MSCs may modulate autophagy in hepatocytes, having beneficial effects in the treatment of liver fibrosis [19–23].

Elevated expression of *BECN1* gene, which is responsible for the synthesis of autophagy-related Beclin-1 protein, is observed in autoreactive T and B lymphocytes of patients suffering from systemic lupus erythematosus (SLE) [58–60]. Similarly, an increased synthesis of antinuclear antibodies and elevated production of inflammatory cytokines were observed in plasma cells and macrophages of SLE patients in which autophagy-related proteins and signaling molecules were altered [60]. T regulatory cells produce immunosuppressive factors that attenuate effector functions of autoreactive T cells, suppress production of auto-antibodies in plasma cells, and inhibit production of inflammatory cytokines in macrophages [61]. MSCs may modulate progression of SLE by affecting autophagy-related pathways in Tregs [62]. Umbilical cord-derived MSCs deliver functional mitochondria to Tregs, improving their longevity which, consequently results in enhanced Treg-dependent suppression of autoreactive T and B cells and with the inhibition of inflammatory macrophages [62].

Beneficial effects of MSCs in the treatment of diabetic rats were attributed to the MSC-dependent modulation of autophagy in insulin-producing pancreatic β-cells [63]. Autophagy plays an important role in the pathogenesis of diabetes mellitus [63]. Autophagy is enhanced in insulin-producing pancreatic β-cells exposed to the high-glucose conditions [63]. Significantly increased generation of autophagosome and autolysosomes, attenuated ROS release and reduced apoptosis were observed in pancreatic β-cells after their co-culture with bone marrow-derived MSCs [63]. When autophagy inhibitors (3-MA) or chloroquine) were added to the cell co-culture, viability of pancreatic β-cells was reduced and the suppressive effects of MSCs on ROS levels were abrogated, confirming important role of autophagy for beneficial effects of MSCs in the treatment of diabetes mellitus [63].

MSC-derived interleukin (IL)-15 induces autophagy in mucosal-associated invariant T (MAIT) cells, innate-like unconventional T cells that play important role in the defense against bacterial and viral pathogens [64]. By activating autophagy in MAIT, MSC-sourced IL-15 induced increased synthesis of inflammatory cytokines (tumor necrosis factor alpha (TNF-α), interferon gamma (IFN-γ)), and cytotoxic granzyme B, significantly enhancing antimicrobial properties of MAIT [64]. Inhibition of autophagy in MAIT completely diminished immunostimulatory effects of MSC-derived IL-15, confirming that IL-15-dependent modulation of autophagy was crucially responsible for MSC-dependent alteration of MAIT's phenotype and function [64].

Modulation of Autophagy as New Approach for Enhancement of MSCs' Immunosuppressive Properties

Modulation of autophagy in MSCs has been shown to be a novel technique for improving MSC-based therapeutic benefits in the therapy of immune cell-mediated diseases (Fig. 9.2) [65–68]. Autophagy improves MSC-dependent suppression of CD4+T cells by increasing production of immunosuppressive TGF-β [65]. It is well known that MSC-derived TGF-β inhibits the activation of the Janus kinase-Signal transducer and activator of transcription (JAK-STAT) signaling pathway in T cells, resulting in G1 cell cycle arrest of T lymphocytes [66]. Pretreatment with rapamycin dramatically increased MSCs' ability to release TGF-β and suppress CD4+T lymphocyte proliferation in TGF-β-dependent manner [65]. In an analogy, 3-MA-induced inhibition of autophagy considerably reduced TGF-β-dependent suppression of T cells by MSCs [65]. The addition of recombinant TGF-β completely restored the immunosuppressive potential of 3-MA-pretreated MSCs, whereas blocking of TGF-β-signaling diminished immunosuppressive capacity of rapamycin-pretreated MSCs, suggesting important role of autophagy for TGF-β-driven suppression of T cells by MSCs [65].

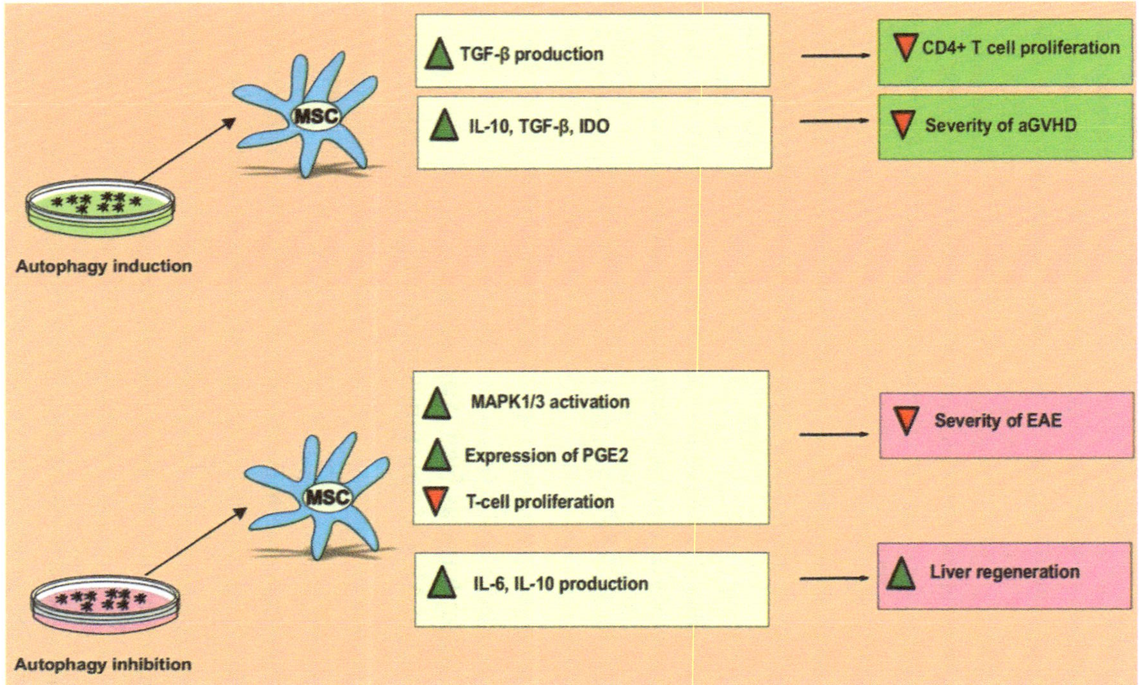

Fig. 9.2 Autophagy regulates immunomodulatory properties of MSCs. Activation of autophagy induces increased synthesis of immunosuppressive factors (transforming growth factor beta (TGF-β), interleukin (IL)-10 and indoleamine 2, 3-dioxygenase (IDO)) in MSCs which results in suppression of CD4+T cells and leads to the attenuation of acute Graft-versus-Host Disease (aGVHD). Inhibition of autophagy in MSCs induces activation of mitogen-activated protein kinases (MAPK1/3), enhances synthesis of prostaglandin E2 (PGE2) which results in suppression of CD4+T cells and attenuates experimental autoimmune encephalomyelitis. Inhibition of autophagy in MSCs induces increased production of IL-6 and IL-10 and enhances liver regeneration

MSC-based therapy was effective in the treatment of acute Graft-versus-Host Disease (aGVHD), a potentially fatal consequence of allogeneic bone marrow transplantation [67]. Rapamycin-induced activation of autophagy significantly increased production of immunosuppressive factors (TGF-β, IL-10, and indoleamine 2,3-dioxygenase (IDO)) in MSCs [19, 66, 67]. Accordingly, rapamycin-treated MSCs showed better therapeutic effects in the treatment of mice with aGVHD when compared with aGVHD mice that received rapamycin-untreated MSCs [67]. MSCs in IL-10-dependent manner inhibit maturation of dendritic cells (DCs) and suppress DC-dependent T cell activation, while MSC-derived IDO induce expansion of Tregs by preventing their reprogramming into inflammatory, IL-17-producing effector Th17 cells [66]. Accordingly, increased production of IL-10 and IDO in rapamycin-treated MSCs was associated with greater expansion of Tregs and decreased presence of inflammatory Th17 cells in mice that received rapamycin-treated MSCs [67], indicating that induction of autophagy substantially increased immunomodulatory properties of MSCs. Opposite to these findings are results obtained by Dang and colleagues who analyzed the effects of autophagy on MSC-based therapy of experimental autoimmune encephalomyelitis (EAE), well-established animal model of multiple sclerosis [68]. They found that inhibiting rather than activating autophagy boosted MSC-mediated suppression of CD4+T cell-driven inflammation in the central nervous system (CNS) of experimental mice [68]. Inflammatory cytokines (TNF-α and IFN-γ), produced by autoreactive, myelin-specific CD4+T cells, induced autophagy in transplanted MSCs by promoting Beclin 1 expression [68]. Beclin 1 deletion, in turn, enhanced MSC therapeutic benefits in EAE by suppressing activation of autoreactive CD4+T cells in CNS [68]. Inhibition of autophagy increased MAPK1/3 activation in MSCs, resulting in increased production of prostaglandin E2 (PGE2), which inhibited IL-2 receptor: JAK3 signaling pathway in activated T cells and suppressed their proliferation [68]. The beneficial effects of MSCs in the therapy of immune cell-mediated acute and chronic liver disorders mostly relied on the immunosuppressive effects of MSCs [12]. Autophagy inhibition in MSCs via downregulation of ATG7 boosted MSC survival and enhanced their therapeutic effects by increasing production of immunosuppressive and hepatoprotective IL-10 [19, 69].

Autophagy Improves the Pro-angiogenic Properties of MSCs

Although MSCs are widely used in the treatment of cardiovascular diseases, their therapeutic efficacy is limited due to their low pro-angiogenic ability in the pathological milieu [70, 71]. Several lines of evidence suggested that activation of autophagy in MSCs could be an effective method for enhancement of MSCs' pro-angiogenic potential [ref]. Rapamycin treatment or over-expression of *BECN-1* gene significantly enhanced the production of basic fibroblast growth factor (bFGF) and angiopoietin in MSCs [72]. Hypoxia activates adenosine monophosphate-activated protein kinase (AMPK)/mTOR signaling pathway and autophagy in MSCs, improving their survival in ischemic environment [72]. Accordingly, after engraftment in hypoxic microenvironment, MSCs increased production of pro-angiogenic factors (bFGF, angiopoietin, and vascular endothelial growth factor (VEGF)), resulting in increased capillary formation of MSC-treated ischemic tissues [73, 74].

Xia and colleagues isolated extracellular vesicles from MSCs which were previously derived from human induced pluripotent stem cells (iPSC-MSC-EVs) [75]. Infusion of iPSC-MSC-EVs significantly reduced infarct volume, improved angiogenesis, and alleviated long-term neurological impairments in mice with ischemic brain injury [75]. Molecular mechanisms responsible for beneficial effects of iPSC-MSC-EVs relied on STAT3-dependent suppression of autophagy in ischemic neural tissue [75]. Administration of iPSC-MSC-EVs increased synthesis of STAT-3 which, in turn, suppressed autophagy, improved migration of endothelial cells and enhanced generation of new blood vessels in ischemic brains of experimental animals [75].

Conclusions

Autophagy regulates stemness, viability, potential for differentiation. immunomodulatory and pro-angiogenic properties of MSCs [19–23]. Modulation of autophagy in MSCs prior to their transplantation represents potentially new approach which could improve therapeutic potential of MSCs in the treatment of degenerative, inflammatory, and ischemic diseases [19–23].

Acknowledgements None.

Disclosure of Interests All authors declare they have no conflict of interest.

Compliance with Ethical Standards This article does not contain any studies with human participants and animals performed by any of the authors.

References

1. Friedenstein AJ, Petrakova KV, Kurolesova AI, Frolova GP (1968) Heterotopic of bone marrow. Analysis of precursor cells for osteogenic and hematopoietic tissues. Transplantation 6:230–47
2. Friedenstein AJ, Chailakhjan RK, Lalykina KS (1970) The development of fibroblast colonies in monolayer cultures of guinea-pig bone marrow and spleen cells. Cell Tissue Kinet 3:393–403
3. Dominici M, Le Blanc K, Mueller I, Slaper-Cortenbach I, Marini F, Krause D, Deans R, Keating A, Prockop Dj, Horwitz E (2006) Minimal criteria for defining multipotent mesenchymal stromal cells. The International Society for Cellular Therapy position statement. Cytotherapy 8:315–7
4. Katz AJ, Tholpady A, Tholpady SS, Shang H, Ogle RC (2005) Cell surface and transcriptional characterization of human adipose-derived adherent stromal (hADAS) cells. Stem Cells 23:412–423

5. Kassis I, Zangi L, Rivkin R, Levdansky L, Samuel S, Marx G, Gorodetsky R (2006) Isolation of mesenchymal stem cells from G-CSF-mobilized human peripheral blood using fibrin microbeads. Bone Marrow Transplant 37:967–976
6. Agha-Hosseini F, Jahani MA, Jahani M, Mirzaii-Dizgah I, Ali-Moghaddam K (2010) In vitro isolation of stem cells derived from human dental pulp. Clin Transplant 24:E23–E28
7. Roubelakis MG, Pappa KI, Bitsika V, Zagoura D, Vlahou A, Papadaki HA, Antsaklis A, Anagnou NP (2007) Molecular and proteomic characterization of human mesenchymal stem cells derived from amniotic fluid: comparison to bone marrow mesenchymal stem cells. Stem Cells Dev. 16:931–952
8. Vellasamy S, Sandrasaigaran P, Vidyadaran S, George E, Ramasamy R (2012) Isolation and characterisation of mesenchymal stem cells derived from human placenta tissue. World J Stem Cells 4:53–61
9. Girdlestone J, Limbani VA, Cutler AJ, Navarrete CV (2009) Efficient expansion of mesenchymal stromal cells from umbilical cord under low serum conditions. Cytotherapy 11:738–748
10. Batsali AK, Pontikoglou C, Koutroulakis D, Pavlaki KI, Damianaki A, Mavroudi I, Alpantaki K, Kouvidi E, Kontakis G, Papadaki HA (2017) Differential expression of cell cycle and WNT pathway-related genes accounts for differences in the growth and differentiation potential of Wharton's jelly and bone marrow-derived mesenchymal stem cells. Stem Cell Res Ther 8:102
11. Gazdic M, Volarevic V, Harrell CR, Fellabaum C, Jovicic N, Arsenijevic N, Stojkovic M (2018) Stem cells therapy for spinal cord injury. Int J Mol Sci 19:1039
12. Volarevic V, Nurkovic J, Arsenijevic N, Stojkovic M (2014) Concise review: therapeutic potential of mesenchymal stem cells for the treatment of acute liver failure and cirrhosis. Stem Cells 32:2818–2823
13. Harrell CR, Volarevic A, Djonov V, Volarevic V (2021) Mesenchymal stem cell-derived exosomes as new remedy for the treatment of neurocognitive disorders. Int J Mol Sci 22:1433
14. Harrell CR, Djonov V, Volarevic V (2021) The cross-talk between mesenchymal stem cells and immune cells in tissue repair and regeneration. Int J Mol Sci 22:2472
15. Harrell CR, Fellabaum C, Jovicic N, Djonov V, Arsenijevic N, Volarevic V (2019) Molecular mechanisms responsible for therapeutic potential of mesenchymal stem cell-derived secretome. Cells 8:467
16. Naji A, Eitoku M, Favier B, Deschaseaux F, Rouas-Freiss N, Suganuma N (2019) Biological functions of mesenchymal stem cells and clinical implications. Cell Mol Life Sci 76:3323–3348
17. Volarevic V, Markovic BS, Gazdic M, Volarevic A, Jovicic N, Arsenijevic N, Armstrong L, Djonov V, Lako M, Stojkovic M (2018) Ethical and safety issues of stem cell-based therapy. Int J Med Sci 15:36–45
18. Harrell CR, Volarevic V (2021) Apoptosis: a friend or foe in mesenchymal stem cell-based immunosuppression. Adv Protein Chem Struct Biol 126:39–62
19. Jakovljevic J, Harrell CR, Fellabaum C, Arsenijevic A, Jovicic N, Volarevic V (2018) Modulation of autophagy as new approach in mesenchymal stem cell-based therapy. Biomed Pharmacother 104:404–410
20. Qin C, Bai L, Li Y, Wang K (2022) The functional mechanism of bone marrow-derived mesenchymal stem cells in the treatment of animal models with Alzheimer's disease: crosstalk between autophagy and apoptosis. Stem Cell Res Ther 13:90
21. Menshikov M, Zubkova E, Stafeev I, Parfyonova Y (2021) Autophagy, mesenchymal stem cell differentiation, and secretion. Biomedicines 9:1178
22. He J, Liu J, Huang Y, Tang X, Xiao H, Hu Z (2021) Oxidative stress, inflammation, and autophagy: potential targets of mesenchymal stem cells-based therapies in ischemic stroke. Front Neurosci 15:641157
23. Deng J, Zhong L, Zhou Z, Gu C, Huang X, Shen L, Cao S, Ren Z, Zuo Z, Deng J, Yu S (2021) Autophagy: a promising therapeutic target for improving mesenchymal stem cell biological functions. Mol Cell Biochem 476:1135–1149
24. Denton D, Xu T, Kumar S (2015) Autophagy as a pro-death pathway. Immunol Cell Biol 93:35–42

25. Kim KH, Lee MS (2014) Autophagy–a key player in cellular and body metabolism. Nat Rev Endocrinol 10:322–337
26. Feng Y, He D, Yao Z, Klionsky DJ (2014) The machinery of macroautophagy. Cell Res 24:24–41
27. Wong AS, Cheung ZH, Ip NY (2011) Molecular machinery of macroautophagy and its deregulation in diseases. Biochim Biophys Acta 1812:1490–1497
28. Li WW, Li J, Bao JK (2012) Microautophagy: lesser-known self-eating. Cell Mol Life Sci 69:1125–1136
29. Cuervo AM (2010) Chaperone-mediated autophagy: selectivity pays off. Trends Endocrinol Metab 21:142–150
30. Rubinsztein DC, Mariño G, Kroemer G (2011) Autophagy and aging. Cell 146:682–695
31. Nuschke A, Rodrigues M, Stolz DB, Chu CT, Griffith L, Wells A (2014) Human mesenchymal stem cells/multipotent stromal cells consume accumulated autophagosomes early in differentiation. Stem Cell Res Ther 5:140
32. Chang TC, Hsu MF, Wu KK (2015) High glucose induces bone marrow-derived mesenchymal stem cell senescence by upregulating autophagy. PLoS ONE 10:e0126537
33. Capasso S, Alessio N, Squillaro T, Di Bernardo G, Melone MA, Cipollaro M, Peluso G, Galderisi U (2015) Changes in autophagy, proteasome activity and metabolism to determine a specific signature for acute and chronic senescent mesenchymal stromal cells. Oncotarget 6:39457–39468
34. Molaei S, Roudkenar MH, Amiri F, Harati MD, Bahadori M, Jaleh F, Jalili MA, Mohammadi RA (2015) Down-regulation of the autophagy gene, ATG7, protects bone marrow-derived mesenchymal stem cells from stressful conditions. Blood Res 50:80–86
35. Sbrana FV, Cortini M, Avnet S, Perut F, Columbaro M, De Milito A, Baldini N (2016) The role of autophagy in the maintenance of stemness and differentiation of mesenchymal stem cells. Stem Cell Rev Rep. 12:621–633
36. Xiang X, Zhao J, Xu G, Li Y, Zhang W (2011) mTOR and the differentiation of mesenchymal stem cells. Acta Biochim Biophys Sin (Shanghai) 43:501–510
37. Dong W, Zhang P, Fu Y, Ge J, Cheng J, Yuan H, Jiang H (2015) Roles of SATB2 in site-specific stemness, autophagy and senescence of bone marrow mesenchymal stem cells. J Cell Physiol 230:680–690
38. Wang L, Fan J, Lin YS, Guo YS, Gao B, Shi QY, Wei BY, Chen L, Yang L, Liu J, Luo ZJ (2015) Glucocorticoids induce autophagy in rat bone marrow mesenchymal stem cells. Mol Med Rep 11:2711–2716
39. Whittier X, Saag KG (2016) Glucocorticoid-induced Osteoporosis. Rheum Dis Clin North Am 42:177–189
40. Park S, Choi Y, Jung N, Kim J, Oh S, Yu Y, Ahn JH, Jo I, Choi BO, Jung SC (2017) Autophagy induction in the skeletal myogenic differentiation of human tonsil-derived mesenchymal stem cells. Int J Mol Med 39:831–840
41. Russo FP, Parola M (2012) Stem cells in liver failure. Best Pract Res Clin Gastroenterol 26:35–45
42. Ugland H, Naderi S, Brech A, Collas P, Blomhoff HK (2011) cAMP induces autophagy via a novel pathway involving ERK, cyclin E and Beclin 1. Autophagy 7:1199–1211
43. Park M, Kim YH, Woo SY, Lee HJ, Yu Y, Kim HS, Park YS, Jo I, Park JW, Jung SC, Lee H, Jeong B, Ryu KH (2015) Tonsil-derived mesenchymal stem cells ameliorate CCl4-induced liver fibrosis in mice via autophagy activation. Sci Rep 5:8616
44. Sanchez CG, Penfornis P, Oskowitz AZ, Boonjindasup AG, Cai DZ, Dhule SS, Rowan BG, Kelekar A, Krause DS, Pochampally RR (2011) Activation of autophagy in mesenchymal stem cells provides tumor stromal support. Carcinogenesis 32:964–972
45. Hou J, Han ZP, Jing YY, Yang X, Zhang SS, Sun K, Hao C, Meng Y, Yu FH, Liu XQ, Shi YF, Wu MC, Zhang L, Wei LX (2013) Autophagy prevents irradiation injury and maintains stemness through decreasing ROS generation in mesenchymal stem cells. Cell Death Dis 4:e844
46. Liu GY, Jiang XX, Zhu X, He WY, Kuang YL, Ren K, Lin Y, Gou X (2015) ROS activates JNK-mediated autophagy to counteract apoptosis in mouse mesenchymal stem cells in vitro. Acta Pharmacol Sin 36:1473–1479

47. Luo S, Rubinsztein DC (2007) Atg5 and Bcl-2 provide novel insights into the interplay between apoptosis and autophagy. Cell Death Differ 14:1247–1250
48. Maiuri MC, Criollo A, Tasdemir E, Vicencio JM, Tajeddine N, Hickman JA, Geneste O, Kroemer G (2007) BH3-only proteins and BH3 mimetics induce autophagy by competitively disrupting the interaction between Beclin 1 and Bcl-2/Bcl-X(L). Autophagy 3:374–376
49. Levin-Salomon V, Bialik S, Kimchi A (2014) DAP-kinase and autophagy. Apoptosis 19:346–356
50. Miao C, Lei M, Hu W, Han S, Wang Q (2017) A brief review: the therapeutic potential of bone marrow mesenchymal stem cells in myocardial infarction. Stem Cell Res Ther 8:242
51. El Nashar EM, Alghamdi MA, Alasmari WA, Hussein MMA, Hamza E, Taha RI, Ahmed MM, Al-Khater KM, Abdelfattah-Hassan A (2021) Autophagy promotes the survival of adipose mesenchymal stem/stromal cells and enhances their therapeutic effects in cisplatin-induced liver injury via modulating TGF-β1/Smad and PI3K/AKT signaling pathways. Cells 10:2475
52. Wu DJ, Adamopoulos IE (2017) Autophagy and autoimmunity. Clin Immunol 176:55–62
53. Terman A, Brunk UT (2005) Autophagy in cardiac myocyte homeostasis, aging, and pathology. Cardiovasc Res 68:355–365
54. Yin XM, Ding WX, Gao W (2008) Autophagy in the liver. Hepatology 47:1773–1785
55. Patel AS, Lin L, Geyer A, Haspel JA, An CH, Cao J, Rosas IO, Morse D (2012) Autophagy in idiopathic pulmonary fibrosis. PLoS ONE 7:1–9
56. Millecamps S, Julien JP (2013) Axonal transport deficits and neurodegenerative diseases. Nat Rev Neurosci 14:161–176
57. Shin JY, Park HJ, Kim HN, Oh SH, Bae JS, Ha HJ, Lee PH (2014) Mesenchymal stem cells enhance autophagy and increase β-amyloid clearance in Alzheimer disease models. Autophagy 10:32–44
58. Gergely P Jr, Grossman C, Niland B, Puskas F, Neupane H, Allam F, Banki K, Phillips PE, Perl A (2002) Mitochondrial hyperpolarization and ATP depletion in patients with systemic lupus erythematosus. Arthritis Rheum 46:175–190
59. Caza TN, Fernandez DR, Talaber G, Oaks Z, Haas M, Madaio MP, Lai ZW, Miklossy G, Singh RR, Chudakov DM, Malorni W, Middleton F, Banki K, Perl A (2014) HRES-1/Rab4-mediated depletion of Drp1 impairs mitochondrial homeostasis and represents a target for treatment in SLE. Ann Rheum Dis 73:1888–1897
60. Martinez J, Cunha LD, Park S, Yang M, Lu Q, Orchard R, Li QZ, Yan M, Janke L, Guy C, Linkermann A, Virgin HW, Green DR (2016) Noncanonical autophagy inhibits the autoinflammatory, lupus-like response to dying cells. Nature 533:115–119
61. Goswami TK, Singh M, Dhawan M, Mitra S, Emran TB, Rabaan AA, Mutair AA, Alawi ZA, Alhumaid S, Dhama K (2022) Regulatory T cells (Tregs) and their therapeutic potential against autoimmune disorders-advances and challenges. Hum Vaccin Immunother 18:2035117
62. Chen J, Wang Q, Feng X, Zhang Z, Geng L, Xu T, Wang D, Sun L (2016) Umbilical cord-derived mesenchymal stem cells suppress autophagy of T cells in patients with systemic lupus erythematosus via transfer of mitochondria. Stem Cells Int 2016:4062789
63. Zhao K, Hao H, Liu J, Tong C, Cheng Y, Xie Z, Zang L, Mu Y, Han W (2015) Bone marrow-derived mesenchymal stem cells ameliorate chronic high glucose-induced β-cell injury through modulation of autophagy. Cell Death Dis 17(6):e1885
64. Ye G, Wang P, Xie Z, Cao Q, Li J, Zheng G, Wang S, Li M, Liu W, Cen S, Li Z, Yu W, Wu Y, Shen H (2021) Autophagy-mediated activation of mucosal-associated invariant T cells driven by mesenchymal stem cell-derived IL-15. Stem Cell Rep 16:926–939
65. Gao L, Cen S, Wang P, Xie Z, Liu Z, Deng W, Su H, Wu X, Wang S, Li J, Ouyang Y, Wu Y, Shen H (2016) Autophagy improves the immunosuppression of CD4+T cells by mesenchymal stem cells through transforming growth Factor-β1. Stem Cells Transl Med 5:1496–1505
66. Volarevic V, Gazdic M, Simovic Markovic B, Jovicic N, Djonov V, Arsenijevic N (2017) Mesenchymal stem cell-derived factors: immuno-modulatory effects and therapeutic potential. BioFactors 43:633–644
67. Kim KW, Moon SJ, Park MJ, Kim BM, Kim EK, Lee SH, Lee EJ, Chung BH, Yang CW, Cho ML (2015) Optimization of adipose tissue-derived mesenchymal stem cells by rapamycin in a murine model of acute graft-versus-host disease. Stem Cell Res Ther 6:202

68. Dang S, Xu H, Xu C, Cai W, Li Q, Cheng Y, Jin M, Wang RX, Peng Y, Zhang Y, Wu C, He X, Wan B, Zhang Y (2014) Autophagy regulates the therapeutic potential of mesenchymal stem cells in experimental autoimmune encephalomyelitis. Autophagy 10:1301–1315
69. Amiri F, Molaei S, Bahadori M, Nasiri F, Deyhim MR, Jalili MA, Nourani MR, Habibi RM (2016) Autophagy-modulated human bone marrow-derived mesenchymal stem cells accelerate liver restoration in mouse models of acute liver failure. Iran Biomed J 20:135–144
70. Joshi J, Kothapalli CR (2022) Role of inflammatory niche and adult cardiomyocyte coculture on differentiation, matrix synthesis, and secretome release by human bone marrow mesenchymal stem cells. Appl Biochem Biotechnol 194:1938–1954
71. Hare JM (2009) Translational development of mesenchymal stem cell therapy for cardiovascular diseases. Tex Heart Inst J 36:145–147
72. Wei W, An Y, An Y, Fei D, Wang Q (2018) Activation of autophagy in periodontal ligament mesenchymal stem cells promotes angiogenesis in periodontitis. J Periodontol 89:718–727
73. Liu J, Hao H, Huang H, Tong C, Ti D, Dong L, Chen D, Zhao Y, Liu H, Han W, Fu X (2015) Hypoxia regulates the therapeutic potential of mesenchymal stem cells through enhanced autophagy. Int J Low Extrem Wounds 14:63–72
74. Boyette LB, Creasey OA, Guzik L, Lozito T, Tuan RS (2014) Human bone marrow-derived mesenchymal stem cells display enhanced clonogenicity but impaired differentiation with hypoxic preconditioning. Stem Cells Transl Med 3:241–254
75. Xia Y, Ling X, Hu G, Zhu Q, Zhang J, Li Q, Zhao B, Wang Y, Deng Z (2020) Small extracellular vesicles secreted by human iPSC-derived MSC enhance angiogenesis through inhibiting STAT3-dependent autophagy in ischemic stroke. Stem Cell Res Ther 11:313

Chapter 10
Autophagic Control of Stem Cells Differentiation into Osteogenic Lineage—Implications in Bone Disorders

Shalmoli Bhattacharyya and Aditi Mahajan

Abstract Autophagy is a degradative metabolic process involved in sustaining the cellular homeostasis by regulating the turnover of cytosolic components. Being an essential cell quality control system, the process of autophagy regulates various properties of all types of stem cells such as their self-renewal, homeostasis, stemness and most importantly, their lineage commitment. Undifferentiated stem cells accumulate autophagosomes due to arrested autophagy which is resumed only during differentiation into specific lineages. In terms of differentiation, autophagy is crucially involved in maintaining the balance between osteoblastogenesis and adipogenesis in Wnt/β-catenin mediated manner. Enhanced autophagy is seen in the case of osteogenic differentiation of stem and progenitor cells, and defects in the autophagic process have been implicated in the development of various bone disorders such as osteoporosis, osteopetrosis, Paget's disease and arthritis. This book chapter describes the role of autophagy in the regulation of stem cells characteristics, including the regulation of embryonic stem cells during embryogenesis and in skeletal tissue formation with special emphasis on the recent evidence supporting the autophagic regulation of osteogenic differentiation of MSCs and the consequent implications in bone pathology.

Keywords Autophagy · Stem cells · Embryogenesis · Bone disorders · Osteogenic differentiation · Skeletal development

---

S. Bhattacharyya (✉) · A. Mahajan
Department of Biophysics, Post Graduate Institute of Medical Education and Research (PGIMER), Chandigarh, India
e-mail: shalmoli2007@yahoo.co.in

B. V. Shravage and K. Turksen (eds.), *Autophagy in Stem Cell Maintenance and Differentiation*, Stem Cell Biology and Regenerative Medicine 73,
https://doi.org/10.1007/978-3-031-17362-2_10

Abbreviations

3-MA	3-Methyladenine
AMBRA1	Autophagy and beclin regulator-1
AMPK	Adenosine monophosphate activated protein kinase
ATG	Autophagy-related gene
ATP	Adenosine tri-phosphate
BMPs	Bone morphogenetic proteins
BMSCs	Bone marrow-derived mesenchymal stem cells
C/EBPβ	CCAAT enhancer binding protein beta
CEP55	Centrosomal protein (55KDa)
Dlx5	Distal-less homeobox 5
ESCs	Embryonic stem cells FABP3fatty acid binding protein 3
FOXO3	Forkhead box O-3
GATA-1	GATA binding protein 1
GFER	Growth factor, augmenter of liver regeneration
GSK3β	Glycogen synthase kinase 3 beta
HSCs	Hematopoietic stem cells
IGF-1	Insulin-like growth factor-1
IGFBP-2	Insulin-like growth factor binding protein
2 IL-1β	Interleukin 1 beta
iPSCs	Induced pluripotent stem cells
JNK	C-Jun N-terminal kinase
Klf 1/2	Kruppel-like factor-1/2
LC3	Microtubule associated proteins 1A/1B light chain 3B
MAPK	Mitogen activated protein kinase
miRNA	Micro-RNA
MMP-13	Matrix metalloproteinase 13
MSCs	Mesenchymal stem cells
Msx2	Msh homeobox 2
mTORC1	Mammalian target of rapamycin complex 1
NBR1	Neighbor of BRCA1 gene 1
NFkB	Nuclear factor kappa light chain enhancer of B cells
Nrf2	Nuclear factor erythroid-related factor 2
NuRD	Nucleosome remodeling and deacetylase
NuRD	Nucleosome remodeling and deacetylase
OA	Osteoarthritis
Oct4	Octamer binding transcription factor
4OPG	Osteoprotegerin
OVX	Osteoporotic mice
P13K	Phosphoinositide 3-kinase
PDGF	Platelet-derived growth factor
POU5F1	POU class 5 homeobox 1
ROS	Reactive oxygen species

Runx2	Runt-related transcription factor 2
SmadA	Suppressor of mothers against decapentaplegic
Sox2	SRY box 2
Spp1	Secreted phosphoprotein 1
SQSTM1	Sequestosome-1
TIMP-1A	Tissue inhibitor of metalloprotease1
TSC1	Tuberous sclerosis complex subunit 1
ULK	UNC-51-like kinases
Wnt	Wingless-related integration site

Introduction

Autophagy is one of the major catabolic mechanisms for maintaining cell-bioenergetic homeostasis by controlling the molecular degradation and recycling and organelle turnover. The process is active at baseline levels in all mammalian cells and targets the potentially harmful and toxic organelles and misfolded or dysfunctional proteins to degradation by lysosomal hydrolases. There are three types of autophagy—microautophagy, chaperone mediated autophagy and macroautophagy, each of which differs in the mechanism of degradation of cytosolic products [1, 2]. The classical 'macroautophagy' is induced under severe conditions such as nutrient starvation, hypoxia and metabolic and oxidative stress which results in the degradation of bulk cytosolic material to support cell survival and function. The process is mainly regulated by two major protein complexes: mammalian target of rapamycin complex (mTORC1) and adenosine monophosphate activated protein kinase (AMPK). Under nutrient deprivation, mTORC1 is inactive while AMPK is activated which induces phagophore formation by activating ULK1/2 (UNC-51-like kinases) complex. This protein complex activates Beclin-1 which, in turn, recruits VSP34 to synthesize phosphatidyl inositol 3-phosphates on the phagophore membrane to mark a signal for the formation of autophagosomes. ATG protein complexes are recruited to the site of autophagosome formation and result in the lipidation of LC3 protein. LC3 protein is the major mediator of autophagosome membrane elongation, target recognition and fusion of autophagosomes with lysosomes. The final degraded products are then translocated to the cytosol where they are utilized for use in new metabolic and physiological reactions to sustain cell homeostasis and growth [3, 4].

Since the process of autophagy supports such critical functions of a cell, alterations in the process have been implicated in several pathological conditions such as neurodegeneration, autoimmune disorders, cancer and cardiovascular and bone metabolic diseases due to modulation of tissue-specific cells such as neurons, tumor cells, osteoblasts and osteocytes. Emerging evidence over the past years have also uncovered the fundamentals of autophagy in the fate determination of stem cells and in the modulation of their potency and functions, which is proposed to have an impact on the regenerative properties of these cells. Stem cells are distinct types of cells that persist throughout the lifespan of an organism from the stage of embryogenesis up to old age with varying abilities to replicate and differentiate into different cell lineages. While embryonic stem cells (ESCs) are considered pluripotent with indefinite self-renewal and proliferation ability, the adult somatic stem cells, particularly those residing quiescently in tissues, are multipotent or unipotent and are usually activated in response to tissue injury. Various regulatory mechanisms act in a concerted manner to maintain quiescence and homeostasis in these stem cells, particularly by means of proteostatic and metabolic regulation. One major outcome of cellular metabolism is the formation and accumulation of free radicals which can cause macromolecules and organelle damage. Thus, healthy stem cells need to maintain a balance between their metabolic demands and cytoprotective measures to prevent the build-up of reactive oxygen species (ROS) and damaged cytosolic material which is achieved by the means of cell quality control mechanisms including autophagy [5]. Moreover, during the process of differentiation, stem cells undergo remodeling which requires the elimination of existing cellular components which are no longer required which is also achieved through the process of autophagy. Overall, growing evidence suggest the critical involvement of the autophagic process in maintaining several crucial functions and properties of both embryonic and adult stem cells. This chapter focusses on autophagic regulation and maintenance of stem cells' homeostasis, potency and their multi-lineage differentiation with a special focus on the osteogenic fate of adult stem cells and its implications in bone disorders.

Autophagic Regulation of Embryonic Development

Autophagy During Embryogenesis

The embryonic development requires a series of morphological and physiological remodeling events which are crucial for differentiation into functional tissues and organ development. These events, which occur at the cellular level, are associated with simultaneous changes in cytoskeleton, cell membrane, extracellular matrix composition and organelles. These changes are under the control of various regulatory mechanisms including autophagy which act in a concerted manner to ultimately determine cell fate and organ development during the initial stages of embryogenesis. After fertilization, the zygote undergoes a series of cell divisions and is reprogrammed to form pluripotent ESCs which requires induction of pluripotency genes, clearance of maternal and paternal genomes, epigenetic modifications and induction of proteolytic systems such as autophagy and ubiquitin-proteasome degradation system. Autophagy plays a major role during both the development and differentiation of an early embryo by providing the cytosolic products for recycling into new macromolecules as well as by clearing out old structures by means of programmed cell death. Since all the structures of an embryo are formed from ESCs derived from the inner cell mass, it is plausible to assume that autophagy plays a critical role in the normal development and functioning of an embryo. It has been established that autophagy is predominantly involved in maintaining the metabolic equilibrium during early and post-implantation stages, with several in vivo studies confirming that mice lacking *atg* genes not only display postnatal defects but also show induction of several compensatory mechanisms including ubiquitin-proteasome system suggesting that autophagy, although necessary for optimal embryogenesis, is not critical for the survival of the organism [6]. Autophagy majorly regulates the initial stages of embryogenesis in mTOR and PI-3K signaling-dependent manner. Different molecular mechanisms have been documented which induce autophagy during early embryonic reprogramming. For example, the earliest surge of autophagy is induced just after fertilization due to calcium oscillations and is reported to be independent of mTORC1 activity [7]. As the embryo reaches the 4-8 cell stage, downregulation of mTOR activity by Sox2 and NuRD complex becomes indispensable for the activation of the autophagic process.

Overactivation of autophagy at this stage also accelerates the formation of the blastocyst [8]. Most of our current knowledge about autophagy during early development comes from the autophagic regulation of ESCs cultured outside the blastocyst. Studies have shown that autophagy deficient ESCs are able to form undifferentiated cell aggregates, embryoid bodies which show defects in ATP production, primitive endoderm and an inner core of ectodermal cells. This inner core, however, fails to undergo the cavitation event which is necessary for clearing out apoptotic products during early embryogenesis [9]. Moreover, oocytes with conditional knockdown of Atg5 gene and fertilized by Atg5 null sperm fail to proceed beyond the 4–8 cell stage [7]. However, the role of autophagy in the later stages of embryogenesis remains ambiguous as mice with null mutations in autophagic genes *atg3*, *atg5*, *atg7* and *atg9* have been reported to produce phenotypically normal pups without any anatomical abnormalities. However, these pups are reported to die within 1–2 days of birth probably due to dysregulated neurological development [10]. These results indicate that autophagy is not essential for the survival of the embryo at later stages but appear to be indispensable for its optimal development. Since autophagy also involves the removal of damaged cellular components, it is presumed that the absence of autophagy causes the build-up of toxic by-products which contribute to such defects which are manifested later in the embryo development. However, the precise cellular and tissue defects arising due to defective autophagy at later stages of embryogenesis are yet to be identified.

Autophagy in ESCs

ESCs are pluripotent cells derived from the inner cell mass of pre-implantation blastocysts in the early embryo with the capacity of indefinite self-renewal and differentiation into the lineages of all the three germ layers—ectoderm, endoderm and mesoderm [10]. ESCs have been shown to have higher basal levels of autophagy on the basis of the presence of LC3 puncta. Moreover, the early stages of spontaneous differentiation of ESCs are associated with enhanced autophagic levels, which possibly contribute to the cell reprogramming events during the transition from undifferentiated to differentiated state of these cells [11, 12]. Inhibition of autophagy also leads to greater accumulation of pluripotency associated proteins in both cytoplasm and nucleus, confirming the involvement of autophagy in maintaining the pluripotency of ESCs [13, 14]. Several studies have shown the role of mTOR signaling in regulating the fate decision in ESCs. Induction of autophagy by inhibiting mTOR leads to considerably reduced pluripotency transcription factors such as POU5F1, SOX2 and OCT4 in ESCs and promotes their mesodermal and endodermal activities with arrested proliferation [14, 15]. mTOR interacts with extrinsic pluripotency supporting factors in the microenvironment of ESCs and suppresses the transcription of growth and developmental inhibitory genes, thereby maintaining ESCs pluripotency [14].

The first-ever detailed molecular mechanism of autophagy in a mammalian cell setting was done using ESCs which showed that Atg5 deficient ESCs had a lower protein turnover rate but normal growth rate and clonogenic potential [16]. Another study also reported similar findings that beclin-1 knockdown did not induce any growth defect in mouse ESCs [17]. However, a later study showed that these ESCs were incapable of forming embryoid bodies [18]. Additionally, functional deficiency of AMBRA1, which regulates Beclin-1 mediated autophagy, has been shown to impede ESCs proliferation and degradation of ubiquitinated proteins leading to excessive cell death [10]. During embryogenesis, the presence of primary cilia plays an important role in sensing external signals and mediating signal transduction via Hedhegog, Wnt and PDGF signaling pathways. The process of ciliogenesis has been associated with the activation of autophagy in ESCs prior to their differentiation. In comparison to ESCs, the neuroectodermal and mesodermal differentiated cells show upregulated autophagy, and the inactivation of genes involved in ciliogenesis drastically reduces the autophagic flux in these cells [19, 20].

Other than macroautophagy, mitochondria selective autophagy or mitophagy has also been implicated in the quality control of ESCs. There are only a fewer mitochondria in ESCs which also have poorly formed cristae. GFER protein which is present in inter membranous space of mitochondria maintains the integrity of mitochondria. Loss of function of GFER is associated with degradation of mitochondria and removal of the fragmented mitochondria by mitophagy. GFER deficient ESCs also show decreased expression of pluripotent markers which suggests that the structural and functional integrity of mitochondria is crucial to ESCs stemness. Furthermore, defective GFER does not lead to alterations in mitochondria of ESCs-derived differentiated cells, suggesting that mitophagy and GFER function are restricted to ESCs only [21].

Autophagy in Skeletogenesis

The process of autophagy has also been implicated in the formation and development of the vertebral skeletal system. Since the early developmental stage, autophagy is involved in the formation, differentiation and functional maintenance of key skeletal tissues such as bone, cartilage and associated connective tissues via modulation of mesenchymal stem cells (MSCs) or bone progenitor cells and native tissue cells such as osteoblasts (bone forming cells), osteocytes (bone residing cells), osteoclasts (bone resorbing cells) and chondrocytes (cartilage forming cells) (Fig. 10.1). These cells are all derived from MSCs (except osteoclasts) via mesenchymal condensations followed by their differentiation at the site of skeletal tissue formation. Several studies have indicated the role of autophagy in the differentiation of MSCs toward osteogenic and chondrogenic lineages. Most of the human skeleton (except for the craniofacial skeleton) develops with the deposition of an unmineralized collagen-rich cartilage matrix by MSCs-derived chondrocytes. These cells survive the hypoxic conditions present in the cartilaginous matrix and proliferate until they become hypertrophic. Several *in vitro* studies have emphasized the role of autophagy during the proliferation and differentiation of chondrocytes. Autophagy enables the survival of chondrocytes in hypoxic and nutrient-depleted conditions in the middle of the growth plate. This is also corroborated by several studies which have shown that depletion of ATG5 or ATG7 causes chondrocyte cell death in *in vitro* and *ex vivo* cultures of bone growth plate tissues [22, 23]. Moreover, autophagy deficient chondrocytes show accumulation of glycogen granules, suggesting that autophagy actively participates in the breakdown of glycogen to meet the glucose demands in avascular growth plates [24]. Once the chondrocytes grow in size and become hypertrophic in nature, they allow the vascular invasion of the extracellular matrix and then undergo apoptosis. What's interesting is that these terminal chondrocytes also exhibit elevated autophagic levels. Thus, autophagy is initially activated during skeletogenesis to support the survival and growth of chondrocytes but later switches to induce apoptosis for the removal of these cells [25, 26]. Chondrocytic apoptosis triggers the resorption of cartilage from the matrix and recruitment of MSCs-derived osteoblasts to these regions which form intracellular mineralized aggregates that are secreted onto the osteoid matrix, thereby mineralizing it. Autophagy is activated in both MSCs and osteoblasts which supports their survival under a hypoxic environment during bone formation [27]. The first evidence of autophagy in the role of osteoblastic mineralization was given in 2014, in which a study showed that primary osteoblasts isolated from mice calvaria contained mineral aggregates inside the autophagic vesicles, and inhibition of autophagy dramatically reduced the mineralization capacity of these osteoblasts [28]. In addition to this, silencing of key autophagic mediators ATG5, ATG7 and Beclin-1 which are involved in autophagosome formation in osteoblasts resulted in decreased bone mineralization and bone mass in vivo [27, 29]. Moreover, autophagy is also actively engaged in pathways critical to osteoblasts formation. For example, insulin-like growth factor-1 (IGF-1) and insulin-like growth factor binding protein-2 (IGFBP-2) are involved in osteogenic differentiation in osteoblasts, the functions of which are partly achieved by the activation of AMPK signaling and autophagy [30]. Defective autophagy has also been reported to impede skeletal tissue formation as Atg7 conditional knockout mice showed severely reduced osteoblast formation, matrix mineralization and bone mass accompanied by increased osteoclast function at the developmental stage. Remarkably, this Atg7 deficiency correlated with severe endoplasmic reticulum stress which resulted in the accumulation of misfolded proteins in the osteoblasts [31].

Once osteoblasts lay down the mineralized matrix, they either undergo apoptosis or become embedded in the matrix as osteocytes. The transition of osteoblasts into osteocytes is accompanied by profound changes in cell morphology and composition which requires active recycling of organelles. Moreover, due to limited vascular perfusion in the bone matrix, osteocytes encounter more oxidative stress and hypoxia and thus require stringent nutrient preservation. In comparison to osteoblasts, osteocytes show accumulated levels of LC3 puncta, suggesting enhanced basal autophagy in these cells which is likely to be a survival mechanism in these cells [29]. The deficiency of autophagy has been associated with defective osteoblast differentiation into osteocytes, further demonstrating the important role of autophagy in bone formation and function [32]. *In vivo* studies investigating the role of autophagy during skeletogenesis have shown that osteocytes deficient in Atg7 cause significant loss in bone mass, cancellous and cortical bone thickness and decrease bone formation rate [33]. Since autophagy also controls the mineralization process, the levels of autophagy are controlled by EphrinB2 signaling molecule in osteocytes which limits secondary mineralization and consequent formation of brittle bones. Loss of EphrinB2 in osteocytes causes hyperactivation of autophagy which accelerates the process of secondary mineral formation and causes the brittle bone phenotype [34].

Other than bone forming cells, bone resorbing cells called as osteoclasts which originate from the lineage of hematopoietic stem cells (HSCs) are essential in maintaining bone homeostasis by constantly remodeling the old bone. Osteoclasts secrete a variety of hydrolytic enzymes to resorb the bone minerals and matrix and recycle these degraded products via endocytosis. Autophagy plays a dual role in regulating osteoclast formation and function. The products released during the autophagic degradative pathway have been implicated in Wnt and NFkB signaling critical to osteoclasts formation and function [35]. Autophagy also controls the spatial localization of protein complexes involved in protein synthesis and secretion in osteoclasts. Loss of ATG7 in HSCs in mice causes severe genomic damage and results in their defective differentiation into osteoclasts [36]. Contrary to this, a few studies have reported that OPG-induced autophagy suppresses the osteoclasts mediated bone resorption via AKT/mTOR/ULK1 and AMPK/mTOR/p70S6K signaling axes [37, 38].

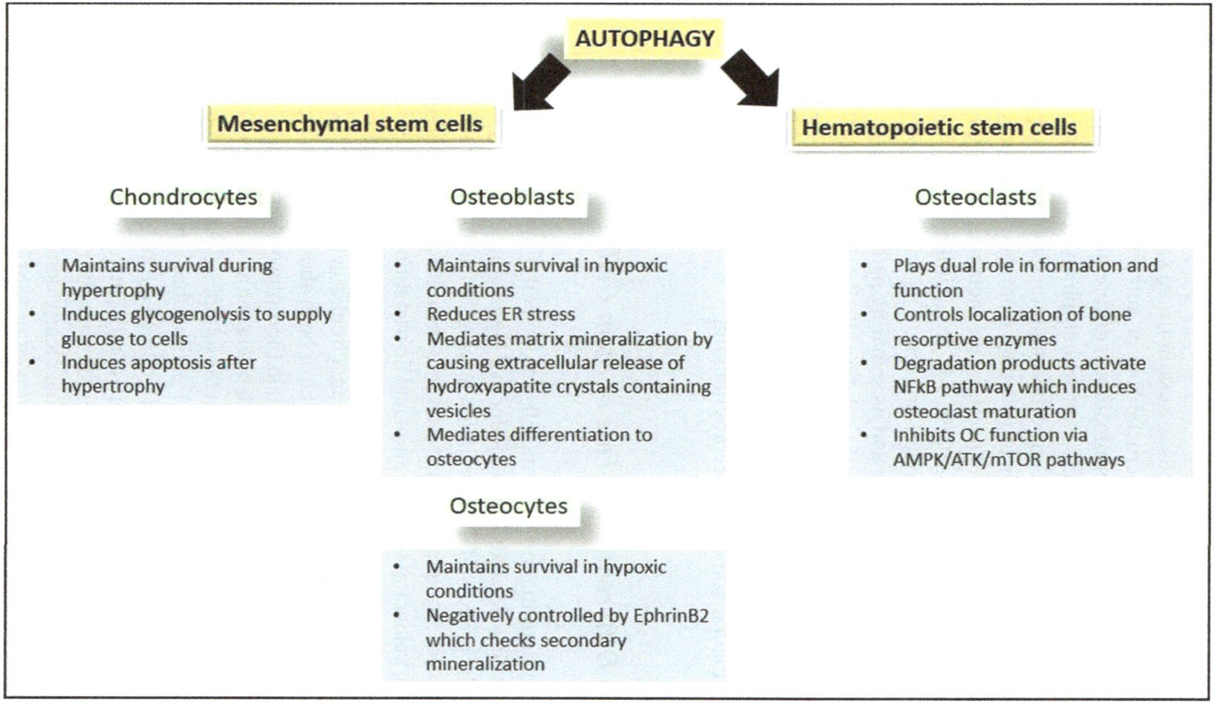

Fig. 10.1 Summarized role of autophagy on cells of bone tissue: Autophagy regulates the differentiation of MSCs into chondrocytes and osteoblasts and of HSCs into osteoclasts. Autophagy then maintains or inhibits the functions of chondrocytes, osteoblasts, osteoclasts and osteocytes which are the major cells governing the process of skeletal tissue formation

Regulation of Stem Cells Characteristics by Autophagy

Since most of the stem cell pool is required to maintain its functionality throughout the entire lifetime of an organism, stringent cellular regulatory mechanisms are required to preserve the capability of these cells to sustain tissue regeneration. Over the past years, evidence have suggested the crucial role of autophagy in the survival, self-renewal, metabolism and lineage determination of different types of stem cells. MSCs have also been shown to constitutively express autophagy-related proteins at basal levels, which are upregulated during early differentiation. Since undifferentiated stem cells are phenotypically and metabolically distinct from their differentiated counterparts, the implication of autophagy in governing the metabolic features of these stem cells is well reported in the literature [39, 40]. In addition to this, the process of autophagy has also been shown to confer protection against MSCs apoptosis under starvation [41]. Moreover, autophagy and its signaling pathway are also shown to be activated during the reprogramming of somatic cells into iPSCs [42, 43]. All these emerging evidence suggest the crucial role of autophagy in producing energy precursors in the form of non-degraded cytosolic products in MSCs and other stem cells for their utilization at the time of stressed metabolic demands pertaining to starvation, hypoxia, differentiation and cellular reprogramming.

Autophagy in Differentiation Fate of Stem Cells

The process of differentiation of stem cells into specific lineages is a multi-stage, complex process that depends on the inductive signals present in their microenvironment. Since the process of differentiation involves changes in cellular phenotype, autophagy is presumed to play an important role in orchestrating metabolic equilibrium by carefully balancing the formation and degradation of new cellular products. The morphological changes induced during the process of differentiation are also regulated by autophagy. Midbodies, which are circular cytoplasmic bridges formed after cytokinesis to separate daughter cells, have been shown to selectively accumulate in undifferentiated stem cells and maintain their pluripotency state. Midbodies' release into the extracellular fluid or their degradation is a characteristic feature of differentiating stem cells and occurs through the autophagic degradation via interaction between the autophagic receptor NBR1 and midbody protein CEP55 [44, 45]. The presence of fewer mitochondria in undifferentiated MSCs as compared to differentiated MSCs has suggested the possible cross-talk between mitophagy and MSCs differentiation. Although mitophagy has been shown to be important in ESCs maintenance, there still occurs a lack of evidence that clearly demonstrates the mechanistic link between mitophagy and lineage commitment of MSCs and it still remains to be elucidated if both macroautophagy and mitophagy act concertedly or independently to promote or inhibit MSCs differentiation [1, 46].

Although autophagy is shown to upregulate at the onset of the differentiation process, several studies have indicated that this upregulation is selective to the lineage specification. For example, after induced lineage commitment, ESCs develop primary cilium which activates autophagy. Autophagic mediated degradation of p62 and consequent inactivation of nuclear factor erythroid-related factor 2 (Nrf2) directs the ESCs fate toward neuroectodermal lineage. This does not happen when ESCs are destined for mesodermal lineage, indicating that autophagy is required only for specific lineage fates [19, 39]. Just like ESCs, induced pluripotent stem cells (iPSCs) also show elevated basal levels of autophagy, and inhibition of the autophagic process induces apoptosis in these cells. Conversely, activation of autophagy is associated with spontaneous embryoid body formation and active differentiation into all the three germ layers [47]. Other than osteogenic differentiation of MSCs which is described in detail in the later section, autophagy has been shown to influence trans-differentiation of MSCs into several other lineages such as adipogenic, chondrogenic, neural and hepatogenic. Knockdown of *atg5* and *atg7* gene in adipose progenitor cells downregulated the expression of adipogenic differentiation factors and prevented lipid accumulation and led to lean body mass *in vivo* [48]. C/EBPβ, which is an important transcription factor for adipogenesis, also targets ATG4b which is critical for LC3 maturation and activation of autophagy. ATG4b results in ubiquitination and degradation of two adipogenic inhibitory factors, Klf2 and Klf3 via SQSTM1/p62 mediated pathway [49]. Similarly, autophagy also regulates the chondrogenic differentiation of stem and progenitor cells. Several studies have indicated autophagy deficient MSCs display impaired chondrogenic differentiation [24, 50]. Synovium-derived MSCs which show superior chondrogenic potential show reduced chondrogenesis when autophagy is inhibited. Pro-inflammatory cytokines such as IL-1β have shown to inhibit autophagy by inhibiting autophagosome formation and also result in the reduced chondrogenic potential of synovium MSCs by decreasing Sox9, aggrecan and collagen type II expression [51]. It is observed that autophagy regulates osteogenic, chondrogenic and adipogenic differentiation of MSCs by regulating Wnt/β-catenin signaling. In fact, both autophagy and Wnt signaling have a direct role in the maintenance of stem cells' homeostasis, pluripotency and their differentiation into specific lineages. Induction of autophagy and Wnt signaling together promotes chondrogenic and osteogenic differentiation while inhibiting adipogenic differentiation of MSCs (Fig. 10.2) [52, 53]. Not only this, but autophagy also regulates the differentiation of MSCs into other lineages as well. Activation of autophagy improves the neural differentiation of bone marrow MSCs in the presence of neuronal inductive signals, and inhibition of the autophagic process reduces the neural differentiation of these MSCs [54]. Similar results have been reported in the case of neural commitment of human placental MSCs [55]. It has been reported that mTOR expression declines during neural differentiation of MSCs which results in the upregulation of autophagy. However, excessive activation of autophagy inhibits the differentiation process, suggesting that a certain level of mTOR signaling is required for optimum differentiation of MSCs [56].

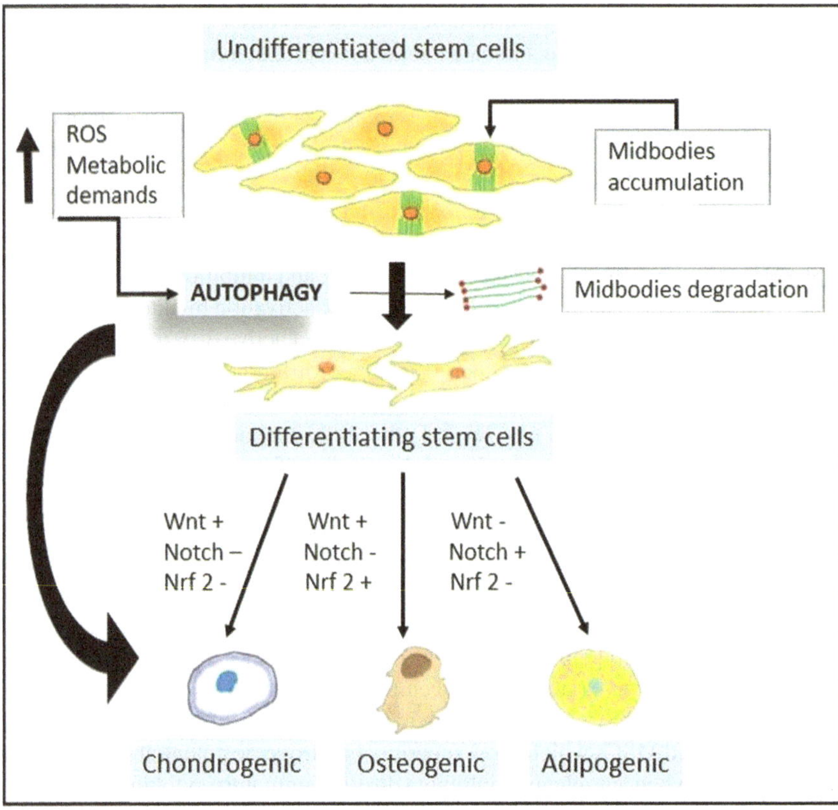

Fig. 10.2 Autophagic regulation of differentiation of stem cells: Undifferentiated stem cells accumulate midbodies and have high metabolic demands which increase ROS production in these cells. Autophagy is induced during the transition to differentiating cells which results in the degradation or extracellular release of midbodies via binding of CEP55 by the autophagic receptor NBR1. Induction of autophagy with other signaling axes such as Wnt, Notch and Nrf2 ultimately regulates the osteogenic, chondrogenic and adipogenic fate of stem cells

In the case of hematopoietic stem cells (HSCs) which reside in hypoxic conditions inside bone marrow, the energy demands during the quiescence state are quite low. As such, HSCs appear to evade oxidative phosphorylation for meeting their energy demands which result in reduced production of free radicals in the niche. The transition from quiescence to differentiation state requires increased metabolic demands and ROS levels and is regulated by mTOR activity. Moreover, this increase in ROS levels is also accompanied by an increase in oxygen levels which is proposed to promote the differentiation of HSCs toward the myeloid lineage [10, 57]. Interestingly, this state of myeloid differentiation is considered as a hallmark of autophagy deficient hematopoiesis. Autophagy is also shown to affect the terminal differentiation of HSCs into different blood cell types. Interestingly, the master regulator of hematopoiesis, GATA-1 has been shown to activate autophagy by transcriptional activation of LC3B and genes involved in the biogenesis of lysosomes. GATA-1 is reported to utilize FOXO3 protein for the activation of autophagy genes, which then regulates not only the self-renewal of HSCs but also acts to control their terminal lineage commitment [39, 58].

Autophagic Regulation of Osteogenic Differentiation of Stem Cells

The role of autophagy in osteogenic differentiation of MSCs is well documented in the literature. Although the exact mechanism by which autophagy regulates the osteogenic differentiation process is still not clear, several studies have indicated that upregulation of autophagy is indispensable for driving the differentiation process. Undifferentiated MSCs show autophagic arrest and accumulate autophagosomes with little turnover. In fact, this arrested autophagy is considered as a hallmark of undifferentiated MSCs and provides the advantage of supplying substrates for the resynthesis of various cytosolic metabolites during stressed conditions. MSCs have been shown to accumulate autophagosomes in their stem state which are then delivered to lysosomes at the onset of osteogenic differentiation [11]. The role of autophagy in spontaneous osteogenic differentiation of MSCs is still unclear, but several studies have indicated the role of autophagy in switching between osteogenic and adipogenic differentiation. Transient inhibition of autophagy at the onset of differentiation favors their adipogenic fate while induction of autophagy stimulates the osteogenic differentiation of MSCs. It is noteworthy that only the transient inhibition of autophagy at the early stages of differentiation results in adipogenesis, while permanent blocking of autophagy impairs adipogenic differentiation [11]. MSCs also show transitory activation of AMPK during early stages of differentiation, which is then downregulated at later stages. Several studies have indicated that AMPK activation during the initial stages of differentiation is required for mTOR inhibition and autophagy induction, which controls the osteoblastic differentiation via regulation of Wnt/β-catenin and Smad1/5/8-Dlx5-Runx2 signaling axes [59–61]. Late AMPK/Akt mediated activation of mTOR has been shown to be essential for optimal osteogenic differentiation of dental pulp MSCs [59]. Furthermore, downregulation of AMPK is assumed to favor glycolysis which is necessary for later stages of differentiation. However, constitutive expression of AMPK has been shown to cause defective terminal osteogenic differentiation, suggesting that the progressive downregulation of autophagy to the basal level is imperative to achieve full differentiation [62, 30]. Forkhead box class O (FOXO) proteins are important in regulating the survival, self-renewal and osteogenic differentiation of MSCs as well as osteoclast differentiation of HSCs. In fact, FOXO proteins are reported to be essential to activate autophagy during osteogenic differentiation of MSCs. The process of osteogenic differentiation of MSCs results in shift in cellular metabolism from glycolysis to mitochondrial respiration to allow more energy production to sustain the differentiation process. As mitochondrial respiration is associated with the production of free radicals, it leads to the activation of the autophagic process where it is presumed to confer protection against ROS-induced intrinsic apoptosis as well as to produce free amino acids for increased protein synthesis demands during the differentiation process [63]. In this context, Puerto et al showed that endogenous ROS production led to phosphorylation of FOXO3 via MAPK/JNK activity and resulted in the induction of autophagy to regulate the elevated ROS levels. Downregulation of autophagy proteins resulted in impaired osteoblast differentiation of BMSCs [64]. Another study showed that FOXO3 levels are upregulated in MSCs which result in increased LC3/LC1 ratio and upregulation of autophagy during osteogenic differentiation. Furthermore, this osteogenic differentiation is inhibited by miR-223-3p which targets FOXO3 and downregulates autophagic proteins [65]. Recently, it was demonstrated that TSC1 depletion in bone marrow MSCs resulted in the suppression of autophagy and an increase in Notch1 protein which resulted in GSK3β independent degradation of β-catenin. This led to the inhibition of osteogenic differentiation with the increase in adipocyte formation, further confirming that autophagy plays a pivotal role in maintaining a balance between osteoblast and adipocyte differentiation of MSCs [66]. Genetic ablation of mTOR inhibitor, Tsc1, in bone marrow MSCs resulted in MSCs hyper-proliferation and increased bone width and mass but also resulted in a significant decrease in bone length and matrix mineralization suggesting defective osteogenic and chondrogenic differentiation of these MSCs [67]. The functional autophagic process has been shown to be essential for BMPs-induced osteogenic differentiation of MSCs. BMP-9 upregulates multiple autophagy-related genes *atg3*, *atg5*, *atg8*, *atg9a*, *atg10*, *atg14*, *atg101*, *fip200* and *ulk* in MSCs undergoing osteogenic differentiation. Moreover, inhibition of autophagy effectively prevents matrix mineralization and osteogenesis in MSCs. This is also attributed *in vivo*, when atg5 silenced MSCs were shown to fail to ectopically form bone as compared to wild-type MSCs further confirming that activated autophagy is essential to drive the osteogenic differentiation and bone formation by MSCs [68]. Figure 10.3 outlines the events associated with autophagy mediated induction of early stages of osteogenic differentiation of MSCs.

Aged bone marrow MSCs have been shown to have reduced levels of autophagy which dramatically reduces their ability to differentiate into osteoblasts in comparison to young bone marrow MSCs which have high levels of autophagy and enhanced osteogenic differentiation ability. A study depicted that treating the aged MSCs with autophagy activator rapamycin restored their osteogenic ability, and contrariwise the treatment with autophagy inhibitor 3-methyladenine (3-MA) reduced the osteogenic ability of young MSCs and promoted their adipogenic differentiation ability [69]. Adding to this, another study showed that bone marrow MSCs derived from osteoporotic patients displayed senescent phenotype and downregulated autophagy compared to the MSCs from healthy individuals. These MSCs showed reduced capacity to differentiate into the osteogenic lineage, which was drastically improved upon autophagy induction by rapamycin. Moreover, MSCs treated with osteogenic induction in the presence of autophagy activator rapamycin showed larger ectopic bone formation and more osteoid tissue as compared to MSCs induced in

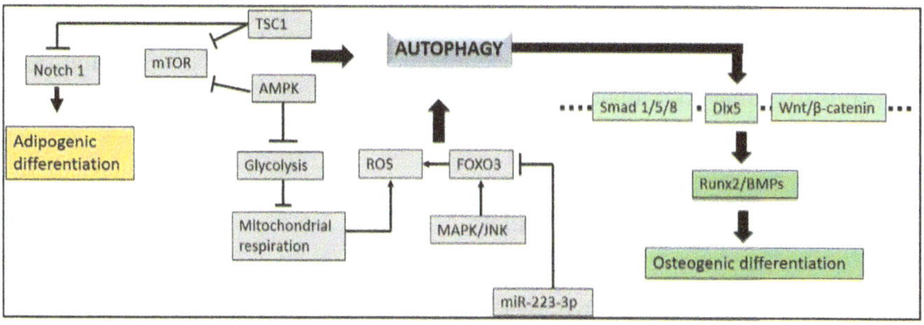

Fig. 10.3 Autophagic regulation of early stages of osteogenic differentiation: Induction of autophagy by AMPK/TSC1/ROS/FOXO3 proteins activates the lysosomal degradation of accumulated autophagosomes in undifferentiated MSCs, thus triggering their osteogenic differentiation. Autophagy-related proteins then activate various signaling pathways, Wnt/β-catenin, Smad1/5/8, Dlx5 that result in upregulation of Runx2/BMP signaling network which results in differentiation of MSCs into osteoblasts. Downregulation of AMPK and activation of mTOR inhibit autophagy during the late stages of osteogenic differentiation

the presence of autophagy inhibitor 3-MA, suggesting how autophagy modulates the osteogenic differentiation capacity of MSCs [70]. Autophagy also accelerates the osteogenic differentiation of MSCs derived from human gingiva. In fact, autophagy inducer resveratrol is shown to have a synergistic effect with osteogenic factors by inducing AMPK-BECLIN1-pro-autophagic pathway during osteogenic differentiation of MSCs [62]. Beclin-1 depletion with concomitant inhibition of autophagy prevents osteogenic differentiation, further demonstrating that the osteogenic fate of gingival MSCs is strictly dependent on the Beclin-1 mediated autophagic pathway. Since mechanical strains are considered to have a stimulatory effect on osteogenesis, a study evaluated the osteogenic differentiation of circulating MSCs on 22 individuals before and after physical exercise and found elevated levels of *runx2*, *msx2* and *spp1* gene levels and increased BMP2/6 protein levels in MSCs suggesting the induction of their osteogenic differentiation. What's interesting was that these osteogenic markers correlated positively with autophagy markers *atg3* and *ulk1*, further demonstrating how autophagy regulates MSCs osteogenic fate under different conditions [71]. Exercise induced muscle myokine, and irisin works in a similar manner. It is observed that irisin induces autophagy by elevating the expression of Atg12-Atg5-Atg16L protein complex, which promotes the osteogenic differentiation of bone marrow MSCs by upregulating Wnt/β- catenin signaling [72]. The autophagic regulation of MSCs during osteogenic differentiation is not only limited to human origin MSCs. Equine adipose MSCs have also been reported to show elevated autophagic flux during osteogenic differentiation. Induction of autophagy by its activator, rapamycin, also upregulated osteogenic markers Runx2 and BMP2 in ESCs, particularly by inhibiting mTOR and stimulating BMP/Smad signaling [73]. However, in another study, rapamycin impeded osteogenic differentiation of bone marrow MSCs induced by dexamethasone [74]. Therefore, no general statements can be made regarding the role of rapamycin in inducing or inhibiting osteogenic differentiation of MSCs. Regardless of this, it is well established that the dynamics of the autophagic process drive the osteoblastic differentiation of MSCs and other stem cells in a time coordinated manner.

Implications in Bone Disorders

The complex regulation of functions of osteoblasts and osteoclasts and the process of bone remodeling is crucial to maintain normal bone homeostasis and bone mineral density. Since autophagy is critical to MSCs differentiation into skeletal cells, HSCs differentiation into osteoclasts and in the formation and maintenance of skeletal tissue, defects pertaining to the autophagic process have pathogenic implications on bone homeostasis. Although autophagy has been shown to contribute to several bone disorders such as osteoporosis, osteopenia, osteopetrosis, Paget's disease and osteoarthritis, only two bone pathological conditions viz. osteoporosis and arthritis have been shown to have defective autophagy regulation of MSCs osteogenic differentiation.

Autophagic Dysregulation in Osteoporosis

It is a metabolic disorder characterized by significant bone loss due to increased osteoclast and reduced osteoblast activity. The disease is caused as a result of the complex interplay of various factors such as age, estrogen deficiency, increased oxidative stress and secondary medical conditions. Abnormal lineage commitment of bone marrow MSCs resulting in increased adipogenesis and reduced osteogenesis has also been implicated as one of the causes of osteoporosis. In this context, a study showed that autophagy levels in bone marrow MSCs derived from osteoporotic mice (OVX) were significantly reduced as evident by decreased levels of Beclin-1 and LC3II-LC3I conversion. Induction of autophagy restored the osteogenic commitment of MSCs and augmented bone formation in OVX mice. They also showed that autophagy did not have any profound effect on the survival of MSCs but rather in restoring the balance between osteogenesis and adipogenesis, suggesting that autophagy works to maintain the function of MSCs to prevent osteoporosis [75]. In general, MSCs derived from osteoporotic patients show reduced osteogenic capability due to suppressed autophagy as compared to MSCs from healthy controls [70]. Another study showed that the autophagy receptor, optineurin, decreases in MSCs with age which compromises the degradation of fatty acid binding protein 3 (FABP3) via selective autophagy and contributes to senile osteoporosis [51]. Another study reported a direct relationship between osteoporotic phenotype and autophagy pathway by genome-wide association study of wrist ultradistal radius [76]. The study, however, did not functionally evaluate the mechanism of autophagy in regulating the wrist osteoporosis.

Autophagic Dysregulation in Arthritis

Arthritis is a disorder that results in the inflammation of the joints. Osteoarthritis is the most common form of arthritis, which results in the gradual loss of articular cartilage tissue. During the initial stages of the development of the disease, autophagy is upregulated as an adaptive response to protect chondrocytes in the inflammatory environment and also results in the regulation of expression of several osteoarthritis associated genes by modulating ROS levels and apoptosis [54, 77] However, the process declines with the progression of the disease resulting in chondrocytic death and tearing of the cartilage tissue. In an experimental model of rats, induction of autophagy via the blocking of mTOR signaling was shown to abrogate the clinical manifestations of inflammatory arthritis. Moreover, mTOR activation has also been observed in the synovial membrane of arthritis patients, suggesting that decreased autophagy contributes, in part, to the disease phenotype [78]. Recently, an interesting observation was made in a study that showed that MSCs exerted their protective effects by means of secreted factors (called as secretome) in osteoarthritis (OA) settings. The study showed that autophagic flux was remarkably reduced in OA rats and the levels of autophagic proteins Beclin-1 and LC3 were restored upon transplantation of MSCs secretome. Enhanced autophagy correlated with reduced MMP-13/TIMP-1 ratio and rescued the subchondral bone architecture from degradation [79]. Adding to this, another study reported that adipose-derived MSCs secreted exosomes also conferred protection to chondrocytes and enhanced matrix synthesis by inhibiting mTOR via miR100-5p and upregulating the autophagy pathway [80]. Adipose MSCs have also been shown to alleviate cartilage destruction and protect chondrocytes against apoptosis by inducing autophagy in these cells [81]. While these studies suggest the protective role of autophagy during osteoarthritis, further studies are warranted to fully decipher the potential of autophagic modulators as preventive measures for the disease.

Conclusion and Final Remarks

Autophagy is a complex catabolic process that is involved in maintaining cellular homeostasis under stressed conditions. In general, autophagy is considered a dynamic process that can act as a double edge sword depending on the metabolic and physiological needs of a cell. The autophagic process contributes to maintaining the self-renewal, survival, stemness and differentiation of all kinds of stem cells such as ESCs, iPSCs, MSCs and HSCs. Accumulating evidence suggests that autophagy directly regulates the osteogenic commitment of MSCs and prevents their adipogenic differentiation. It also contributes to the survival and function of MSCs and MSCs-derived skeletal cells like chondrocytes, osteoblasts and osteocytes. Moreover, autophagy also regulates the myeloid lineage commitment of HSCs into osteoclasts and osteoclasts function, thereby orchestrating bone tissue formation and remodeling. Perturbations in the autophagic process contribute to several bone pathologies like osteoporosis, osteopetrosis, Paget's disease and osteoarthritis. However, there are only limited studies exploring the involvement of autophagy in the osteogenic fate of stem cells, and therefore, further investigations are required on uncovering the pleiotropic effects of autophagy in the modulation of bone physiology.

Acknowledgements None

Conflict of Interest The authors declare no competing interests in this study.

References

1. Ceccariglia S, Cargnoni A, Silini AR, Parolini O (2019) Autophagy: a potential key contributor to the therapeutic action of mesenchymal stem cells. 16(1):28–37. https://doi.org/10.1080/15548627.2019.1630223
2. Wirawan E, Berghe TV, Lippens S, Agostinis P, Vandenabeele P (2011) Autophagy: for better or for worse. Cell Res 22(1):43–61. https://doi.org/10.1038/cr.2011.152
3. Parzych KR, Klionsky DJ (2014) An overview of autophagy: morphology, mechanism, and regulation. 20(3): 460–73. https://doi.org/10.1089/ARS.2013.5371. https://Home.Liebertpub.Com/Ars
4. Tanida I (2011) Autophagosome formation and molecular mechanism of autophagy. 14(11):2201–2214. https://doi.org/10.1089/ARS.2010.3482. https://Home.Liebertpub.Com/Ars
5. Hou J, Han ZP, Jing YY, Yang X, Zhang SS, Sun K, Hao C, et al (2013) Autophagy prevents irradiation injury and maintains stemness through decreasing ROS generation in mesenchymal stem cells. Cell Death Dis 4(10):e844–e844. https://doi.org/10.1038/cddis.2013.338
6. Rodolfo C, Di Bartolomeo S, Cecconi F (2016) Autophagy in stem and progenitor cells. Cell Mol Life Sci CMLS 73(3):475–496. https://doi.org/10.1007/S00018-015-2071-3
7. Tsukamoto S, Kuma A, Murakami M, Kishi C, Yamamoto A, Mizushima N (2008) Autophagy is essential for preimplantation development of mouse embryos. Science (New York, N.Y.) 321(5885):117–120. https://doi.org/10.1126/SCIENCE.1154822
8. Wang S, Xia P, Ye B, Huang G, Liu J, Fan Z (2013) Transient activation of autophagy via Sox2-mediated suppression of MTOR is an important early step in reprogramming to pluripotency. Cell Stem Cell 13(5):617–625. https://doi.org/10.1016/J.STEM.2013.10.005
9. Sbrana FV, Cortini M, Avnet S, Perut F, Columbaro M, De Milito A, Baldini N (2016) The role of autophagy in the maintenance of stemness and differentiation of mesenchymal stem cells. Stem Cell Rev Rep 12(6):621–633. https://doi.org/10.1007/S12015-016-9690-4
10. Guan JL, Simon AK, Prescott M, Menendez JA, Liu F, Wang F, Wang C, Wolvetang E, Vazquez-Martin A, Zhang J (2013) Autophagy in stem cells. Autophagy 9(6):830. https://doi.org/10.4161/AUTO.24132
11. Nuschke A, Rodrigues M, Stolz DB, Chu CT, Griffith L, Wells A (2014) Human mesenchymal stem cells/multipotent stromal cells consume accumulated autophagosomes early in differentiation. Stem Cell Res Ther 5(6). https://doi.org/10.1186/SCRT530
12. Tra T, Gong L, Kao LP, Li XL, Grandela C, Devenish RJ, Wolvetang E, Prescott M (2011) Autophagy in human embryonic stem cells. PLoS One 6(11). https://doi.org/10.1371/JOURNAL.PONE.0027485

13. Cho YH, Han KM, Kim D, Lee J, Lee SH, Choi KW, Kim J, Han YM (2014) Autophagy regulates homeostasis of pluripotency-associated proteins in HESCs. Stem Cells 32(2):424–435. https://doi.org/10.1002/STEM.1589
14. Zhou J, Pei S, Wang L, Chen J, Zimmermann M, Genbacev O, Afonja O et al (2009) MTOR supports long-term self-renewal and suppresses mesoderm and endoderm activities of human embryonic stem cells. Proc Natl Acad Sci USA 106(19):7840–7845. https://doi.org/10.1073/PNAS.0901854106
15. Murakami M, Ichisaka T, Maeda M, Oshiro N, Hara K, Edenhofer F, Kiyama H, Yonezawa K, Yamanaka S (2004) MTOR is essential for growth and proliferation in early mouse embryos and embryonic stem cells. Mol Cell Biol 24(15):6710–6718. https://doi.org/10.1128/MCB.24.15.6710-6718.2004/ASSET/7D1197D7-2E81-4470-B499-CAFCF3545C8B/ASSETS/GRAPHIC/ZMB0150442570007.JPEG
16. Mizushima N, Yamamoto A, Hatano M, Kobayashi Y, Kabey Y, Suzuki K, Tokuhis T, Ohsumi Y, Yoshimori T (2001) Dissection of autophagosome formation using Apg5-deficient mouse embryonic stem cells. J Cell Biol 152(4):657. https://doi.org/10.1083/JCB.152.4.657
17. Qu X, Jie Y, Bhagat G, Furuya N, Hibshoosh H, Troxel A, Rosen J et al (2003) Promotion of tumorigenesis by heterozygous disruption of the beclin 1 autophagy gene. J Clin Investig 112(12):1809–1820. https://doi.org/10.1172/JCI20039
18. Qu X, Zou Z, Sun Q, Luby-Phelps K, Cheng P, Hogan RN, Gilpin C, Levine B (2007) Autophagy gene-dependent clearance of apoptotic cells during embryonic development. Cell 128(5):931–946. https://doi.org/10.1016/J.CELL.2006.12.044
19. Jang J, Wang Y, Lalli MA, Guzman E, Godshalk SE, Zhou H, Kosik KS (2016) Primary Cilium-Autophagy-Nrf2 (PAN) axis activation commits human embryonic stem cells to a neuroectoderm fate. Cell 165(2):410–420. https://doi.org/10.1016/J.CELL.2016.02.014
20. Offei EB, Yang X, Brand-Saberi B (2018) The role of autophagy in morphogenesis and stem cell maintenance. Histochem Cell Biol 150(6):721–732. https://doi.org/10.1007/S00418-018-1751-0/FIGURES/4
21. Todd LR, Gomathinayagam R, Sankar U (2010) A novel Gfer-Drp1 link in preserving mitochondrial dynamics and function in pluripotent stem cells. 6(6):821–822. https://doi.org/10.4161/AUTO.6.6.12625
22. Srinivas V, Bohensky J, Shapiro IM (2009) Autophagy: a new phase in the maturation of growth plate chondrocytes is regulated by HIF, MTOR and AMP kinase. Cells Tissues Organs 189(1–4):88–92. https://doi.org/10.1159/000151428
23. Vuppalapati KK, Bouderlique T, Newton PT, Kaminskyy VO, Wehtje H, Ohlsson C, Zhivotovsky B, Chagin AS (2015) Targeted deletion of autophagy genes Atg5 or Atg7 in the chondrocytes promotes caspase-dependent cell death and leads to mild growth retardation. J Bone Min Res 30(12):2249–2261. https://doi.org/10.1002/JBMR.2575
24. Horigome Y, Ida-Yonemochi H, Waguri S, Shibata S, Endo N, Komatsu M (2020) Loss of autophagy in chondrocytes causes severe growth retardation. Autophagy 16(3):501–511. https://doi.org/10.1080/15548627.2019.1628541/SUPPL_FILE/KAUP_A_1628541_SM8468.ZIP
25. Di Bartolomeo S, Nazio F, Cecconi F (2010) The role of autophagy during development in higher Eukaryotes. Traffic 11(10):1280–1289. https://doi.org/10.1111/J.1600-0854.2010.01103.X
26. Shapiro IM, Adams CS, Freeman T, Srinivas V (2005) Fate of the hypertrophic chondrocyte: microenvironmental perspectives on apoptosis and survival in the epiphyseal growth plate. Birth Defects Res Part C Embryo Today Rev 75(4):330–39. https://doi.org/10.1002/BDRC.20057
27. Moss JJ, Hammond CL, Lane JD (2020) Zebrafish as a model to study autophagy and its role in skeletal development and disease. Histochem Cell Biol 154(5):549–564. https://doi.org/10.1007/S00418-020-01917-2/FIGURES/4
28. Nollet M, Santucci-Darmanin S, Breuil V, Al-Sahlanee R, Cros C, Topi M, Momier D et al (2014) Autophagy in osteoblasts is involved in mineralization and bone homeostasis. Autophagy 10(11):1965. https://doi.org/10.4161/AUTO.36182

29. Zahm AM, Bohensky J, Adams CS, Shapiro IM, Srinivas V (2011) Bone cell autophagy is regulated by environmental factors. Cells Tissues Organs 194(2–4):274. https://doi.org/10.1159/000324647
30. Xi G, Rosen CJ, Clemmons DR (2016) IGF-I and IGFBP-2 stimulate AMPK activation and autophagy, which are required for osteoblast differentiation. Endocrinology 157(1):268–281. https://doi.org/10.1210/EN.2015-1690
31. Li H, Li D, Ma Z, Qian Z, Kang X, Jin X, Li F et al (2018) Defective autophagy in osteoblasts induces endoplasmic reticulum stress and causes remarkable bone loss. Autophagy 14(10):1726–1741. https://doi.org/10.1080/15548627.2018.1483807
32. Liu F, Fang F, Yuan H, Yang D, Chen Y, Williams L, Goldstein SA, Krebsbach PH, Guan JL (2013) Suppression of autophagy by FIP200 deletion leads to osteopenia in mice through the inhibition of osteoblast terminal differentiation. J Bone Min Res Offic J Am Soc Bone Min Res 28(11):2414. https://doi.org/10.1002/JBMR.1971
33. Onal M, Piemontese M, Xiong J, Wang Y, Han L, Ye S, Komatsu M et al (2013) Suppression of autophagy in osteocytes mimics skeletal aging. J Biol Chem 288(24):17432–17440. https://doi.org/10.1074/JBC.M112.444190
34. Vrahnas C, Blank M, Dite TA, Tatarczuch L, Ansari N, Crimeen-Irwin B, Nguyen H, et al (2019) Increased autophagy in EphrinB2-deficient osteocytes is associated with elevated secondary mineralization and brittle bone. Nat Commun 10(1):1–16. https://doi.org/10.1038/s41467-019-11373-9
35. Yin X, Zhou C, Li J, Liu R, Shi B, Yuan Q, Zou S (2019) Autophagy in bone homeostasis and the onset of osteoporosis. Bone Res 7(1):1–16. https://doi.org/10.1038/s41413-019-0058-7
36. Mortensen M, Soilleux EJ, Djordjevic G, Tripp R, Lutteropp M, Sadighi-Akha E, Stranks AJ et al (2011) The autophagy protein Atg7 is essential for hematopoietic stem cell maintenance. J Exp Med 208(3):455–467. https://doi.org/10.1084/JEM.20101145
37. Tong X, Jianhong G, Song R, Wang D, Sun Z, Sui C, Zhang C, Liu X, Bian J, Liu Z (2018) Osteoprotegerin inhibit osteoclast differentiation and bone resorption by enhancing autophagy via AMPK/MTOR/P70S6K signaling pathway in vitro. J Cell Biochem 120(2):1630–1642. https://doi.org/10.1002/JCB.27468
38. Zhao H, Sun Z, Ma Y, Song R, Yuan Y, Bian J, Jianhong G, Liu Z (2020) Antiosteoclastic bone resorption activity of osteoprotegerin via enhanced AKT/MTOR/ULK1-mediated autophagic pathway. J Cell Physiol 235(3):3002–3012. https://doi.org/10.1002/JCP.29205
39. Boya P, Codogno P, Rodriguez-Muela N (2018) Autophagy in stem cells: repair, remodelling and metabolic reprogramming. Development 145(4). https://doi.org/10.1242/DEV.146506
40. Shyh-Chang N, Daley GQ, Cantley LC (2013) Stem cell metabolism in tissue development and aging. Development 140(12):2535–2547. https://doi.org/10.1242/DEV.091777
41. Zhang Q, Yang YJ, Wang H, Dong QT, Wang TJ, Qian HY, Hui X (2012) Autophagy activation: a novel mechanism of atorvastatin to protect mesenchymal stem cells from hypoxia and serum deprivation via AMP-activated protein kinase/mammalian target of rapamycin pathway. Stem Cells Dev 21(8):1321. https://doi.org/10.1089/SCD.2011.0684
42. Chen T, Shen L, Jie Y, Wan H, Guo A, Chen J, Long Y, Zhao J, Pei G (2011) Rapamycin and other longevity-promoting compounds enhance the generation of mouse induced pluripotent stem cells. Aging Cell 10(5):908–911. https://doi.org/10.1111/J.1474-9726.2011.00722.X
43. Menendez JA, Vellon L, Oliveras-Ferraros C, Cufí S, Vazquez-Martin A (2011) MTOR-regulated senescence and autophagy during reprogramming of somatic cells to pluripotency: a roadmap from energy metabolism to stem cell renewal and aging. Cell Cycle 10(21):3658–3677. https://doi.org/10.4161/CC.10.21.18128
44. Ettinger AW, Wilsch-Bräuninger M, Marzesco AM, Bickle M, Lohmann A, Maliga Z, Karbanová J, Corbeil D, Hyman AA, Huttner WB (2011) Proliferating versus differentiating stem and cancer cells exhibit distinct midbody-release behaviour. Nat Commun 2(1):1–12. https://doi.org/10.1038/ncomms1511
45. Kuo TC, Chen CT, Baron D, Onder TT, Loewer S, Almeida S, Weismann CM, et al (2011) Midbody accumulation through evasion of autophagy contributes to cellular reprogramming and tumorigenicity. Nat Cell Biol 13(10):1214–1223. https://doi.org/10.1038/ncb2332

46. Li Q, Gao Z, Chen Y, Guan MX (2017) The role of mitochondria in osteogenic, adipogenic and chondrogenic differentiation of mesenchymal stem cells. Protein Cell 8(6):439–445. https://doi.org/10.1007/S13238-017-0385-7/FIGURES/1
47. Sotthibundhu A, McDonagh K, Von Kriegsheim A, Munoz A, Klawiter A, Thompson K, Chauhan KD et al (2016) Rapamycin regulates autophagy and cell adhesion in induced pluripotent stem cells. Stem Cell Res Ther 7(1):1–16. https://doi.org/10.1186/S13287-016-0425-X/FIGURES/9
48. Singh R, Xiang Y, Wang Y, Baikati K, Cuervo AM, Luu YK, Tang Y, Pessin JE, Schwartz GJ, Czaja MJ (2009) Autophagy regulates adipose mass and differentiation in mice. J Clin Investig 119(11):3329–3339. https://doi.org/10.1172/JCI39228
49. Guo L, Huang J-X, Liu Y, Li X, Zhou S-R, Qian S-W, Liu Y et al (2013) Transactivation of Atg4b by C/EBPβ promotes autophagy to facilitate adipogenesis. Mol Cell Biol 33(16):3180–3190. https://doi.org/10.1128/MCB.00193-13/ASSET/31E508DC-C169-44EA-99D4-33793BE5D5F8/ASSETS/GRAPHIC/ZMB9991000710008.JPEG
50. Menshikov M, Zubkova E, Stafeev I, Parfyonova Y (2021) Autophagy, mesenchymal stem cell differentiation, and secretion. Biomedicines 9 (9). https://doi.org/10.3390/BIOMEDICINES9091178
51. Liu W, Luo H, Wang R, Kang Y, Liao W, Sun Y, Chen G, Shao L (2020) Rapamycin-induced autophagy promotes the chondrogenic differentiation of synovium-derived mesenchymal stem cells in the temporomandibular joint in response to IL-1 β. Biomed Res Int https://doi.org/10.1155/2020/4035306
52. Kang S, Bennett CN, Gerin I, Rapp LA, Hankenson KD, MacDougald OA (2007) Wnt signaling stimulates osteoblastogenesis of mesenchymal precursors by suppressing CCAAT/enhancer-binding protein α and peroxisome proliferator-activated receptor γ *. J Biol Chem 282(19):14515–14524. https://doi.org/10.1074/JBC.M700030200
53. Kirton JP, Crofts NJ, George SJ, Brennan K, Canfield AE (2007) Wnt/β-catenin signaling stimulates chondrogenic and inhibits adipogenic differentiation of pericytes: potential relevance to vascular disease? Circ Res 101(6):581–589. https://doi.org/10.1161/CIRCRESAHA.107.156372
54. Li B, Duan P, Li C, Jing Y, Han X, Yan W, Xing Y (2016) Role of autophagy on bone marrow mesenchymal stem-cell proliferation and differentiation into neurons. Mol Med Rep 13(2):1413–1419. https://doi.org/10.3892/MMR.2015.4673/HTML
55. Sotthibundhu A, Muangchan P, Phonchai R, Promjantuek W, Chaicharoenaudomrung N, Kunhorm P, Noisa P (2021) Autophagy promoted neural differentiation of human placenta-derived mesenchymal stem cells. In Vivo 35(5):2609–2620. https://doi.org/10.21873/INVIVO.12543
56. Li Y, Wang C, Zhang G, Wang X, Duan R, Gao H, Peng T, Teng J, Jia Y (2014) Role of autophagy and MTOR signaling in neural differentiation of bone marrow mesenchymal stem cells. Cell Biol Int 38(11):1337–1343. https://doi.org/10.1002/CBIN.10320
57. Owusu-Ansah E, Banerjee U (2009) Reactive oxygen species prime Drosophila haematopoietic progenitors for differentiation. Nature 461(7263):537–541. https://doi.org/10.1038/nature08313
58. Kang Y-A, Sanalkumar R, O'Geen H, Linnemann AK, Chang C-J, Bouhassira EE, Farnham PJ, Keles S, Bresnick EH (2012) Autophagy driven by a master regulator of hematopoiesis. Mol Cell Biol 32(1):226–239. https://doi.org/10.1128/MCB.06166-11/ASSET/018FE648-0369-4A61-BD2C-E58B89E029BC/ASSETS/GRAPHIC/ZMB9991093280011.JPEG
59. Pantovic A, Krstic A, Janjetovic K, Kocic J, Trajkovic L, Bugarski D, Trajkovic V (2013) Coordinated time-dependent modulation of AMPK/Akt/MTOR signaling and autophagy controls osteogenic differentiation of human mesenchymal stem cells. Bone 52(1):524–531. https://doi.org/10.1016/J.BONE.2012.10.024
60. Takatani T, Minagawa M, Takatani R, Kinoshita K, Kohno Y (2011) AMP-activated protein kinase attenuates Wnt/β-Catenin signaling in human osteoblastic Saos-2 cells. Mol Cell Endocrinol 339(1–2):114–119. https://doi.org/10.1016/J.MCE.2011.04.003

61. Jang WG, Kim EJ, Lee KN, Son HJ, Koh JT (2011) AMP-activated protein kinase (AMPK) positively regulates osteoblast differentiation via induction of Dlx5-dependent Runx2 expression in MC3T3E1 cells. Biochem Biophys Res Commun 404(4):1004–1009. https://doi.org/10.1016/J.BBRC.2010.12.099
62. Vidoni C, Ferraresi A, Secomandi E, Vallino L, Gardin C, Zavan B, Mortellaro C, Isidoro C (2019) Autophagy drives osteogenic differentiation of human gingival mesenchymal stem cells. Cell Commun Signal CCS 17(1). https://doi.org/10.1186/S12964-019-0414-7
63. Marycz K, Kornicka K, Marędziak M, Golonka P, Nicpoń J (2016) Equine metabolic syndrome impairs adipose stem cells osteogenic differentiation by predominance of autophagy over selective mitophagy. J Cell Mol Med 20(12):2384–2404. https://doi.org/10.1111/JCMM.12932
64. Gómez-Puerto MC, Verhagen LP, Braat AK, Lam EWF, Coffer PJ, Lorenowicz MJ (2016) Activation of autophagy by FOXO3 regulates redox homeostasis during osteogenic differentiation. Autophagy 12(10):1804–1816. https://doi.org/10.1080/15548627.2016.1203484/SUPPL_FILE/KAUP_A_1203484_SM0295.ZIP
65. Long C, Cen S, Zhong Z, Zhou C, Zhong G (2021) FOXO3 is targeted by MiR-223-3p and promotes osteogenic differentiation of bone marrow mesenchymal stem cells by enhancing autophagy. Hum Cell 34(1):14–27. https://doi.org/10.1007/S13577-020-00421-Y
66. Choi HK, Yuan H, Fang F, Xiaoxi Wei L, Liu QL, Guan JL, Liu F (2018) Tsc1 regulates the balance between osteoblast and adipocyte differentiation through autophagy/Notch1/β-Catenin cascade. J Bone Miner Res 33(11):2021–2034. https://doi.org/10.1002/JBMR.3530
67. Wu H, Zhixiang W, Li P, Cong Q, Chen R, Wenrui X, Biswas S et al (2017) Bone size and quality regulation: concerted actions of MTOR in mesenchymal stromal cells and osteoclasts. Stem Cell Rep 8(6):1600–1616. https://doi.org/10.1016/J.STEMCR.2017.04.005
68. Zhao X, Huang B, Wang H, Ni N, He F, Liu Q, Shi D et al (2021) A functional autophagy pathway is essential for BMP9-induced osteogenic differentiation of mesenchymal stem cells (MSCs). Am J Transl Res 13(5):4233
69. Ma Y, Qi M, An Y, Zhang L, Yang R, Doro DH, Liu W, Jin Y (2018) Autophagy controls mesenchymal stem cell properties and senescence during bone aging. Aging Cell 17(1). https://doi.org/10.1111/ACEL.12709
70. Wan Y, Zhuo N, Li Y, Zhao W, Jiang D (2017) Autophagy promotes osteogenic differentiation of human bone marrow mesenchymal stem cell derived from osteoporotic vertebrae. Biochem Biophys Res Commun 488(1):46–52. https://doi.org/10.1016/J.BBRC.2017.05.004
71. Dalle Carbonare L, Mottes M, Cheri S, Deiana M, Zamboni F, Gabbiani D, Schena F, Salvagno GL, Lippi G, Valenti MT (2019) Increased gene expression of RUNX2 and SOX9 in mesenchymal circulating progenitors is associated with autophagy during physical activity. Oxid Med Cell Longev https://doi.org/10.1155/2019/8426259
72. Chen X, Sun K, Zhao S, Geng T, Fan X, Sun S, Zheng M, Jin Q (2020) Irisin promotes osteogenic differentiation of bone marrow mesenchymal stem cells by activating autophagy via the Wnt//β-Catenin signal pathway. Cytokine 136:155292. https://doi.org/10.1016/J.CYTO.2020.155292
73. Lee KW, Yook JY, Son MY, Kim MJ, Koo DB, Han YM, Cho YS (2010) Rapamycin promotes the osteoblastic differentiation of human embryonic stem cells by blocking the MTOR pathway and stimulating the BMP/Smad pathway. Stem Cells Dev 19(4):557–568. https://doi.org/10.1089/SCD.2009.0147
74. Isomoto S, Hattori K, Ohgushi H, Nakajima H, Tanaka Y, Takakura Y (2007) Rapamycin as an inhibitor of osteogenic differentiation in bone marrow-derived mesenchymal stem cells. J Orthop Sci 12(1):83–88. https://doi.org/10.1007/S00776-006-1079-9
75. Qi M, Zhang L, Ma Y, Shuai Y, Li L, Luo K, Liu W, Jin Y (2017) Autophagy maintains the function of bone marrow mesenchymal stem cells to prevent estrogen deficiency-induced osteoporosis. Theranostics 7(18):4498–4516. https://doi.org/10.7150/THNO.17949
76. Zhang L, Guo YF, Liu YZ, Liu YJ, Xiong DH, Liu XG, Wang L et al (2010) Pathway-based genome-wide association analysis identified the importance of regulation-of-autophagy pathway for ultradistal radius BMD. J Bone Min Res 25(7):1572–1580. https://doi.org/10.1002/JBMR.36

77. Sasaki H, Takayama K, Matsushita T, Ishida K, Kubo S, Matsumoto T, Fujita N, Oka S, Kurosaka M, Kuroda R (2012) Autophagy modulates osteoarthritis-related gene expression in human chondrocytes. Arthritis Rheum 64(6):1920–1928. https://doi.org/10.1002/ART.34323
78. Cejka D, Hayer S, Niederreiter B, Sieghart W, Fuereder T, Zwerina J, Schett G (2010) Mammalian target of Rapamycin signaling is crucial for joint destruction in experimental arthritis and is activated in osteoclasts from patients with rheumatoid arthritis. Arthritis Rheum 62(8):2294–2302. https://doi.org/10.1002/ART.27504
79. Chen W, Sun Y, Xueping G, Hao Y, Liu X, Lin J, Chen J, Chen S (2019) Conditioned medium of mesenchymal stem cells delays osteoarthritis progression in a rat model by protecting subchondral bone, maintaining matrix homeostasis, and enhancing autophagy. J Tissue Eng Regen Med 13(9):1618–1628. https://doi.org/10.1002/TERM.2916
80. Wu J, Kuang L, Chen C, Yang J, Zeng WN, Li T, Chen H et al (2019) MiR-100-5p-abundant exosomes derived from infrapatellar fat pad MSCs protect articular cartilage and ameliorate gait abnormalities via inhibition of MTOR in osteoarthritis. Biomaterials 206:87–100. https://doi.org/10.1016/J.BIOMATERIALS.2019.03.022
81. Zhou J, Wang Y, Liu Y, Zeng H, Hanshi X, Lian F (2019) Adipose derived mesenchymal stem cells alleviated osteoarthritis and chondrocyte apoptosis through autophagy inducing. J Cell Biochem 120(2):2198–2212. https://doi.org/10.1002/JCB.27530
82. Sheng LY, Zhang FJ, Zeng C, Luo W, Xiao WF, Gao SG, Lei GH (2016) Autophagy in osteoarthritis. Joint Bone Spine 83(2):143–148. https://doi.org/10.1016/J.JBSPIN.2015.06.009
83. Liu ZZ, Hong CG, Hu WB, Chen ML, Duan R, Li HM, Yue T, et al (2020) Autophagy receptor OPTN (Optineurin) regulates mesenchymal stem cell fate and bone-fat balance during aging by clearing FABP3. 17(10):2766–2782. https://doi.org/10.1080/15548627.2020.1839286

Chapter 11
Autophagy in Cancer Metastasis

Ruhi Polara, Daphni van Rinsum, and Nirmal Robinson

Abstract During cancer development, tumour cells are exposed to various intrinsic and extrinsic stresses such as nutrient deficiency, lack of oxygen, DNA damage, and growth factor deprivation that regulate cell growth and homeostasis. In response to these stresses, tumour cells, unlike healthy cells, develop adaptive strategies to grow and migrate successfully. One of the key mechanisms that cancer cells utilize to circumvent cellular stresses is autophagy, which is a catabolic process that facilitates the degradation and recycling of damaged organelles, thereby reducing cellular stress and promoting cell survival. Emerging studies have shown a vital role of autophagy in cancer metastasis, which is the major cause of cancer-associated deaths. However, the role of autophagy in metastasis is multidimensional and involves both metastasis-promoting and suppressing roles dependent on the demands of tumour cells during the metastatic process. As novel compounds targeting autophagy emerge, it will be crucial to consider the stage of metastatic progression at which autophagy is being targeted to efficiently overcome metastasis.

Keywords Cancer · Metastasis · Autophagy · Cell stress · Tumour microenvironment

Abbreviations

APP	Amyloid precursor protein
ARHI	Aplasia Ras homolog member I
ATF	Activating transcription factor
ATG	Autophagy-related gene
CAF	Cancer-associated fibroblasts
CDCP1	CUB domain-containing protein-1

R. Polara · D. van Rinsum · N. Robinson (✉)
Centre for Cancer Biology, University of South Australia and SA Pathology, Adelaide, Australia
e-mail: nirmal.robinson@unisa.edu.au

B. V. Shravage and K. Turksen (eds.), *Autophagy in Stem Cell Maintenance and Differentiation*, Stem Cell Biology and Regenerative Medicine 73,
https://doi.org/10.1007/978-3-031-17362-2_11

CSF-1	Colony stimulating factor 1
CTC	Circulating tumour cell
CXC	CXC-chemokine ligand
DIAPH3	Diaphanous homologue 3
EC	Endothelial cell
ECM	Extracellular matrix
EGF	Epidermal growth factor
EMT	Epithelial-mesenchymal transition
ER	Endoplasmic reticulum
ERK	Extracellular-regulated kinase
FAK	Focal adhesion kinase
FAS	FS-7-associated surface antigen
HGF	Hepatocyte growth factor
IGF	Insulin-like growth factor
LC3	Microtubule-associated protein 1 light chain 3
MAPK	Mitogen-activated protein kinase
MET	Mesenchymal-epithelial transition
MHC	Major histocompatibility complex
MICA/B	MHC-I polypeptide-related sequence A/B
MLCK	Myosin light chain kinase
MMP	Matrix metalloproteinase
mTOR	Mechanistic target of rapamycin
NF-κB	Nuclear factor kappa B
NK	Natural killer
NKG2D	Natural killer group 2D
NRF2	NF-E2-related factor 2
PAR	Protease-activated receptor
PERK	Protein kinase R-like ER kinase
PI3K	Phosphoinositide 3 kinase
PTEN	Phosphate and tensin homolog
ROCK	RhoA-Rho-associated protein kinase
ROS	Reactive oxygen species
TAM	Tumour-associated macrophage
TGFβ	Transforming growth factor β
TKI	Tyrosine kinase inhibitor
TME	Tumour microenvironment
TNF	Tumour necrosis factor
uPA	Urokinase-type plasminogen activator
VCAM	Vascular cell adhesion molecule
VEGF	Vascular endothelial growth factor
ZEB1	Zinc finger E-box binding homeobox 1

Introduction

Cancer metastasis refers to the spread and colonization of primary tumour cells to a secondary organ and is one of the leading causes of death, which accounts for 90% of cancer-associated deaths globally [1]. During tumour development, cancer cells undergo genetic mutations, adopt their microenvironment, and stimulate angiogenesis that potentially facilitates metastasis [2]. Metastatic cascade of solid tumours can be divided into five major steps: (1) invasion of the basement membrane; (2) intravasation into the surrounding vasculature or lymphatic system (3) survival during circulation; (4) extravasation from vasculature to secondary tissue; and (5) colonization at secondary tumour sites (Fig. 11.1) [2].

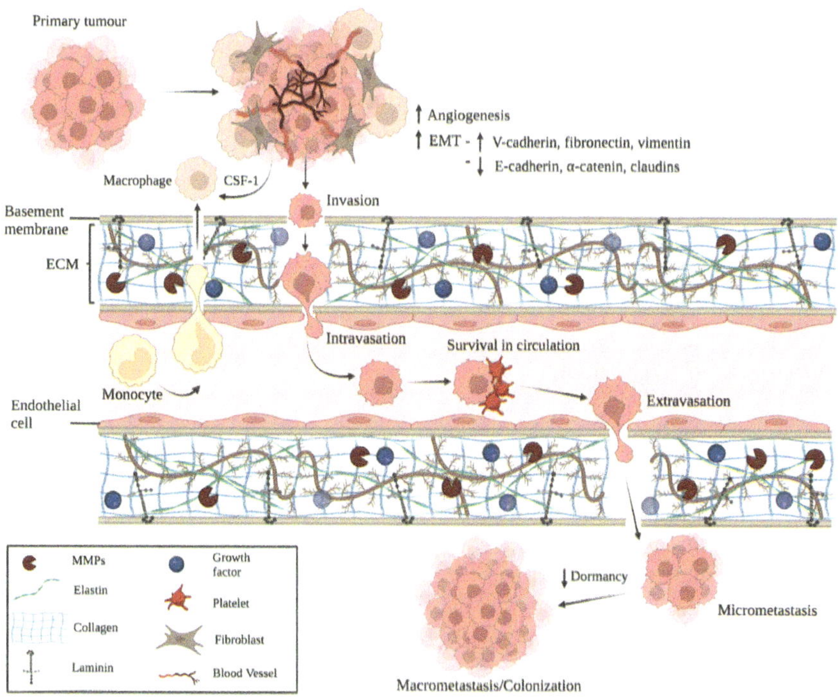

Fig. 11.1 Overview of the five key steps in cancer metastasis. Invasion is initiated by growth factors, MMPs, and EMT, followed by intravasation involving the degradation of endothelial gap junctions. Following intravasation, tumour cell survival in the circulation is facilitated by many factors including tumour cell coating by platelets. Subsequent extravasation occurs as a result of tumour cell impingement and breakage of blood vessels or via the recruitment of an active machinery. Ultimately tumour colonization at the secondary site is mediated by the development of pre-metastatic niche including suppression of the immune system. Abbreviations: MMPS, Matrix Metalloproteinases; ECM, Extracellular matrix; CFS-1, colony stimulating factor-1; EMT, epithelial-mesenchymal transition

Given the intratumour heterogeneity and the dynamic nature of metastasis, certain cancer cell subpopulations may outperform others to successfully metastasize [2]. Furthermore, synergies may exist between cancer cell subpopulations, which enable different subpopulations to collectively complete the cascade [2]. Overall, metastasis is known to be a highly inefficient process that requires orchestration of multiple complex events to prevent the elimination of migrating tumour cells during the process [1].

Autophagy is a highly conserved self-degradative process that is widely known to play a key role in maintaining cell survival under various cellular stress conditions [3]. Recent studies have shown that autophagy is critical during the progression of numerous cancers and is identified to play a crucial role at every phase of the metastatic cascade [3]. Specifically, it has been shown to be implicated in regulating tumour cell motility and invasion, cancer stem cell viability and differentiation, resistance to programmed cell death (anoikis), epithelial-to-mesenchymal transition, tumour cell dormancy and escape from immune detection, with emerging roles in establishing pre-metastatic niche [3]. This chapter focuses on discussing how autophagy regulates metastasis in carcinomas, which constitute 90% of all cancers and potential therapies that can target autophagy to inhibit metastasis for disease management.

Cancer Metastasis

Invasion and Migration

Metastasis is initiated upon invasion and migration of primary tumour cells that permeate the extracellular matrix (ECM) with the help of stromal cells such as fibroblasts, adipocytes, and endothelial cells (ECs), which are reprogrammed to promote tumour cell invasion [1, 4]. There are two main components of the ECM: basement membrane that separates epithelial cells from the stroma which forms connective tissue; and the interstitial matrix that forms a three-dimensional scaffold to support tissue architecture, provide cell attachment, and separate tissue compartments [5]. Additionally, the ECM also sequesters growth factors and cytokines that determine the fate of cells surrounding the matrix [5]. The biochemical interactions between healthy cells and the ECM are frequently altered in cancer, which influence tumour cell proliferation and invasion. Tumour cell infiltration through the ECM relies on a number of factors such as cell motility, remodelling of the specific tissue matrix, release of chemotactic factors to recruit stromal cells, and loss of cell–cell interactions [6].

In order to acquire motility and invasiveness, cancer cells must shed their epithelial phenotype, which tightly holds the cells together and gain mesenchymal (stem cell-like) phenotype with enhanced migratory and invasive capacity [7]. Initiation of epithelial to mesenchymal transition (EMT) that commonly occurs during embryogenesis and wound healing is dependent on the loss of cell–cell adhesions, activation of transcription factors, modifications in the expression of specific cell-surface proteins, and production of ECM degradation enzymes [7, 8]. Epithelial cells are normally closely associated by cell–cell junctions in the form of tight junctions, adherens junctions, gap junctions, desmosomes and integrins [9]. The absence of these intercellular connections, which is identified by the loss of E-cadherin, α-catenin, and claudins, and acquisition of mesenchymal markers such as N-cadherin, fibronectin, and vimentin, is critical during EMT to facilitate physical detachment of tumour cells from primary tumour [9]. Several studies have demonstrated that loss of E-cadherin alone that often coincides with gain of N-cadherin is a key hallmark of EMT in many cancers [10]. E-cadherin at the cell membrane typically interacts with β-catenin intracellularly that is linked to actin-cytoskeleton through α-catenin, which is essential in maintaining epithelial cell–cell adhesion [1, 11]. However, upon EMT induction and loss of E-cadherin, β-catenin detaches from α-catenin and translocates to the nucleus where it drives transcription of tumour invasive genes such as cyclin D1, c-Myc, Matrix Metalloproteinase (MMP)-7 and Membrane Type (MT)-1-MMP [12–15].

Accumulating evidence suggests that EMT-inducing signals are partially initiated by growth factors such as hepatocyte growth factor (HGF), epidermal growth factor (EGF), insulin-like growth factor (IGF), and transforming growth factor β (TGFβ) [6, 16]. These stimulate a signalling cascade leading to the activation of numerous EMT-inducing transcription factors, namely, Snail, Slug, Twist, and zinc finger E-box binding homeobox 1 (ZEB1), some of which including Snail and Slug are crucial in repressing E-cadherin [17, 18]. Importantly, macrophages and stromal cells are known to play a vital role in regulating EMT [19]. These cells are recruited by tumour cells via the release of various chemokines such as colony-stimulating factor 1 (CSF-1), C–C motif ligand 5 (CCL5), and C–C motif ligand 2 (CCL2) [20, 21]. In addition to inducing and maintaining EMT through the secretion of growth factors like HGF, EGF, and IGF, these cells also cleave ECM by producing hydrolytic enzymes such as MMPs, for example, MT-1-MMP and MMP-9 [22]. This creates space in the interstitial matrix to enable tumour cell movement and further promotes the release of ECM-tethered growth factors, which leads to the formation of a positive feedback loop that enhances tumour progression into the matrix [22].

Upon ECM degradation, mesenchymal-like tumour cells, in response to a stimulus such as growth factors, commence extending actin-rich cytoplasmic projections termed lamellipodia and filopodia that form a leading edge at the front end of the cell [23]. Following protrusion, the leading edge adheres to ECM fibres via integrins present at the plasma membrane of the migratory cell [23, 24]. Integrin molecules couple to actin cytoskeleton intracellularly and induce mechanical forces and integrin clustering [23]. Clustered integrins recruit further actin filaments forming contractile structures comprised of actin myofilaments that are linked to myosin II molecules called stress fibres [23]. This integrin/cytoskeleton complex matures into a focal adhesion after the recruitment of various additional proteins including paxillin, zyxin, and tyrosine kinase focal adhesion kinase (FAK) [25]. The assembly of these proteins stimulates downstream signalling cascades that activate mitogen-activated protein kinase (MAPK)/extracellular regulated kinase (ERK) pathway, resulting in the transcriptional modulation of cell cycle and differentiation [26]. Furthermore, activation of these signalling cascades also causes phosphorylation of small GTPases of the Rho family, which induce actomyosin contraction leading to rear-end (trailing edge) retraction and forward movement of the cell body and nucleus [27, 28]. Importantly, several studies have reported that members of the Rho family such as RhoA and Rac1 are aberrantly expressed in various cancers, which is associated with enhanced tumour cell migration and advanced disease stage [29–31].

Intravasation

Following invasion of ECM, tumour cells can intravasate (enter) into lymphatic or blood vessels, with the majority of entry occurring through blood vessel routes [32]. While this process can occur passively as a result of tumour expansion, which leads to impingement and breakage of tumour generated angiogenic blood vessels lacking intact endothelial cell–cell interaction, increasing evidence suggests requirement of an active cell migration machinery [33]. Active tumour cell intravasation can occur paracellularly involving tumour cell entrance through EC junctions or transcellularly, whereby tumour cells migrate directly through the endothelial cell body [34]. Paracellular transmigration that is more commonly employed by tumour cells proceeds with the involvement of tumour-associated macrophages that are recruited by cancer cells secreting CSF-1 [35]. These macrophages reciprocate by producing vascular endothelial growth factor (VEGF), EGF and tumour necrosis factor (TNF)-1α, which induce retraction of EC junctions promoting tumour cell migration [35, 36]. A multitude of other factors including stromal cell-mediated secretion of TGF-β, CXCL12/CXCR4, MMPs such as MMP-1, -2 and -9 also play a crucial role in regulating intravasation [32, 36, 37]. Mechanistically, TGF-β and VEGF disrupt vascular endothelial (VE)-cadherin and β-catenin complexes in ECs that tightly conjugate ECs together, which generates gaps in EC junctions enabling tumour cell entrance in the circulation [38, 39]. However, recent findings also indicate that prolonged exposure of TGF-β facilitates endothelial proliferation leading to inhibition of tumour cell migration, which illustrates a paradoxical role of TGF-β in intravasation [36].

Furthermore, MMP-12/ADAM12 expression on tumour vasculature that induces ectodomain cleavage of VE-cadherin and angiopoietin 1 receptor TIE2 selectively expressed on ECs is proposed to damage EC junctions, implying a potential role of ADAM12 in cancer cell intravasation [40]. Notch signalling in tumour cells mediated by the interaction of Notch ligands on ECs and Notch receptors on tumour cells is also suggested to promote breast and colon cancer cell intravasation [41, 42]. Moreover, secretion of a serine protease Urokinase-type Plasminogen Activator (uPA) by stromal cells and macrophages that acts upon protease-activated receptor (PAR) on tumour cells has been shown to facilitate tumour cell transmigration through endothelial barrier [43, 44]. Binding of uPA to PAR that is promoted by MMP-1 can stimulate cleavage of plasminogen into plasmin that subsequently cleaves CUB domain-containing protein-1 (CDCP1) in tumour cells [32]. This results in the formation of a complex between retained CDCP1 and β_1-integrin that activates FAK/phosphoinositide 3-kinase (PI3K) pathway, which ultimately results in the transcription of various genes involved in enhancing tumour cell motility and migration through blood vessels [32, 45]. A recent study also identified that cancer cells can intravasate independently of stroma invasion following angiogenesis and development of intratumour blood vessels in the core of the tumour in the presence of EGF receptor (EGFR) [46].

During transcellular intravasation, inflammatory mediators commonly found in cancer microenvironment stimulate calcium influx in ECs [47]. This leads to calcium ion interaction with calmodulin, a calcium receptor responsible for the regulation of several fundamental cellular processes [48]. This binding activates myosin light chain kinase (MLCK) at the tumour cell attachment site inducing phosphorylation of MLC and actomyosin contraction ultimately causing rapid cytoskeletal and membrane remodelling, which generates a transitory pore-like structure for tumour cell movement across the EC body [47, 48].

Survival in the Circulation

Upon intravasation, tumour cells must overcome several barriers within the bloodstream including shear stress of the blood flow, ECM detachment-induced apoptosis (i.e., anoikis), and immune attack to survive in the circulation [49]. One of the mechanisms through which tumour cells resist mechanical forces in circulation is by coating themselves with platelets, which facilitates immune evasion and provides the required structure to circumvent mechanical stresses of the blood flow [50]. These platelets also contribute to the maintenance of EMT in tumour cells while in circulation, which is required to confer tumour cell resistance to apoptosis and senescence [51, 52]. Circulating tumour cells (CTCs) also lose anchorage-dependent adhesion to ECM, which is mediated via integrins that play a crucial role in maintaining tumour cell survival [53]. Typically, loss of cell adhesion induces anoikis, a form of cell death triggered by apoptosis that can be initiated by various extrinsic and intrinsic signals, all of which lead to the activation of caspases that ultimately stimulate endonucleases, DNA fragmentation, and cell death [54]. The extrinsic pathway is triggered in response to the activation of cell death receptors such as FAS and TNF receptor 1 (TNFR1), which are upregulated upon ECM detachment [55]. Conversely, the intrinsic pathway involving mitochondria is activated upon induction of DNA damage and endoplasmic reticulum (ER) stress—identified by the accumulation of unfolded/misfolded proteins in the ER, both of which occur in response to the loss of ECM anchorage [54]. Importantly, while a majority of tumour cells succumb to anoikis, some tumour cells develop adaptive pathways to overcome this process. These mechanisms include acquisition of mutations in anti-apoptotic/pro-survival pathways such as repression of tumour suppressor PTEN (phosphatase and tensin homolog), and activation of receptor tyrosine kinases such as insulin-like growth factor 1 receptor (IGF-1R) that stimulates PI3K/AKT pro-survival signalling that counteracts anoikis and enhances tumour cell survival [56, 57]. Furthermore, activating transcription factor 4 (ATF 4) that is activated in the absence of ECM-cell attachment induces a cytoprotective autophagy program that represses anoikis and promotes metastasis [49].

In addition to forming tumour cell-platelet microaggregates, cancer cells also downregulate MHC-I, and overexpress immune checkpoints such as CD47, and PD-L1 to inhibit immune responses and escape from immune-mediated destruction [58]. Reduction in the expression of MHC-I polypeptide-related sequence A (MICA) and B (MICB), which serve as ligands for Natural Killer Group 2D (NKG2D) receptor on Natural Killer (NK) cells for subsequent activation of NK cells, is also identified in numerous cancers [59]. Tumour cells also overexpress FAS ligand to induce T-cell killing and downregulate FAS expression to escape apoptosis and achieve immune escape [60].

Extravasation

Tumour cell migration from the circulation into secondary organs (extravasation) can occur as a result of tumour cell growth and entrapment in capillaries that causes rupturing of capillaries and invasion of tumour cells into tissues [61]. Alternatively, tumour cells can also directly and actively transmigrate through the endothelium as single cells by making changes to cellular components and disrupting inter-endothelial cell–cell junctions [49]. The latter method is found to be the predominant mode of extravasation in most cancers [49]. This process is proposed to be initiated by tumour cell rolling on the endothelium, which is mediated by various ligand-receptor interactions including the interaction of endothelial selectin (E-selectin) with its ligands sialyl Lewis x (sLe^x) or its isomer sialyl Lewis a (sLe^a) and CD44 expressed on tumour cells [62–66]. Stable adhesion of tumour cells to ECs is subsequently achieved by adhesion receptors including the binding of integrins such as $\alpha 4\beta 1$ on cancer cells to vascular cell adhesion molecule (VCAM) on ECs [67–71]. Upon attachment, tumour cells breakdown EC junctions to generate gaps between ECs, which enables tumour cell entry into tissue. One of the mechanisms by which ECs junctions are disrupted is through the induction of programmed necrosis (necroptosis) in ECs, which is facilitated by the ligation between membrane-bound amyloid precursor protein (APP) on tumour cells and endothelial death receptor 6 (DR6) on ECs [72]. Tumour cells also exploit platelets to secrete ATP, which acts on $P2Y_2$ receptor on EC leading to cytoskeleton rearrangement in ECs and opening of EC junctions [73]. Additionally, cancer cells also recruit monocytes that differentiate into macrophages within the underlying tissue, which release VEGF and increase vascular permeability [74, 75].

Furthermore, tumour cells also stimulate Rac1, RhoA-Rho-associated protein kinase (ROCK) and/or p38 MAPK in ECs via E-selectin interaction, which induces phosphorylation of MLC, formation of stress fibres and actomyosin-mediated contraction of EC junctions [76–78]. Importantly, expression of Rac1 in tumour cells, in addition to other molecules including β1 integrin, FAK, and protein diaphanous homologue 3 (DIAPH3), drives the formation of filopodium-like extensions that promote cancer cell movement across the endothelium and into the tissue upon EC junction disassembly [79].

It is well established that many tumours selectively metastasize to specific organs, which is defined by the 'seed and soil' hypothesis where CTCs (seeds) colonize an organ that provides sufficient nutrients (soil) for tumours to grow [80]. For example, lung cancers metastasize frequently to the bone, liver, and brain; breast cancers metastasize to the lung, liver, bone, and brain; and prostate cancers metastasize to the bone [81]. The site of extravasation and colonization is determined by a number of factors including the interaction between specific chemokines, exosomes, and growth factors with their respective receptors/ligands expressed by cells of the target organ, tumour cells or the target endothelium [82]. For example, CXC-chemokine ligand 12 (CXC12) expressed in the liver attracts breast cancer cells expressing its complementary receptor, CXCR4, thus promoting breast cancer cell extravasation and migration into the liver [83, 84]. Moreover, secreted growth factors such as IGF-1 released from the bone marrow also contribute to the recruitment of primary tumour cells expressing IGF-1 receptor such as breast and prostate cancer [85, 86]. However, several studies have also reported that various tumours commonly metastasize to the liver and bone, which is attributed to the increased permeability of blood vessels within these organs [87].

Exosomes are membrane-bound extracellular vesicles that comprise of proteins, RNA, DNA, and lipids that can be transferred to other cells to promote the expression of specific proteins and lipids within the recipient cell [88]. Increasing evidence suggests a vital role of exosomes in organ-specific metastasis. Breast cancer cell-released exosomes expressing α6β1 are associated with lung metastasis [89]. These exosomes interact with lung fibroblasts and activate Src, which upregulates S100A4 protein expression [89]. This can eventually lead to the production of VEGF-A, and other proteases, which results in vascular leakiness in the lung endothelium and formation of a pre-metastatic niche that facilitates breast cancer metastasis to the lung [89–91]. These reprogrammed fibroblasts also secrete MMPs to degrade local ECM and further promote the formation of a permissive pre-metastatic niche for tumour engraftment.

Colonization

Despite the release of a large number of tumour cells in the circulation, only a small population of cells develop into micro or macrometastases posing colonization as the rate-limiting step of the metastatic cascade [61]. This is attributed to a number of factors such as exposure of disseminated tumour cells (CTCs lodged in secondary tissue) to a largely functional immune system that eliminates tumour cells, and insufficient vascularization that suppresses tumour growth beyond 1–2 mm in diameter [19]. Thus, following extravasation, a majority of disseminated tumour cells often enter a state of dormancy (quiescence) with no active growth which can extend for up to several years. Factors that contribute to tumour dormancy include activated stress signals present in the new microenvironment (e.g., hypoxia), lack of growth factors, and expression of tumour suppressor genes such as kisspeptin1 (KISS1) [92–94]. Some of these stimuli play a vital role in inducing autophagy that helps tumour cells maintain a quiescent state (Discussed in Sect. Autophagy in Regulating Tumour Cell Dormancy and Therapy Resistance) [94].

Increasing evidence suggests that upon invasion of secondary tissues, disseminating tumour cells undergo mesenchymal-epithelial transition (MET) [95]. Though the detailed mechanisms by which MET is induced are not well understood, several pathways are proposed to have a crucial role in this process. For example, lack of EGFR signalling is found to trigger MET as demonstrated by increased E-cadherin levels in prostate cancer cells cocultured with hepatocytes [96]. In breast cancer, blockade of PI3K/AKT signalling was associated with MET, suggesting a potential involvement of this pathway in regulating MET [97]. It is also proposed that MET is induced simply due to the absence of EMT-inducing signals, however, the detailed mechanism/s through which this occurs remains to be elucidated [95].

After MET, tumour cells can enter G_0 quiescent state or form micrometastases and continue to remain dormant due to lack of blood vessels and oxygen deprivation or immune surveillance, which generates an equilibrium between tumour cell proliferation and apoptosis preventing tumour outgrowth. In order to successfully colonize and grow progressively, tumour cells must overcome dormancy. Activation of inhibitor of DNA binding 1 (ID1) transcription factor, which is implicated in the regulation of cell growth, senescence, and differentiation, has been reported to promote the recruitment of endothelial progenitor cells, which facilitate the formation of new vasculature in micrometastases [98]. Tumour cells also commence the recruitment of bone marrow-derived cells by secreting factors such as osteopontin (OPN) and stromal cell-derived factor 1 (SDF-1), which create a favourable environment to support tumour growth [99–101]. In order to circumvent immune-mediated cell death, these tumour cells also secrete a plethora of anti-inflammatory cytokines such as TGF-β and VEGF that facilitate immune evasion [19]. Several transcription factors such as Snail and Twist that confer self-renewal capacity to disseminated tumour cells have also been shown to be crucial in promoting tumour outgrowth in secondary tissues [99]. A small proportion of tumour cells overcome multiple barriers and stimulate self-renewal abilities to eventually develop into large colonies that are clinically detectable as macrometastases that often impede with organ function.

Autophagy in Cancer Metastasis

Autophagy is a cell-autonomous mechanism that destroys dysfunctional organelles or protein aggregates. It regulates the fundamental metabolic functions inside the cells and is implicated in various diseases such as cancer, neurodegenerative and lysosomal disorders. Autophagy regulates cellular homeostasis and stress responses and is triggered by various stimuli including nutrient deprivation, oxidative stress, and toxic molecules. Autophagy involves the formation of an intermediate organelle called the autophagosome, which is generated by an isolation membrane, or phagophore that sequesters a small portion of the cytoplasm. The autophagosome subsequently fuses with lysosomes to form an autolysosome, which degrades and recycles the materials contained within it.

Autophagy is considered to be a tumour-suppressive mechanism during tumour initiation and malignant transformation [102]. By removing damaged cells and organelles, autophagy limits cell proliferation and genomic instability in cancer. Gain of function mutation in tumour-suppressor protein p53 counteracts with the autophagic process through the AKT/mTOR pathway. p53, thus, shows a relationship with autophagy, thereby regulating cancer progression [103]. Epidermal growth factor receptor (EGFR) inhibits autophagy by binding to Beclin1, which allows cancer cells to survive against stress conditions [102]. Deactivated EGFR favours the upregulation of Beclin1, thereby favouring autophagy in cancer cells [102].

On the contrary, studies have suggested a protective role of autophagy in cancer cells. For example, autophagy plays a vital role in producing essential cellular metabolites to meet the metabolic and energy demands of tumour cells [102]. Furthermore, autophagic flux was found to be increased in cancer-associated adipocytes and cancer-associated fibroblasts in response to hypoxia and ROS production by malignant cells when compared to normal counterparts [104]. Cancer cells also exhibit higher glutamine utilization, leading to autophagy induction through inactivation of mTOR which supports tumour cell survival in harsh microenvironmental conditions [105]. The deprived nutrient condition, limited energy, and hypoxia of the tumour microenvironment are some of the factors that induce autophagy through different pathways leading to tumour development and metastasis [102].

Autophagy plays a significant role in every phase of the metastatic pathway. It is involved in modulating tumour cell motility and invasion, cancer stem cell viability and differentiation, resistance to anoikis, epithelial-to-mesenchymal transition (EMT), tumour cell dormancy, and escape from immune surveillance (Fig. 11.2) [106]. Environmental stresses such as hypoxia, or nutrient deprivation that promote metastasis induce autophagic flux resulting in dissemination of tumour cells [107], and detachment from the extracellular matrix (ECM) [99, 108–110]. Autophagy flux cannot be measured directly; therefore the microtubule-associated light chain B (LC3B) is used as a marker for autophagic flux. An increase in punctate LC3B staining was generally associated with metastasis, a poorer outcome, and with a more aggressive and invasive phenotype in human breast cancer [111, 112], melanoma

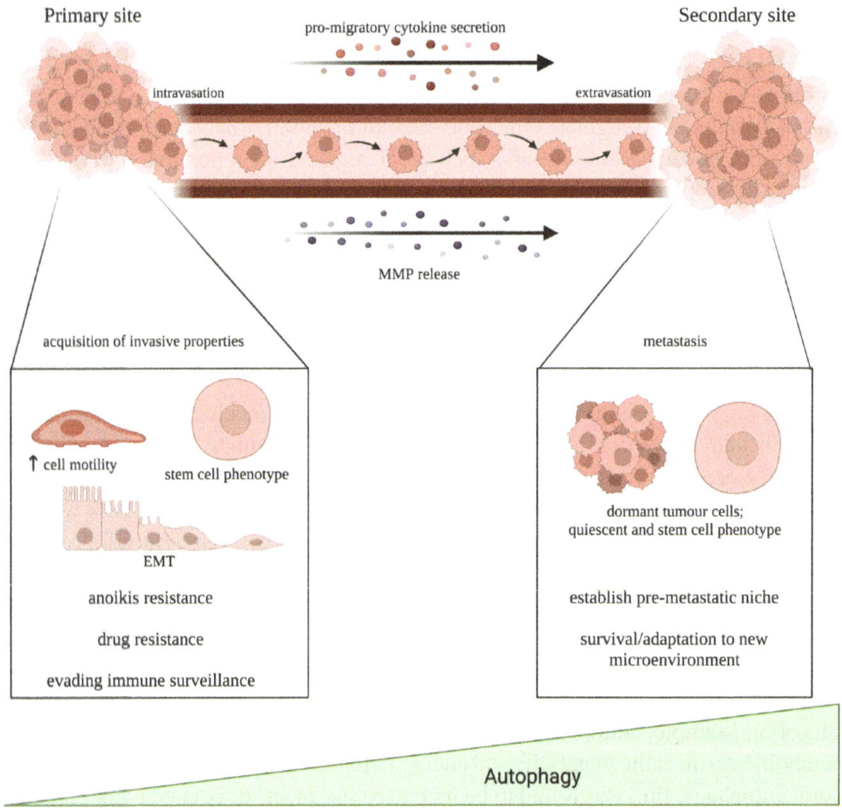

Fig. 11.2 Role of autophagy in Cancer Metastasis. As tumour cells progress to invasiveness, autophagy increases. This increase in autophagy correlates with increased cell motility, EMT, secretion of pro-migratory factors, release of matrix metalloproteinases (MMPs), drug resistance, and escape from immune surveillance at the primary tumour site. Autophagy is required at the secondary tumour site to maintain tumour cells in a dormant state through its ability to promote quiescence and stemness

[111, 113, 114], hepatocellular carcinoma [115, 116], and in human glioblastoma [117].

Role of Autophagy in Regulating EMT

For tumour cells to progress to invasiveness and metastasis, they undergo EMT as stated in Sect. Invasion and Migration [106, 118]. Importantly, hypoxia and TGF-β, inducers of EMT can also activate autophagy [119, 120]. Whether EMT and autophagy act directly or indirectly to promote cancer invasiveness and metastasis remains unknown. However, one study reported that in hepatocellular carcinoma cell lines, autophagy is required for TGF-β-induced EMT [120]. Additionally, ULK2, a protein that induces autophagy through phosphorylation of Beclin1-containing initiation complex, stimulates EMT, thereby downregulating E-cadherin and increasing invasiveness in vitro [121].

Another autophagy protein, p62, which acts as a cargo receptor, has been shown to bind to the EMT regulator Twist, thereby preventing proteasomal degradation and increasing EMT, invasiveness in vitro, and metastasis in vivo [122]. Furthermore, EMT and autophagy both promote cancer stem cells (CSCs) state [123–125], and CSCs have been hypothesized to drive metastasis due to their motile and plastic phenotype, as well as for their ability to propagate de novo CSCs and tumour heterogeneity at secondary tumour sites [108, 126, 127]. On the contrary, several studies have shown autophagy indirectly inhibits EMT by limiting p62 accumulation, and thus preventing tumour cell migration instead of promoting tumour cell motility [111, 128–132]. These studies have shown a definite relation between autophagy and EMT, however, both a tumour suppressing and tumour promoting role have been described. Therefore, further research is required to understand the precise role of autophagy in regulating EMT.

Autophagy In Inducing Resistance To Anoikis

Matrix detachment or direct inhibition of integrins has been shown to induce autophagy in epithelial cells, while autophagy inhibition increases apoptosis upon detachment [133]. It is, therefore, hypothesized that autophagy plays a key role in preventing anoikis and supporting the survival of detached tumour cells during metastasis [134]. A previous study has shown that autophagy induction in response to matrix detachment or integrin blockade is associated with reactive oxygen species (ROS)-dependent activation of the ER-stress kinase, protein kinase R-like ER kinase (PERK1), in mammary tumour models [109]. Inhibition of PERK or autophagy during matrix detachment or integrin signalling blockade promoted apoptosis and diminished clonogenic recovery, suggesting a role for PERK-induced autophagy in mammary tumour cell survival during ECM detachment [109, 110]. However, how PERK activates autophagy downstream of integrin signalling blockade is less understood. PERK is known to activate eukaryotic initiation factor 2 (eIF-2α), which suppresses general protein translation but permits selective translation of the transcription factor ATF4 [109]. ATF4 is known to induce autophagy-related gene 5 (ATG5) and LC3B expression [135]. Moreover, PERK is also known to induce the NF-E2-related factor 2 (NRF2) detoxification pathway that activates LKB1-AMPK signalling downstream of the integrin signalling blockade, which inhibits mechanistic target of rapamycin (mTOR) signalling to relieve mTOR complex 1 (mTORC1)-mediated autophagy inhibition [136]. However, another study has also shown that while mTORC1 signalling is reduced during detachment in mammary epithelial cells, mTORC1 reactivation does not inhibit autophagy, suggesting that mTORC1 inhibition is not necessary for autophagy induction during matrix detachment [137]. Alternatively, the study showed that inhibition of IκB kinase activity stimulates autophagy upon nutrient deprivation [138], which is essential for autophagy induction upon integrin signalling blockade or matrix detachment in mammary epithelial cells [137]. Furthermore, elevated levels of ROS induced by matrix detachment also contribute to autophagy induction through direct ATG4 activation [139]. Autophagy promoting survival following matrix detachment has also been shown in hepatocellular carcinoma and melanoma [115, 116, 140]. Thus, matrix detachment is likely to induce autophagy downstream of the integrin signalling blockade via multiple signalling pathways.

Role of Autophagy in Governing Tumour Cell Motility

The increased motility of tumour cells is important for metastasis for escaping the primary tumour site and to colonize a secondary tumour site [125, 126, 141]. Multiple studies have shown that autophagy has a direct role in essential aspects of tumour cell motility and invasion [132, 142–144] through, for example, modulation of the tumour cell secretome [129], turnover of components of the cell migration mechanism [130, 131], and through ECM proteins [145]. Autophagy has been linked to the function of the Rho family members, which are key regulators of cell motility. One study showed that the production of Rho1-induced cell protrusions and cell spreading of hemocytes was dependent on *Atg1* and on the *Drosophila* homolog of p62, and that the inhibition of autophagy prevented migration of blood cells to larval wound sites [142]. Concomitantly, knockdown of *ULK1* and *Beclin1* involved in the autophagic process, prevented cell spreading in mouse macrophages [142]. Furthermore, p62 has also been shown to play a role in targeting active RhoA, mammalian homolog of *Drosophila* Rho, to the autophagosome for degradation [146]. Inhibition of autophagy through ATG5 knockdown resulted in aberrant accumulation of RhoA, causing cytokinesis defects, multinucleation, and aneuploidy [146], illustrating the consequences of autophagy deficiency in tumour cells. Conversely, starvation-induced autophagy has been shown to be regulated by the downstream effector of Rho called Rho-associated kinase-1 (ROCK1) [147]. A subsequent study showed that ROCK1 activation upon amino acid deprivation promotes autophagy by directly phosphorylating *Beclin1* [148]. Furthermore, the interaction between autophagy and cell migration has also been shown through the identification of the coordinated control of a small GTPase involved in autophagosome and lysosomal fusion, named Rab7 [149], and Rac1, a Rho family GTPase required for lamellipodia formation and cell motility [150]. Autophagy also has a direct role in focal adhesion dynamics [130, 131, 144]. Focal adhesion kinase (FAK)-interacting protein of 200 kD (FIP200–required for autophagosome formation) binds and inhibits FAK kinase activity [151] but is also the mammalian homolog of yeast autophagy gene

Atg17 and an essential component of the autophagy pre-initiation/Ulk1 complex in mammals [152].

Autophagy has also been linked to increased tumour cell migration and invasion. It has been demonstrated that autophagy regulates the production of pro-migratory factors IL-6, MMP2, and WNT-5A during tumour cell invasion [129].

Autophagy in Regulating Tumour Cell Dormancy and Therapy Resistance

Cancer cell dormancy is induced upon cellular stress, whereby cells enter a state of quiescence as mentioned earlier. Quiescence is regulated by metabolism, specifically through autophagy [153, 154]. This happens in response to cellular stress as a way for the cell to continue to survive on low resources by limiting its energy expenditure. The use of autophagy to maintain survival in a stressful environment has been shown to induce quiescence. Autophagy may support the metabolic needs of tumour cells by maintaining amino acid levels, ATP production, and preventing energetic catastrophe [154–156]. Autophagy induction, G1 arrest, and cell survival are coordinated downstream of LKB1-AMPK activation [157]. In addition to stimulating ULK1-dependent autophagy [158], LKB1-AMPK activates $p27^{kip1}$-dependent G1 growth arrest. The absence of p27/KIP1 activates LKB1-AMPK signalling under nutrient stress resulting in apoptosis [157], suggesting a mechanism linking autophagy induction, growth arrest and apoptosis inhibition.

The tumour suppressor aplasia Ras homolog member I (ARHI) has also been linked in tumour cell dormancy through autophagy induction [153]. AHRI inhibits the PI3K/AKT growth factor signalling pathway, thereby inducing autophagy [159]. Additionally, re-expression of ARHI in ARHI-deficient SKOv3 ovarian cancer cells in vitro induces autophagy, suggesting a role of autophagy in regulating tumour cell dormancy [153].

Involvement of Autophagy During Tumour Immune Surveillance

In regulating immne responses, autophagy is partially responsible for the degradation of intracellular pathogens, secretion of immune modulatory cytokines and proteases, generation of antigen peptides for MHC-II presentation, and for the modulation of pro-inflammatory signalling [160–162]. At the primary tumour site and throughout the metastatic dissemination and colonization, tumour cells evolve mechanisms to evade immune surveillance by cytotoxic T lymphocytes (CTLs) and natural killer (NK) cells [163]. Tumour cell intrinsic autophagy modulates the ability of cancer cells to evade immune defence [160]. Michaud et al. have shown autophagy-dependent recruitment of dendritic cells and T cells to subcutaneous CT26 murine colon carcinoma in response to chemotherapy, whereas autophagy inhibition inhibited immune cell recruitment to the tumour cells in vivo [164]. Similar response has also been shown by Ladoire et al. who showed strong punctate LC3B staining in human breast cancer cells which was correlated with increased tumour infiltration by $CD8^+$ CTLs and reduced infiltration by $FoxP3^+$ T-regulatory (T_{reg}) cells consistent with increased tumour surveillance promoted by tumour cell autophagy [165]. Similar results have also been shown in lung cancer model [166].

Besides promoting immune surveillance, autophagy has also been demonstrated to block immune surveillance. Acquisition of the mesenchymal phenotype has been shown to be associated with an increase in autophagic flux and *Beclin1*-dependent resistance to CTL-mediated cell lysis in MCF-7 breast cancer cells induced to undergo EMT, suggesting that the pro-metastatic behaviour is linked through autophagy with resistance to immune surveillance [167]. This has also been demonstrated in MMTV-PyMT mouse mammary tumour models in which FIP200 deletion increased the CXXL10-dependent infiltration of CTLs into primary tumours, resulting in reduced primary tumour growth and decreased metastatic dissemination [168]. Additionally, hypoxic tumour cell resistance to CTLs and NK cells is promoted by autophagy through autophagy-dependent granzyme B degradation in breast cancer cells [169], and destabilization of the immune synapse and other functions in melanoma cells [170, 171], as well as a described feedback loop wherein autophagy inhibition sensitizes renal cell carcinoma cells to NK cell-mediated lysis [172]. The pro- and anti-inflammatory surveillance functions regulated by autophagy may be subjective to different tissues or differences in stress responses, such as hypoxia, nutrient deprivation, or chemotherapy, that regulate tumour cell interactions with the immune system.

Role of Autophagy in Other Tumour Microenvironment Activities

Autophagy has been suggested to have multiple functions in regulating paracrine signalling between tumour cells and stroma cells in the tumour microenvironment (TME) to regulate tumour cell escape. The fibroblasts and tumour-infiltrating immune cells are essential for pro-migratory growth factors and chemokines in the TME, and are therefore important in regulating the metastatic potential of tumour cells [173]. Stromal cells, macrophages [173], and fibroblasts [174] are essential sources of matrix-degrading enzymes required by the TME to promote invasion as described earlier. It is hypothesized that the release of hydrogen peroxide by tumour cells induces senescence in remodelled cancer-associated fibroblasts (CAFs) resulting in an autophagy-dependent switch to aerobic glycolysis and production of lactic acid, ketone bodies, and free fatty acids, which fuel the growth of tumour cells [175]. Autophagy is part of the endolysosomal system, which suggests that autophagy may also be involved in modulating the release of exosomes by tumour cells, thereby influencing the establishment of a pre-metastatic niche [176, 177].

Autophagy in Cancer Therapeutics

Several studies link autophagy induction with tumour cell dormancy and metastasis, therefore, autophagy inhibition could be seen as a therapeutical option, as dormancy has been hypothesized to underlie the persistence of disease [154]. Autophagy induction has been associated with Imatinib-induced reversible quiescence in gastrointestinal stromal cells [178]. Gupta et al. demonstrated rapid cell death via autophagy inhibition through either knockdown of essential autophagy genes or lysosomotrophic treatment, such as chloroquine [178]. Therefore, therapies combining Imatinib or other tyrosine kinase inhibitors (TKIs) with autophagy inhibitors may increase the plausibility of durable treatment responses [106]. The quiescent CSC state of dormant tumour cells leads to tumour cell resistance to chemotherapeutic agents that target proliferating cells [106]. In vitro, radiation treatment induces autophagy, whereas autophagy inhibition reduced clonogenic survival of several tumour cell lines, including breast, lung, and cervical tumour cell lines following radiation [179]. Similar observations have been made for several chemotherapies, for example, doxorubicin resistance in the HEp3 liver cancer dormancy model is attributed to p38-induced upregulation of ER chaperones and PERK [180], which stimulates autophagy through induction of ATF4 and its downstream targets ATG5 and LC3B [135]. Autophagy inhibition in combination with treatment with alkylating agents suppresses tumour recurrence in a Myc-driven lymphoma model [181]. These studies suggest that autophagy promotes the survival of dormant tumour cells and contributes to therapy resistance. Therefore, the combination of chemotherapy with inhibition of autophagy could favourably eliminate dormant tumour cells and consequently limit metastatic dormancy [106].

Summary

Increasing evidence suggests a key role of autophagy in mediating several processes throughout the cancer metastasis cascade, thus providing the opportunity to target specific steps of the cascade using autophagy inhibitors to stall cancer metastases and ultimately cancer mortality rates. Autophagy has been identified to modulate cancer stem-cell phenotype, cell motility and invasion, tumour dormancy, tumour immune surveillance, drug resistance and several other tumour microenvironment functions. While this provides an insight into the interplay between autophagy and metastasis, further research is required to understand how some of the metastatic processes are regulated mechanistically, given the paradoxical role of autophagy in some steps of the cascade. This will support the development of novel autophagy inhibitors and inducers that can be used in combination with conventional therapies to inhibit cancer dissemination and improve survival outcomes in cancer patients.

Acknowledgements Work in the laboratory of NR is supported by funds from Centre for Cancer Biology, University of South Australia and Neurosurgical Research Foundation (NRF), Adelaide, Australia.

Disclosure of Interests All authors declare they have no conflict of interest.

References

1. van Zijl F, Krupitza G, Mikulits W (2011) Initial steps of metastasis: cell invasion and endothelial transmigration. Mutat Res 728(1–2):23–34. https://doi.org/10.1016/j.mrrev.2011.05.002
2. Hapach LA, Mosier JA, Wang W, Reinhart-King CA (2019) Engineered models to parse apart the metastatic cascade. Npj Precis Oncol 3(1):20. https://doi.org/10.1038/s41698-019-0092-3
3. Mowers EE, Sharifi MN, Macleod KF (2017) Autophagy in cancer metastasis. Oncogene 36(12):1619–1630. https://doi.org/10.1038/onc.2016.333
4. Pachmayr E, Treese C, Stein U (2017) Underlying mechanisms for distant metastasis–molecular biology. Visc Med 33(1):11–20. https://doi.org/10.1159/000454696
5. Kular JK, Basu S, Sharma RI (2014) The extracellular matrix: structure, composition, age-related differences, tools for analysis and applications for tissue engineering. J Tissue Eng 5:2041731414557112. https://doi.org/10.1177/2041731414557112
6. Friedl P, Alexander S (2011) Cancer invasion and the microenvironment: plasticity and reciprocity. Cell 147(5):992–1009. https://doi.org/10.1016/j.cell.2011.11.016
7. Lu W, Kang Y (2019) Epithelial-mesenchymal plasticity in cancer progression and metastasis. Dev Cell 49(3):361–374. https://doi.org/10.1016/j.devcel.2019.04.010
8. Kim DH, Xing T, Yang Z, Dudek R, Lu Q, Chen Y-H (2017) Epithelial mesenchymal transition in embryonic development, tissue repair and cancer: a comprehensive overview. J Clin Med 7(1):1. https://doi.org/10.3390/jcm7010001
9. Ribatti D, Tamma R, Annese T (2020) Epithelial-mesenchymal transition in cancer: a historical overview. Transl Oncol 13(6):100773–100773. https://doi.org/10.1016/j.tranon.2020.100773
10. Cavallaro U, Christofori G (2004) Cell adhesion and signalling by cadherins and Ig-CAMs in cancer. Nat Rev Cancer 4(2):118–132. https://doi.org/10.1038/nrc1276
11. Thiery JP, Acloque H, Huang RYJ, Nieto MA (2009) Epithelial-mesenchymal transitions in development and disease. Cell 139(5):871–890. https://doi.org/10.1016/j.cell.2009.11.007
12. Hlubek F, Spaderna S, Jung A, Kirchner T, Brabletz T (2004) Beta-catenin activates a coordinated expression of the proinvasive factors laminin-5 gamma2 chain and MT1-MMP in colorectal carcinomas. Int J Cancer 108(2):321–326. https://doi.org/10.1002/ijc.11522
13. Kolligs FT, Bommer G, Göke B (2002) Wnt/beta-catenin/tcf signaling: a critical pathway in gastrointestinal tumorigenesis. Dig 66(3):131–144. https://doi.org/10.1159/000066755
14. Li YJ, Wei ZM, Meng YX, Ji XR (2005) Beta-catenin up-regulates the expression of cyclinD1, c-myc and MMP-7 in human pancreatic cancer: relationships with carcinogenesis and metastasis. World J Gastroenterol 11(14):2117–2123. https://doi.org/10.3748/wjg.v11.i14.2117
15. Zucker S, Vacirca J (2004) Role of matrix metalloproteinases (MMPs) in colorectal cancer. Cancer Metastasis Rev 23(1–2):101–117. https://doi.org/10.1023/a:1025867130437
16. Thiery JP (2002) Epithelial–mesenchymal transitions in tumour progression. Nat Rev Cancer 2(6):442–454. https://doi.org/10.1038/nrc822

17. Raja R, Pandey A, Kumar P (2020) Epithelial to mesenchymal plasticity: role in cancer progression. Front Biosci-Landmark 25(5):838–873. https://doi.org/10.2741/4837
18. Cano A, Pérez-Moreno MA, Rodrigo I, Locascio A, Blanco MJ, del Barrio MG, Portillo F, Nieto MA (2000) The transcription factor snail controls epithelial–mesenchymal transitions by repressing E-cadherin expression. Nat Cell Biol 2(2):76–83. https://doi.org/10.1038/35000025
19. Quail DF, Joyce JA (2013) Microenvironmental regulation of tumor progression and metastasis. Nat Med 19(11):1423–1437. https://doi.org/10.1038/nm.3394
20. Kohli K, Pillarisetty VG, Kim TS (2021) Key chemokines direct migration of immune cells in solid tumors. Cancer Gene Ther. https://doi.org/10.1038/s41417-021-00303-x
21. Borsig L, Wolf MJ, Roblek M, Lorentzen A, Heikenwalder M (2014) Inflammatory chemokines and metastasis—tracing the accessory. Oncogene 33(25):3217–3224. https://doi.org/10.1038/onc.2013.272
22. Winkler J, Abisoye-Ogunniyan A, Metcalf KJ, Werb Z (2020) Concepts of extracellular matrix remodelling in tumour progression and metastasis. Nat Commun 11(1):5120. https://doi.org/10.1038/s41467-020-18794-x
23. Friedl P, Wolf K (2003) Tumour-cell invasion and migration: diversity and escape mechanisms. Nat Rev Cancer 3(5):362–374. https://doi.org/10.1038/nrc1075
24. Scales TME, Parsons M (2011) Spatial and temporal regulation of integrin signalling during cell migration. Curr Opin Cell Biol 23(5):562–568. https://doi.org/10.1016/j.ceb.2011.05.008
25. Webb DJ, Donais K, Whitmore LA, Thomas SM, Turner CE, Parsons JT, Horwitz AF (2004) FAK–Src signalling through paxillin, ERK and MLCK regulates adhesion disassembly. Nat Cell Biol 6(2):154–161. https://doi.org/10.1038/ncb1094
26. Slack-Davis JK, Eblen ST, Zecevic M, Boerner SA, Tarcsafalvi A, Diaz HB, Marshall MS, Weber MJ, Parsons JT, Catling AD (2003) PAK1 phosphorylation of MEK1 regulates fibronectin-stimulated MAPK activation. J Cell Biol 162(2):281–291. https://doi.org/10.1083/jcb.200212141
27. Heikenwalder M, Lorentzen A (2019) The role of polarisation of circulating tumour cells in cancer metastasis. Cell Mol Life Sci 76(19):3765–3781. https://doi.org/10.1007/s00018-019-03169-3
28. Friedl P, Wolf K (2009) Proteolytic interstitial cell migration: a five-step process. Cancer Metastasis Rev 28(1):129–135. https://doi.org/10.1007/s10555-008-9174-3
29. Sahai E, Marshall CJ (2002) RHO-GTPases and cancer. Nat Rev Cancer 2(2):133–142. https://doi.org/10.1038/nrc725
30. Kamai T, Arai K, Tsujii T, Honda M, Yoshida K (2001) Overexpression of RhoA mRNA is associated with advanced stage in testicular germ cell tumour. BJU Int 87(3):227–231. https://doi.org/10.1046/j.1464-410x.2001.02030.x
31. Fritz G, Just I, Kaina B (1999) Rho GTPases are over-expressed in human tumors. Int J Cancer 81(5):682–687. https://doi.org/10.1002/(sici)1097-0215(19990531)81:5%3c682::aid-ijc2%3e3.0.co;2-b
32. Chiang SPH, Cabrera RM, Segall JE (2016) Tumor cell intravasation. Am J Physiol Cell Physiol 311(1):C1–C14. https://doi.org/10.1152/ajpcell.00238.2015
33. van Zijl F, Krupitza G, Mikulits W (2011) Initial steps of metastasis: cell invasion and endothelial transmigration. Mutat Res/Rev Mutat Res 728(1):23–34. https://doi.org/10.1016/j.mrrev.2011.05.002
34. Reymond N, d'Água BB, Ridley AJ (2013) Crossing the endothelial barrier during metastasis. Nat Rev Cancer 13(12):858–870. https://doi.org/10.1038/nrc3628
35. Zavyalova MV, Denisov EV, Tashireva LA, Savelieva OE, Kaigorodova EV, Krakhmal NV, Perelmuter VM (2019) Intravasation as a key step in cancer metastasis. Biochem (Mosc) 84(7):762–772. https://doi.org/10.1134/s0006297919070071
36. Guo S, Deng C-X (2018) Effect of stromal cells in tumor microenvironment on metastasis initiation. Int J Biol Sci 14(14):2083–2093. https://doi.org/10.7150/ijbs.25720
37. Boimel PJ, Smirnova T, Zhou ZN, Wyckoff J, Park H, Coniglio SJ, Qian B-Z, Stanley ER, Cox D, Pollard JW, Muller WJ, Condeelis J, Segall JE (2012) Contribution of CXCL12 secretion

to invasion of breast cancer cells. Breast Cancer Res 14(1):R23. https://doi.org/10.1186/bcr3108
38. Hoeben A, Landuyt B, Highley MS, Wildiers H, Van Oosterom AT, De Bruijn EA (2004) Vascular endothelial growth factor and angiogenesis. Pharmacol Rev 56(4):549–580. https://doi.org/10.1124/pr.56.4.3
39. Drabsch Y, ten Dijke P (2011) TGF-β signaling in breast cancer cell invasion and bone metastasis. J Mammary Gland Biol Neoplasia 16(2):97–108. https://doi.org/10.1007/s10911-011-9217-1
40. Fröhlich C, Klitgaard M, Noer JB, Kotzsch A, Nehammer C, Kronqvist P, Berthelsen J, Blobel C, Kveiborg M, Albrechtsen R, Wewer UM (2013) ADAM12 is expressed in the tumour vasculature and mediates ectodomain shedding of several membrane-anchored endothelial proteins. Biochem J 452(1):97–109. https://doi.org/10.1042/bj20121558
41. Chen J, Imanaka N, Chen J, Griffin JD (2010) Hypoxia potentiates notch signaling in breast cancer leading to decreased E-cadherin expression and increased cell migration and invasion. Br J Cancer 102(2):351–360. https://doi.org/10.1038/sj.bjc.6605486
42. Sonoshita M, Aoki M, Fuwa H, Aoki K, Hosogi H, Sakai Y, Hashida H, Takabayashi A, Sasaki M, Robine S, Itoh K, Yoshioka K, Kakizaki F, Kitamura T, Oshima M, Taketo MM (2011) Suppression of colon cancer metastasis by Aes through inhibition of Notch signaling. Cancer Cell 19(1):125–137. https://doi.org/10.1016/j.ccr.2010.11.008
43. Mekkawy AH, Pourgholami MH, Morris DL (2014) Involvement of urokinase-type plasminogen activator system in cancer: an overview. Med Res Rev 34(5):918–956. https://doi.org/10.1002/med.21308
44. Ossowski L (1988) Plasminogen activator dependent pathways in the dissemination of human tumor cells in the chick embryo. Cell 52(3):321–328. https://doi.org/10.1016/s0092-8674(88)80025-4
45. Casar B, Rimann I, Kato H, Shattil SJ, Quigley JP, Deryugina EI (2014) In vivo cleaved CDCP1 promotes early tumor dissemination via complexing with activated β1 integrin and induction of FAK/PI3K/Akt motility signaling. Oncogene 33(2):255–268. https://doi.org/10.1038/onc.2012.547
46. Deryugina EI, Kiosses WB (2017) Intratumoral cancer cell intravasation can occur independent of invasion into the adjacent stroma. Cell Rep 19(3):601–616. https://doi.org/10.1016/j.celrep.2017.03.064
47. Dalal PJ, Muller WA, Sullivan DP (2020) Endothelial cell calcium signaling during barrier function and inflammation. Am J Pathol 190(3):535–542. https://doi.org/10.1016/j.ajpath.2019.11.004
48. Khuon S, Liang L, Dettman RW, Sporn PHS, Wysolmerski RB, Chew T-L (2010) Myosin light chain kinase mediates transcellular intravasation of breast cancer cells through the underlying endothelial cells: a three-dimensional FRET study. J Cell Sci 123(Pt 3):431–440. https://doi.org/10.1242/jcs.053793
49. Strilic B, Offermanns S (2017) Intravascular survival and extravasation of tumor cells. Cancer Cell 32(3):282–293. https://doi.org/10.1016/j.ccell.2017.07.001
50. Fares J, Fares MY, Khachfe HH, Salhab HA, Fares Y (2020) Molecular principles of metastasis: a hallmark of cancer revisited. Signal Transduct Target Ther 5(1):28. https://doi.org/10.1038/s41392-020-0134-x
51. Lou X-L, Sun J, Gong S-Q, Yu X-F, Gong R, Deng H (2015) Interaction between circulating cancer cells and platelets: clinical implication. Chin J Cancer Res 27(5):450–460. https://doi.org/10.3978/j.issn.1000-9604.2015.04.10
52. Leblanc R, Peyruchaud O (2016) Metastasis: new functional implications of platelets and megakaryocytes. Blood 128(1):24–31. https://doi.org/10.1182/blood-2016-01-636399
53. Hamidi H, Ivaska J (2018) Every step of the way: integrins in cancer progression and metastasis. Nat Rev Cancer 18(9):533–548. https://doi.org/10.1038/s41568-018-0038-z
54. Paoli P, Giannoni E, Chiarugi P (2013) Anoikis molecular pathways and its role in cancer progression. Biochim Biophys Acta (BBA)—Mol Cell Res 1833(12):3481–3498. https://doi.org/10.1016/j.bbamcr.2013.06.026

55. Kim Y-N, Koo KH, Sung JY, Yun U-J, Kim H (2012) Anoikis resistance: an essential prerequisite for tumor metastasis. Int J Cell Biol 2012:306879. https://doi.org/10.1155/2012/306879
56. Altomare DA, Testa JR (2005) Perturbations of the AKT signaling pathway in human cancer. Oncogene 24(50):7455–7464. https://doi.org/10.1038/sj.onc.1209085
57. Valentinis B, Morrione A, Peruzzi F, Prisco M, Reiss K, Baserga R (1999) Anti-apoptotic signaling of the IGF-I receptor in fibroblasts following loss of matrix adhesion. Oncogene 18(10):1827–1836. https://doi.org/10.1038/sj.onc.1202471
58. Zhong X, Zhang H, Zhu Y, Liang Y, Yuan Z, Li J, Li J, Li X, Jia Y, He T, Zhu J, Sun Y, Jiang W, Zhang H, Wang C, Ke Z (2020) Circulating tumor cells in cancer patients: developments and clinical applications for immunotherapy. Mol Cancer 19(1):15. https://doi.org/10.1186/s12943-020-1141-9
59. Chitadze G, Lettau M, Bhat J, Wesch D, Steinle A, Fürst D, Mytilineos J, Kalthoff H, Janssen O, Oberg H-H, Kabelitz D (2013) Shedding of endogenous MHC class I-related chain molecules A and B from different human tumor entities: heterogeneous involvement of the "a disintegrin and metalloproteases" 10 and 17. Int J Cancer 133(7):1557–1566. https://doi.org/10.1002/ijc.28174
60. Gordon N, Kleinerman ES (2009) The role of Fas/FasL in the metastatic potential of osteosarcoma and targeting this pathway for the treatment of osteosarcoma lung metastases. Cancer Treat Res 152:497–508. https://doi.org/10.1007/978-1-4419-0284-9_29
61. Massagué J, Obenauf AC (2016) Metastatic colonization by circulating tumour cells. Nature 529(7586):298–306. https://doi.org/10.1038/nature17038
62. Dimitroff CJ, Lechpammer M, Long-Woodward D, Kutok JL (2004) Rolling of human bone−metastatic prostate tumor cells on human bone marrow endothelium under shear flow is mediated by E-selectin. Can Res 64(15):5261. https://doi.org/10.1158/0008-5472.CAN-04-0691
63. Tözeren A, Kleinman HK, Grant DS, Morales D, Mercurio AM, Byers SW (1995) E-selectin-mediated dynamic interactions of breast-and colon-cancer cells with endothelial-cell monolayers. Int J Cancer 60(3):426–431. https://doi.org/10.1002/ijc.2910600326
64. Krause T, Turner GA (1999) Are selectins involved in metastasis? Clin Exp Metastasis 17(3):183–192. https://doi.org/10.1023/a:1006626500852
65. Laferrière J, Houle F, Huot J (2004) Adhesion of HT-29 colon carcinoma cells to endothelial cells requires sequential events involving E-selectin and integrin beta4. Clin Exp Metastasis 21(3):257–264. https://doi.org/10.1023/b:clin.0000037708.09420.9a
66. Takada A, Ohmori K, Yoneda T, Tsuyuoka K, Hasegawa A, Kiso M, Kannagi R (1993) Contribution of carbohydrate antigens sialyl lewis A and sialyl lewis X to adhesion of human cancer cells to vascular endothelium. Cancer Res 53(2):354–361
67. Mattila P, Majuri ML, Renkonen R (1992) VLA-4 integrin on sarcoma cell lines recognizes endothelial VCAM-1. Differential regulation of the VLA-4 avidity on various sarcoma cell lines. Int J Cancer 52(6):918−923. https://doi.org/10.1002/ijc.2910520615
68. Steinbach F, Tanabe K, Alexander J, Edinger M, Tubbs R, Brenner W, Stöckle M, Novick AC, Klein EA (1996) The influence of cytokines on the adhesion of renal cancer cells to endothelium. J Urol 155(2):743–748
69. Taichman DB, Cybulsky MI, Djaffar I, Longenecker BM, Teixidó J, Rice GE, Aruffo A, Bevilacqua MP (1991) Tumor cell surface alpha 4 beta 1 integrin mediates adhesion to vascular endothelium: demonstration of an interaction with the N-terminal domains of INCAM-110/VCAM-1. Cell Regul 2(5):347–355. https://doi.org/10.1091/mbc.2.5.347
70. Tomita Y, Saito T, Saito K, Oite T, Shimizu F, Sato S (1995) Possible significance of VLA-4 (alpha 4 beta 1) for hematogenous metastasis of renal-cell cancer. Int J Cancer 60(6):753–758. https://doi.org/10.1002/ijc.2910600604
71. Klemke M, Weschenfelder T, Konstandin MH, Samstag Y (2007) High affinity interaction of integrin alpha4beta1 (VLA-4) and vascular cell adhesion molecule 1 (VCAM-1) enhances migration of human melanoma cells across activated endothelial cell layers. J Cell Physiol 212(2):368–374. https://doi.org/10.1002/jcp.21029

72. Strilic B, Yang L, Albarrán-Juárez J, Wachsmuth L, Han K, Müller UC, Pasparakis M, Offermanns S (2016) Tumour–cell–induced endothelial cell necroptosis via death receptor 6 promotes metastasis. Nature 536(7615):215–218. https://doi.org/10.1038/nature19076
73. Schumacher D, Strilic B, Sivaraj Kishor K, Wettschureck N, Offermanns S (2013) Platelet-derived nucleotides promote tumor–cell transendothelial migration and metastasis via P2Y2 receptor. Cancer Cell 24(1):130–137. https://doi.org/10.1016/j.ccr.2013.05.008
74. Qian B, Deng Y, Im JH, Muschel RJ, Zou Y, Li J, Lang RA, Pollard JW (2009) A distinct macrophage population mediates metastatic breast cancer cell extravasation establishment and growth. PLoS ONE 4(8):e6562. https://doi.org/10.1371/journal.pone.0006562
75. Qian B-Z, Li J, Zhang H, Kitamura T, Zhang J, Campion LR, Kaiser EA, Snyder LA, Pollard JW (2011) CCL2 recruits inflammatory monocytes to facilitate breast–tumour metastasis. Nature 475(7355):222–225. https://doi.org/10.1038/nature10138
76. Tremblay PL, Auger FA, Huot J (2006) Regulation of transendothelial migration of colon cancer cells by E–selectin–mediated activation of p38 and ERK MAP kinases. Oncogene 25(50):6563–6573. https://doi.org/10.1038/sj.onc.1209664
77. Cain RJ, Vanhaesebroeck B, Ridley AJ (2010) The PI3K p110alpha isoform regulates endothelial adherens junctions via Pyk2 and Rac1. J Cell Biol 188(6):863–876. https://doi.org/10.1083/jcb.200907135
78. Li B, Zhao WD, Tan ZM, Fang WG, Zhu L, Chen YH (2006) Involvement of Rho/ROCK signalling in small cell lung cancer migration through human brain microvascular endothelial cells. FEBS Lett 580(17):4252–4260. https://doi.org/10.1016/j.febslet.2006.06.056
79. Shibue T, Brooks MW, Inan MF, Reinhardt F, Weinberg RA (2012) The outgrowth of micrometastases is enabled by the formation of filopodium-like protrusions. Cancer Discov 2(8):706–721. https://doi.org/10.1158/2159-8290.Cd-11-0239
80. Akhtar M, Haider A, Rashid S, Al-Nabet ADMH (2019) Paget's "Seed and Soil" theory of cancer metastasis: an idea whose time has come. Adv Anat Pathol 26(1)
81. Nguyen DX, Bos PD, Massagué J (2009) Metastasis: from dissemination to organ-specific colonization. Nat Rev Cancer 9(4):274–284. https://doi.org/10.1038/nrc2622
82. Xie H-Y, Shao Z-M, Li D-Q (2017) Tumor microenvironment: driving forces and potential therapeutic targets for breast cancer metastasis. Chin J Cancer 36(1):36. https://doi.org/10.1186/s40880-017-0202-y
83. Müller A, Homey B, Soto H, Ge N, Catron D, Buchanan ME, McClanahan T, Murphy E, Yuan W, Wagner SN, Barrera JL, Mohar A, Verástegui E, Zlotnik A (2001) Involvement of chemokine receptors in breast cancer metastasis. Nature 410(6824):50–56. https://doi.org/10.1038/35065016
84. Teicher BA, Fricker SP (2010) CXCL12 (SDF-1)/CXCR4 pathway in cancer. Clin Cancer Res 16(11):2927. https://doi.org/10.1158/1078-0432.CCR-09-2329
85. Smith HA, Kang Y (2017) Determinants of organotropic metastasis. Annu Rev Cancer Biol 1(1):403–423. https://doi.org/10.1146/annurev-cancerbio-041916-064715
86. Hua H, Kong Q, Yin J, Zhang J, Jiang Y (2020) Insulin-like growth factor receptor signaling in tumorigenesis and drug resistance: a challenge for cancer therapy. J Hematol Oncol 13(1):64. https://doi.org/10.1186/s13045-020-00904-3
87. Fouad YA, Aanei C (2017) Revisiting the hallmarks of cancer. Am J Cancer Res 7(5):1016–1036
88. Zhang Y, Liu Y, Liu H, Tang WH (2019) Exosomes: biogenesis, biologic function and clinical potential. Cell Biosci 9(1):19. https://doi.org/10.1186/s13578-019-0282-2
89. Hoshino A, Costa-Silva B, Shen T-L, Rodrigues G, Hashimoto A, Tesic Mark M, Molina H, Kohsaka S, Di Giannatale A, Ceder S, Singh S, Williams C, Soplop N, Uryu K, Pharmer L, King T, Bojmar L, Davies AE, Ararso Y, Zhang T, Zhang H, Hernandez J, Weiss JM, Dumont-Cole VD, Kramer K, Wexler LH, Narendran A, Schwartz GK, Healey JH, Sandstrom P, Jørgen Labori K, Kure EH, Grandgenett PM, Hollingsworth MA, de Sousa M, Kaur S, Jain M, Mallya K, Batra SK, Jarnagin WR, Brady MS, Fodstad O, Muller V, Pantel K, Minn AJ, Bissell MJ, Garcia BA, Kang Y, Rajasekhar VK, Ghajar CM, Matei I, Peinado H, Bromberg J, Lyden D (2015) Tumour exosome integrins determine organotropic metastasis. Nature 527(7578):329–335. https://doi.org/10.1038/nature15756

90. Connell JT, Sugimoto H, Cooke VG, MacDonald BA, Mehta AI, LeBleu VS, Dewar R, Rocha RM, Brentani RR, Resnick MB, Neilson EG, Zeisberg M, Kalluri R (2011) VEGF-A and Tenascin-C produced by S100A4⁺ stromal cells are important for metastatic colonization. Proc Natl Acad Sci 108(38):16002. https://doi.org/10.1073/pnas.1109493108
91. Chen H, Chengalvala V, Hu H, Sun D (2021) Tumor-derived exosomes: nanovesicles made by cancer cells to promote cancer metastasis. Acta Pharm Sin B 11(8):2136–2149. https://doi.org/10.1016/j.apsb.2021.04.012
92. Aguirre-Ghiso JA (2007) Models, mechanisms and clinical evidence for cancer dormancy. Nat Rev Cancer 7(11):834–846. https://doi.org/10.1038/nrc2256
93. Klein CA (2020) Cancer progression and the invisible phase of metastatic colonization. Nat Rev Cancer 20(11):681–694. https://doi.org/10.1038/s41568-020-00300-6
94. Klein CA (2011) Framework models of tumor dormancy from patient-derived observations. Curr Opin Genet Dev 21(1):42–49. https://doi.org/10.1016/j.gde.2010.10.011
95. Yao D, Dai C, Peng S (2011) Mechanism of the mesenchymal−epithelial transition and its relationship with metastatic tumor formation. Mol Cancer Res 9(12):1608–1620. https://doi.org/10.1158/1541-7786.Mcr-10-0568
96. Yates CC, Shepard CR, Stolz DB, Wells A (2007) Co−culturing human prostate carcinoma cells with hepatocytes leads to increased expression of E-cadherin. Br J Cancer 96(8):1246–1252. https://doi.org/10.1038/sj.bjc.6603700
97. Zhou XD, Agazie YM (2008) Inhibition of SHP2 leads to mesenchymal to epithelial transition in breast cancer cells. Cell Death Differ 15(6):988–996. https://doi.org/10.1038/cdd.2008.54
98. Psaila B, Lyden D (2009) The metastatic niche: adapting the foreign soil. Nat Rev Cancer 9(4):285–293. https://doi.org/10.1038/nrc2621
99. Valastyan S, Weinberg RA (2011) Tumor metastasis: molecular insights and evolving paradigms. Cell 147(2):275–292. https://doi.org/10.1016/j.cell.2011.09.024
100. McAllister SS, Gifford AM, Greiner AL, Kelleher SP, Saelzler MP, Ince TA, Reinhardt F, Harris LN, Hylander BL, Repasky EA, Weinberg RA (2008) Systemic endocrine instigation of indolent tumor growth requires osteopontin. Cell 133(6):994–1005. https://doi.org/10.1016/j.cell.2008.04.045
101. Hiratsuka S, Duda DG, Huang Y, Goel S, Sugiyama T, Nagasawa T, Fukumura D, Jain RK (2011) C-X-C receptor type 4 promotes metastasis by activating p38 mitogen-activated protein kinase in myeloid differentiation antigen (Gr-1)-positive cells. Proc Natl Acad Sci U S A 108(1):302–307. https://doi.org/10.1073/pnas.1016917108
102. Saha S, Panigrahi DP, Patil S, Bhutia SK (2018) Autophagy in health and disease: a comprehensive review. Biomed Pharmacother 104:485–495
103. Cordani M, Butera G, Pacchiana R, Donadelli M (2017) Molecular interplay between mutant p53 proteins and autophagy incancer cells. Biochim Biophys Acta (BBA)-Rev Cancer 1867(1):19−28
104. Rybstein MD, Pedro JMB-S, Kroemer G, Galluzzi L (2018) The autophagic network and cancer. Nat Cell Biol 20:243–251
105. Villar VH, Merhi F, Djavaheri-Mergny M, Durán RV (2015) Glutaminolysis and autophagy in cancer. Autophagy 11(8):1198–1208
106. Mowers EE, Sharifi MN, Macleod KF (2017) Autophagy in cancer metastasis. Oncogene 36:1619–1630
107. Kroemer G, Marino G, Levine B (2010) Autophagy and the integrated stress response. Mol Cell 40(2):280–293
108. Chaffer CL, Weinberg RA (2011) A perspective on cancer cell metastasis. Science 331(6024):1559–1564
109. Avivar-Valderas A, Salas E, Bobrovnikova-Marjon E, Diehl JA, Nagi C, Debnath J, Aguirre-Ghiso JA (2011) PERK integrates autophagy and oxidative stress responses to promote survival during extracellular matrix detachment. Mol Cel Biol 31(17):3616–3629
110. Fung C, Lock R, Gao S, Salas E, Debnath J (2008) Induction of autophagy during extracellular matric detachment promotes cell survival. Mol Biol Cell 19:797–806

111. Lazova R, Camp RL, Klump V, Siddiqui SF, Amaravadi RK, Pawelek JM (2012) Punctate LC3B expression is a common feature of solid tumors and associated with proliferation, metastasis, and poor outcome. Clin Cancer Res 18(2):370–379
112. Zhao H, Yang M, Zhao J, Wang J, Zhang Y, Zhang Q (2013) High expression of LC3B is associated with progression and poor outcome in triple-negative breast cancer. Med Oncol 30:475
113. Lazova R, Klump V, Pawelek J (2010) Autophagy in cutaneous malignant melanoma. J Cutan Pathol 37(2):256–268
114. Han C, Sun B, Wang W, Cai W, Lou D, Sun Y, Zhao X (2011) Overexpression of microtubule-associated protein-1 light chain 3 is associated with melanoma metastasis and vasculogenic mimicry. Tohoku J Exp Med 223(4):243–251
115. Peng Y-F, Shi Y-H, Ding Z-B, Ke A-W, Gu C-Y, Hui B, Zhou J, Qiu S-J, Dai Z, Fan J (2013) Autophagy inhibition suppresses pulmonary metastasis of HCC in mive via impairing anoikis resistance and colonization of HCC cells. Autophagy 9(12):2056–2068
116. Peng Y-F, Shi Y-H, Shen Y-H, Ding Z-B, Ke A-W, Zhou J, Qiu S-J, Fan J (2013) Promoting colonization in metastatic HCC cells by modulation of autophagy. PLoS ONE 8(9):e74407
117. Galavotti S, Bartesaghi S, Faccenda D, Shaked-Rabi M, Sanzone S, McEvoy A, Dinsdale D, Condorelli F, Brandner S, Campanella M, Grose R, Jones C, Salomoni P (2012) The autophagy-associated factors DRAM1 and p62 regulate cell migration and invasion in glioblastoma stem cells. Oncogene 32:699–712
118. Nieto MA, Huang RY-J, Jackson RA, Thiery JP (2016) EMT: 2016. Cell 166(1):21–45
119. Kiyono K, Suzuki HI, Matsuyama H, Morishita Y, Komuro A, Kano M, Sugimoto K, Miyazono K (2009) Autophagy is activated by TGF-B and potentiates TGF-B-mediated growth inhibition in human hepatocellular cancer cells. Can Res 69(23):8844–8852
120. Li J, Yang B, Zhou Q, Wu Y, Shang D, Guo Y, Song Z, Zheng Q, Xiong J (2013) Autophagy promotes hepatocellular carcinoma cell invasion through activation of epithelial-mesenchymal transition. Carcinogenesis 34(6):1343–1351
121. Kim YH, Baek SH, Kim EK, Ha JM, Jin SY, Lee HS, Ha HK, Song SH, Kim SJ, Shin HK, Yong J, Kim D-H, Kim CD, Bae SS (2016) Uncoordinated 51-lile kinase 2 signaling pathway regulates epithelial-mesenchymal transition in A549 lung cancer cells. FEBS Lett 590(9):1365–1374
122. Qiang L, Zhao B, Ming M, Wang N, He T-C, Hwang S, Thorburn A, He Y-Y (2014) Regulation of cell proliferation and migration by p62 through stabilization of twist1. Proc Natl Acad Sci USA 111(25):9241–9246
123. Espina V, Mariani BD, Gallagher RI, Tran K, Banks S, Wiedemann J, Huryk H, Mueller C, Adamo L, Deng J, Petricoin EF III, Pastore L, Zaman S, Menezes G, Mize J, Johal J, Edmiston K, Liotta LA (2010) Malignant precursor cells pre-exist in human breast DCIS and require autophagy for survival. PLoS ONE 5(4):e10240
124. Mani SA, Guo W, Liao M-J, Eaton EN, Ayyanan A, Zhou AY, Brooks M, Reinhard F, Zhang CC, Shipitsin M, Campbell LL, Polyak K, Brisken C, Yang J, Weinberg RA (2008) The epithelial-mesenchymal transition generates cells with properties of stem cells. Cell 133(4):704–715
125. May CD, Sphyris N, Evans KW, Werden SJ, Guo W, Mani SA (2011) Epithelial-mesenchymal transition and cancer stem cells: a dangerously dynamic duo in breast cancer progression. Breast Cancer Res 13(1):202
126. McGowan PM, Kirstein JM, Chambers AF (2009) Micrometastatic disease and metastatic outgrowth: clinical issues and experimental approaches. Futur Med 5(7):1083–1098
127. Chaffer CL, Brueckmann I, Scheel C, Kaestli AJ, Wiggins PA, Rodrigues LO, Brooks M, Reinhardt F, Su Y, Polyak K, Arendt LM, Kupperwasser C, Bierie B, Weinberg RA (2011) Normal and neoplastic nonstem cells can spontaneously convert to a stem-like state. Proc Natl Acad Sci USA 108(19):7950–7955
128. Karsli-Uzunbas G, Yanxiang Guo J, Price S, Teng X, Laddha SV, Khor S, Kalaany NY, Jacks T, Chan CS, Rabinowitz JD, White E (2014) Autophagy is required for glucose homestasis and lung tumor maintenance. Cancer Discov 4(8):914–927

129. Lock R, Kenific CM, Leidal AM, Salas E, Debnath J (2014) Autophagy dependent production of secreted factors facilitates oncogenic RAS-driven invasion. Cancer Discov 4(4):466–479
130. Kenific CM, Stehbens SJ, Goldsmith J, Leidal AM, Faure N, Ye J, Wittmann T, Debnath J (2016) NBR1 enables autophagy-dependent focal adhesion turnover. J Cell Biol 212(5):577–590
131. Sharifi MN, Mowers EE, Drake LE, Chris C, Chen H, Zamora M, Mui S, Macleod KF (2016) Autophagy promotes focal adhesion disassembly and cell motility of metastatic tumor cells through the direct interaction of paxillin with LC3. Cell Rep 15(8):1660–1672
132. Kenific CM, Thorburn A, Debnath J (2010) Autophagy and metastasis: another double-edged sword. Curr Opin Cell Biol 22(2):241–245
133. Fung C, Lock R, Gao S, Salas E, Debnath J (2008) Induction of autophagy during extracellular matrix detachment promotes cell survival. Mol Biol Cell 19(3):797–806
134. Lock R, Debnath J (2008) Extracellular matrix regulation of autophagy. Curr Opin Cell Biol 20(5):538–588
135. Rouschop KMA, van den Beucken T, Dubois L, Niessen H, Bussink J, Savelkouls K, Keulers T, Mujcic H, Landuyt W, Voncken JW, Lambin P, van der Kogel AJ, Koritzinsky M, Wouters BG (2010) The unfolded protein response protects human tumor cells during hypoxia through regulation of the autophagy genes MAP1LC3B and ATG5. J Clin Investig 120(1):127–141
136. Avivar-Valderas A, Bobrovnikova-Marjon E, Diehl JA, Bardeesy N, Debnath J, Aguirre-Ghiso JA (2013) Regulation of autophagy during ECM detachment is linked to a selective inhibition of mTORC1 by PERK. Oncogene 32(41):4932–4940
137. Chen N, Debnath J (2013) IkB kinase complex (IKK) triggers detachment-induced autophagy in mammary epithelial cells independently of the PI3K-AKT-MTORC1 pathway. Autophagy 9(8):1214–1227
138. Criollo A, Senovilla L, Authier H, Maiuri MC, Morselli E, Vitale I, Kepp O, Tasdemir E, Galluzzi L, Shen S, Tailler M, Delahaye N, Tesniere A, De Stefano D, Younes AB, Harper F, Pierron G, Lavandero S, Zitvogel L, Isreal A, Baud V, Kroemer G (2010) The IKK complex contributes to the induction of autophagy. EMBO J 29(3):619–631
139. Scherz-Shouval R, Shvets E, Fass E, Shorer H, Gil L, Elazar Z (2007) Reactive oxygen species are essential for autophagy and specficially regulate the activity of Atg4. EMBO J 26(7):1749–1760
140. Maes H, Kuchnio A, Peric A, Moens S, Nys K, De Bock K, Quaegebeur A, Schoors S, Georgiadou M, Wouters J, Vinckier S, Vankelecom H, Garmyn M, Vion A-C, Radtke F, Boulanger C, Gerhardt H, Dejana E, Dewerchin M, Ghesquière B, Annaert W, Agostinis P, Carmeliet P (2014) Tumor vessel normalization by chloroquine independent of autophagy. Cancer Cell 26(2):190–206
141. Langley RR, Fidler IJ (2011) The seed and soil hypothesis revisited—the role of tumor-stroma interactions in metastasis to different organs. Int J Cancer 128(11):2527–2535
142. Kadandale P, Stender JD, Glass CK, Kiger AA (2010) Conserved role for autophagy in Rho1-mediated cortical remodeling and blood cell recruitment. Proc Natl Acad Sci USA 107(23):10502–10507
143. Macintosh RL, Timpson P, Thorburn J, Anderson KI, Thorburn A, Ryan KM (2012) Inhibition of autophagy impairs tumor cell invasion in an organotypic model. Cell Cycle 11:2022–2029
144. Sandilands E, Serrels B, McEwan DG, Morton JP, Macagno JP, McLeod K, Stevens C, Brunton VG, Langdon WY, Vidal M, Sansom OJ, Dikic I, Wilkinson S, Frame MC (2012) Autophagic targeting of Src promotes cancer cell survival following reduced FAK signalling. Nat Cell Biol 14:51–60
145. Kim SI, Na H-J, Ding Y, Wang Z, Lee SJ, Choi ME (2012) Autophagy promotes intracellular degradation of type i collagen induced by transforming growth factor (TGF)-b1. J Biol Chem 287(15):11677–11688
146. Belaid A, Cerezo M, Chargui A, Corcelle-Termeau E, Pedeutour F, Giuliano S, Ilie M, Rubera I, Tauc M, Barale S, Bertolotto C, Brest P, Vouret-Craviari V, Klionsky DJ, Carle GF, Hofman P, Mograbi B (2013) Autophagy plays a critical role in the degradation of active RHOA, the control of cell cytokinesis and genomic instability. Can Res 73(14):4311–4322

147. Chan EYW, Kir S, Tooze SA (2007) siRNA screening of the kinome identifies ULK1 as a multidomain modulator of autophagy. J Biol Chem 282(35):25464–25474
148. Gurkar AU, Chu K, Raj L, Bouley R, Lee S-H, Kim Y-B, Dunn SE, Mandinova A, Lee SW (2013) Identification of ROCK1 kinase as a critical regulator of Beclin1 mediated autophagy during metabolic stress. Nat Commun 4:2189
149. Gutierrez MG, Munafó DB, Berón W, Colombo MI (2004) Rab7 is required for the normal progression of the autophagic pathway in mammalian cells. J Cell Sci 117(13):2687–2697
150. Mack NA, Whalley HJ, Castillo-Lluva S, Malliri A (2011) The diverse roles of Rac signaling in tumorigenesis. Cell Cycle 10(10):1571–1581
151. Abbi S, Ueda H, Cooper LA, Zhao J, Christopher R, Guan J-L (2002) Regulation of focal adhesion kinase by a novel protein inhibitor FIP200. Mol Biol Cell 13(9):3178–3191
152. Hara T, Takamura A, Kishi C, Iemura S-i, Natsume T, Guan J-L, Mizushima N (2008) FIP200, a ULK-interacting protein, is required for autophagosome formation in mammalian cells. J Cell Biol 181(3):497–510
153. Lu Z, Luo RZ, Lu Y, Zhang X, Yu Q, Khare S, Kondo S, Kondo Y, Yu Y, Mills GB, Liao WS-L, Bast J, Robert C (2008) The tumor suppressor gene ARHI regulates autophagy and tumor dormancy in human ovarian cancer cells. J Clin Investig 118(12):3917–3929
154. Sosa MS, Bragado P, Aguirre-Ghiso JA (2014) Mechanisms of disseminated cancer cell dormancy: an awakening field. Nat Rev Cancer 14(9):611–622
155. Galluzzi L, Pietrocola F, Levine B, Kroemer G (2014) Metabolic control of autophagy. Cell 159(6):1263–1276
156. Lum JJ, Bauer DE, Kong M, Harris MH, Li C, Lindsten T, Thompson CB (2005) Growth factor regulation of autophagy and cell survival in the absence of apoptosis. Cell 120(2):237–248
157. Liang J, Shao SH, Xu Z-X, Hennessy B, Ding Z, Larrea M, Kondo S, Dumont DJ, Gutterman JU, Walker CL, Slingerland JM, Mills GB (2007) The energy sensing LKB1-AMPK pathway regulates p27kip1 phosphorylation mediating the decision to enter autophagy or apoptosis. Nat Cell Biol 9:218–224
158. Egan DF, Shackelford DB, Mihaylova MM, Gelino SR, Kohnz RA, Mair W, Vasquez DS, Joshi A, Gwinn DM, Taylor R, Asara JM, Fitzpatrick J, Dillin A, Viollet B, Kundu M, Hansen M, Shaw RJ (2011) Phosphorylation of ULK1 (hATG1) by AMP-activated protein kinase connects energy sensing to mitophagy. Science 331(6016):456–461
159. Lu Z, Yang H, Sutton MN, Yang M, Clarke CH, Liao WS-L, Bast RC (2014) ARHI (DIRAS3) induces autophagy in ovarian cancer cells by downregulating the epidermal growth factor receptor, inhibiting PI3K and Ras/MAP signaling and activating the FOXo3a-mediated induction of Rab7. Cell Death Differ 21(8):1275–1289
160. Ma Y, Galuzzi L, Zitvogel L, Kroemer G (2013) Autophagy and cellular immune responses. Immunity 39(2):211–227
161. Deretic V, Saitoh T, Akira S (2013) Autophagy in infection, inflammation and immunity. Nat Rev Immunol 13:722–737
162. Zhong Z, Sanchez-Lopez E, Karin M (2016) Autophagy, inflammation and immunity: a troika governing cancer and its treatment. Cell 166(2):288–298
163. Schreiber RD, Old LJ, Smyth MJ (2011) Cancer immunoediting: integrating immunity´s roles in cancer suppression and promotion. Science 331(6024):1565–1570
164. Michaud M, Martins I, Sukkurwala AQ, Adjemian S, Ma Y, Pellegatti P, Shen S, Kepp O, Scoazec M, Mignot G, Rello-Varona S, Tailler M, Menger L, Vacchelli E, Galluzzi L, Ghiringhelli F, Di Virgilio F, Zitvogel L, Kroemer G (2011) Autophagy-dependent anticancer immunre responses induced by chemotherapeutic agents in mice. Science 334(6062):1573–1577
165. Ladoire S, Enot D, Senovilla L, Ghiringhelli F, Poirier-Colame V, Chaba K, Semeraro M, Chaix M, Penualt-Llorca F, Arnould L, Poillot ML, Arveux P, Delaloge S, Andre F, Zitvogel L, Kroemer G (2016) The presence of LC3B puncta and HMGB1 expression in malignant cells correlate with the immune infiltrate in breast cancer. Autophagy 12(5):864–875
166. Rao S, Tortola L, Perlot T, Wirnsberger G, Novatchkova M, Nitsch R, Sykacek P, Frank L, Schramek D, Komnenovic V, Sigl V, Aumayr K, Schmauss G, Fellner N, Handschuh S,

Glösmann M, Pasierbek P, Schlederer M, Resch GP, Ma Y, Popper H, Kenner L, Kroemer G, Penninger JM (2014) A dual role for autophagy in a murine model of lung cancer. Nat Commun 5:3056–3070
167. Akalay I, Janji B, Hasmim M, Noman MZ, André F, De Cremoux P, Bertheau P, Badoual C, Vielh P, Larsen AK, Sabbah M, Tan TZ, Keira JH, Hung NTY, Thiery JP, Mami-Chouaib F, Chouaib S (2013) Epithelial-to-mesenchymal transition and autophagy induction in breast carcinoma promote escape from T-cell-mediated lysis. Can Res 73(8):2418–2427
168. Wei H, Wei S, Gan B, Peng X, Zou W, Gguan J-L (2011) Suppression of autophagy by FIP200 deletion inhibits mammary tumorigenesis. Genes Dev 25(14):1510–1527
169. Baginska J, Viry E, Berchem G, Poli A, Noman MZ, van Moer K, Medves S, Zimmer J, Oudin A, Niclou SP, Bleackley RC, Goping IS, Chouaib S, Janji B (2013) Granzyme B degradation by autophagy decreases tumor cell susceptibility to natural killer-mediated lysis under hypoxia. Proc Natl Acad Sci U S A 110(43):17450–17455
170. Tittarelli A, Janji B, van Moer K, Noman MZ, Chouaib S (2015) The selective degradation of synaptic connexin 43 protein by hypoxia-induced autophagy impairs natural killer cell-mediated tumor cell killing. J Biol Chem 290(39):23670–23679
171. Noman MZ, Janji B, Kaminska B, van Moer K, Pierson S, Przanowski P, Buart S, Berchem G, Romero P, Mami-Chouaib F, Chouaib S (2011) Blocking hypoxia-induced autophagy in tumors restores cytotoxic T-Cell activity and promotes regression. Can Res 71(18):5976–5986
172. Messai Y, Noman MZ, Hasmim M, Janji B, Tittarelli A, Boutet M, Baud V, Viry E, Billot K, Nanbakhsh A, Safta TB, Richon C, Ferlicot S, Donnadieu E, Couve S, Gardie B, Orlanducci F, Albiges L, Thiery J, Olive D, Escudier B, Chouaib S (2014) ITPR1 protects renal cancer cells against natural killer cells by inducing autophagy. Can Res 74(23):6820–6832
173. Joyce JA, Pollard JW (2009) Microenvironmental regulation of metastasis. Nat Rev Cancer 9(4):239–252
174. Wels J, Kaplan RN, Rafii S, Lyden D (2008) Migratory neighbors and distant invaders: tumor-associated niche cells. Genes Dev 22(5):559–574
175. Martinez-Outschoorn UE, Pavlides S, Whitaker-Menezes D, Daumer KM, Milliman JN, Chiavarina B, Migneco G, Witkiewicz AK, Martinez-Cantarin MP, Flomenberg N, Howell A, Pestell RG, Lisanti MP, Sotgia F (2010) Tumor cells induce the cancer associated fibroblast phenotype via caveolin-1 degradation: implications for breast cancer and DCIS therapy with autophagy inhibitors. Cell Cycle 9(12):2423–2433
176. Baixauli F, López-Otín C, Mittelbrunn M (2014) Exosomes and autophagy: coordinated mechanisms for the maintenance of cellular fitness. Front Immunol 5:403
177. Huotari J, Helenius A (2011) Endosome maturation. EMBO J 30(17):3481–3500
178. Gupta A, Roy S, Lazar AJF, Wang W-L, McAuliffe JC, Reynoso D, McMahon J, Tagushi T, Floris G, Debiec-Rychter M, Schöffski P, Trent JA, Debnath J, Rubin BP (2010) Autophagy inhibition and antimalarials promote cell death in gastrointestinal stromal tumor (GIST). Proc Natl Acad Sci USA 107(32):12333–14338
179. Apel A, Herr I, Schwarz H, Rodemann HP, Mayer A (2008) Blocked autophagy sensitizes resistant carcinoma cells to radiation therapy. Can Res 68(5):1485–1494
180. Ranganathan AC, Zhang L, Adam AP, Aguirre-Ghiso JA (2006) Functional coupling of p38-induced up-regulation of BiP and activation of RNA-dependent protein kinase-like endoplasmic reticulum kinase to drug resistance of dormant carcinoma cells. Can Res 66(3):1702–1711
181. Amaravadi RK, Yu D, Lum JJ, Bui T, Christophorou MA, Evan GI, Thomas-Tikhonenko A, Thompson CB (2007) Autophagy inhibition enhances therapy-induced apoptosis in a Myc-induced model of lymphoma. J Clin Investig 117(2):326–336

Index

A

Adult neurogenesis, 194, 195, 197–199, 209, 210
Adult stem cells, 113, 128
Aging, 114, 119, 121, 123, 124, 126, 128
Autophagic cell death, 193, 206, 207, 209, 210
Autophagy, 35, 37–53, 59–77, 85–89, 91–100, 107–128, 137, 143–159, 169, 172–186, 193, 195, 198–204, 206, 207, 209, 210, 219, 221, 223–229, 230, 235, 237–252
Autophagy assays, 1, 3, 19, 26
Autophagy-related genes, 174, 182, 183

B

Bone disorders, 235, 238, 250

C

Cancer, 259, 261, 262, 264–270, 273–275
C. elegans, 85, 86, 88, 90–96, 98–100
Cell stress, 259, 262, 273
Cellular homeostasis, 37, 38, 41, 43, 52, 53
Chaperone Mediated Autophagy (CMA), 86

D

Differentiation, 35, 37, 40, 41, 43, 44, 46–49, 52, 53, 169, 171–174, 180–186
Drosophila, 85, 86, 90–99

E

Embryogenesis, 235, 238–240
Embryonic stem cells, 59, 61

F

Flux, 1, 6, 14, 15, 17–19, 22, 24–26, 29

G

Germline stem cells, 85, 87–91, 93–100

H

Hematological disorders, 116
Hematopoietic stem cells, 107, 108, 113, 117–119, 124

I

Immunomodualtion, 219
Insulin-like IGF-1 Signaling (IIS), 94
Intestine, 171, 174, 185, 186

L

LC3/GABARAP, 6, 15, 23, 28
Lysosome, 195, 201–204, 210

M

Macroautophagy, 85–87
Mesenchymal Stem Cells (MSC), 219, 220, 222, 225

B. V. Shravage and K. Turksen (eds.), *Autophagy in Stem Cell Maintenance and Differentiation*, Stem Cell Biology and Regenerative Medicine 73,
https://doi.org/10.1007/978-3-031-17362-2

Metastasis, 259, 261, 262, 266, 267, 269–272, 275
Microautophagy, 86
Muscle stem cell, 137, 142, 149, 154, 157
Myoblast, 137, 139, 141, 144, 154–158
Myogenesis, 137, 140, 141, 154–157

N
Neo-angiogenesis, 221
Neural Stem Cell (NSC), 59, 61, 62, 69, 70, 72, 193–195, 197, 199, 210
Neurogenesis, 59, 61, 69–73, 75, 77

O
Osteogenic differentiation, 235, 241, 245, 247–250

P
p62, 5–8, 16, 19, 20, 24, 27
Pathologies, 173, 185, 186
P granules, 92
Proliferation, 169, 171–174, 176, 180–186
Protein aggregates, 203, 210

Q
Quiescence, 197–199, 203, 204, 210

R
Regeneration, 221, 225, 226, 228

S
Satellite cell, 137, 139–142, 147, 149–155, 157, 158
Skeletal muscle, 137–140, 143–149, 155
Stem cells, 1, 3, 6–14, 16, 19, 22, 24–29, 35, 37, 40–49, 52, 53, 169, 171–186, 235, 237, 238, 244–246, 250–252

T
Target-of-Rapamycin (TOR), 94–97, 100
Transcription, 59, 61, 66, 68, 69, 72, 73
Tumour microenvironment, 274, 275

The manufacturer's authorised representative in the EU is Springer Nature Customer Service Centre GmbH, Europaplatz 3, 69115 Heidelberg, Germany. If you have any concerns regarding our products, please contact ProductSafety@springernature.com

Printed and bound by CPI Group (UK) Ltd, Croydon, CR0 4YY
30/09/2026
02225239-0010